Stochastic Differential Equations for Science and Engineering

Stochastic Differential Equations for Science and Engineering is aimed at students at the M.Sc. and PhD level. The book describes the mathematical construction of stochastic differential equations with a level of detail suitable to the audience, while also discussing applications to estimation, stability analysis, and control. The book includes numerous examples and challenging exercises. Computational aspects are central to the approach taken in the book, so the text is accompanied by a repository on GitHub containing a toolbox in R which implements algorithms described in the book, code that regenerates all figures, and solutions to exercises.

Features

- Contains numerous exercises, examples, and applications
- Suitable for science and engineering students at Master's or PhD level
- Thorough treatment of the mathematical theory combined with an accessible treatment of motivating examples
- GitHub repository available at: https://github.com/Uffe-H-Thygesen/SDE-book and https://github.com/Uffe-H-Thygesen/SDEtools

Uffe Høgsbro Thygesen received his Ph.D. degree from the Technical University of Denmark in 1999, based on a thesis on stochastic control theory. He worked with applications to marine ecology and fisheries until 2017, where he joined the Department of Applied Mathematics and Computer Science at the same university. His research interests are centered on deterministic and stochastic dynamic systems and involve times series analysis, control, and dynamic games, primarily with applications in life science. In his spare time he teaches sailing and kayaking and learns guitar and photography.

Stochastic Differential Equations for Science and Engineering

Uffe Høgsbro Thygesen
Technical University of Denmark, Denmark

CRC Press
Taylor & Francis Group
Boca Raton London New York

CRC Press is an imprint of the
Taylor & Francis Group, an **informa** business
A CHAPMAN & HALL BOOK

The cover image depicts a sample path of a particle diffusing in a double well system, starting between the two wells. The vertical axis denotes position while the horizontal is time. The smooth curves show the probability density at various points in time; here the horizontal axis depicts the density.

First edition published 2023
by CRC Press
6000 Broken Sound Parkway NW, Suite 300, Boca Raton, FL 33487-2742

and by CRC Press
4 Park Square, Milton Park, Abingdon, Oxon, OX14 4RN

CRC Press is an imprint of Taylor & Francis Group, LLC

ISBN: 978-1-032-23217-1 (hbk)
ISBN: 978-1-032-23406-9 (pbk)
ISBN: 978-1-003-27756-9 (ebk)

DOI: 10.1201/ 9781003277569

Typeset in LM Roman
by KnowledgeWorks Global Ltd.

Publisher's note: This book has been prepared from camera-ready copy provided by the authors

Contents

Section II Stochastic Calculus

Chapter 6 ▪ Stochastic Integrals   117

Chapter 7 ▪ The Stochastic Chain Rule   141

SECTION III   **Applications**

CHAPTER   10 ■ State Estimation   233

Preface

This book has grown from a set of lecture notes written for a course on *Diffusion and Stochastic Differential Equations*, offered at the Technical University of Denmark. This 5 ECTS course is primarily aimed at students in the M.Sc.&Eng. programme, and therefore the book has the same intended audience. These students have a broad background in applied mathematics, science and technology, and although most of them are ultimately motivated by applications, they are well aware that nothing is more practical than a good theory (to paraphrase Kurt Lewin).

Therefore, the book aims to describe the mathematical construction of stochastic differential equations with a fair level of detail, but not with complete rigor, while also describing applications and giving examples. Computational aspects are important, so the book is accompanied by a repository on GitHub which contains a toolbox in R which implements algorithms described in the book, code that regenerates all figures, and solutions to exercises. See `https://github.com/Uffe-H-Thygesen/SDEbook` and `https://github.com/Uffe-H-Thygesen/SDEtools`.

The book assumes that the reader is familiar with ordinary differential equations, is operational in "elementary" probability (i.e., not measure-theoretic), and has been exposed to partial differential equations and to stochastic processes, for example, in the form of Markov chains or time series analysis.

Many students and colleagues have provided feedback and corrections to earlier versions. I am grateful for all of these, which have improved the manuscript. Any remaining errors, of which I am sure there are some, remain my responsibility.

Uffe Høgsbro Thygesen
Lundtofte, October 2022

CHAPTER 1

Introduction

Ars longa, vita brevis.
Hippocrates, c. 400 BC

A stochastic differential equation can, informally, be viewed as a differential equation in which a stochastic "noise" term appears:

$$\frac{dX_t}{dt} = f(X_t) + g(X_t)\,\xi_t, \quad X_0 = x. \qquad (1.1)$$

Here, X_t is the state of a dynamic system at time t, and $X_0 = x$ is the initial state. Typically we want to "solve for X_t" or describe the stochastic process $\{X_t : t \geq 0\}$. The function f describes the dynamics of the system without noise, $\{\xi_t : t \geq 0\}$ is white noise, which we will define later in detail, and the function g describes how sensitive the state dynamics is to noise.

In this introductory chapter, we will outline what the equation means, which questions of analysis we are interested in, and how we go about answering them. A reasonable first question is why we would want to include white noise terms in differential equations. There can be (at least) three reasons:

Analysis: We may want to examine how noise, or uncertain source terms, affect the system. Consider, for example, a wind turbine (Figure 1.1). The wind exerts a force on the turbine, and the strength of this force fluctuates unpredictably. We may model these fluctuations as noise, and ask how they cause the construction to vibrate. To answer this question, we must have a model of how the noise enters the system, i.e., g, as well as how the system dynamics respond, i.e., f. Figure 1.1 shows a simulation from a model with three state variables, i.e., X_t is a 3-vector containing force on the turbine and the position and velocity of the hub. The figure compares a stochastic simulation with a noise-free simulation. Based on such a simulation, or a mathematical analysis of the model, we can get statistics of force, position and velocity. These statistics are important to assess the wear and tear on the turbine, and also affect the regularity of the electrical power which is generated by the turbine.

Time series analysis: We may have a time series of measurements taken from the system. Based on these measurements, we may want to estimate

DOI: 10.1201/9781003277569-1

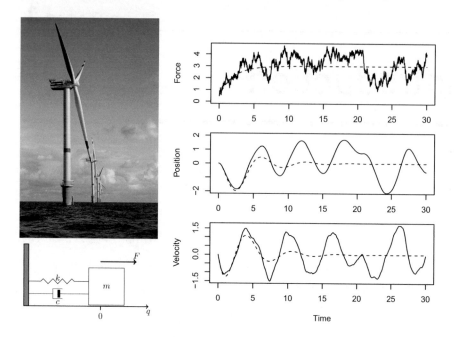

Figure 1.1   A wind turbine is affected by fluctuations in wind speed which causes it to oscillate. We model the fluctuations as low-pass filtered noise and the response of the construction as a linear mass-spring-damper system. *Solid lines:* Simulated force, position and velocity from a stochastic simulation of a dimensionless model. *Dashed lines:* Noise-free simulation. The details of this model are given in Exercise 5.5. Photo credit: CC BY-SA 4.0.

parameters in the differential equation, i.e., in f; we may want to know how large loads the structure has been exposed to, and we may want to predict the future production of electrical power. To answer these questions, we must perform a statistical analysis on the time series. When we base time series analysis on stochastic differential equations, we can use insight in the system dynamics to fix the structure of f and maybe of g. The framework lets us treat statistical errors correctly when estimating unknown parameters and when assessing the accuracy with which we can estimate and predict.

Optimization and control: We may want to design a control system that dampens the fluctuations that come from the wind. On the larger scale of the electrical grid, we may want a control system to ensure that the power supply meets the demand and so that voltages and frequencies are kept at the correct values. To design such control systems optimally, we need to take into account the nature of the disturbances that the control system should compensate for.

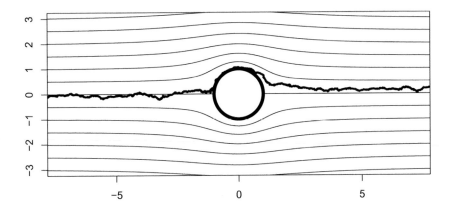

Figure 1.2 A small particle embedded in the two-dimensional flow around a cylinder. The thin lines are *streamlines*, i.e., the paths a water molecule follows, neglecting diffusion. The thick black line is a simulated random trajectory of a particle, which is transported with the flow, but at the same time subjected to molecular diffusion.

Motion of a Particle Embedded in a Fluid Flow

Let us examine in some more detail the origin and form of the noise term ξ_t in (1.1). Figure 1.2 displays water flowing in two dimensions around a cylinder.[1] In absence of diffusion, water molecules will follow the streamlines. A small particle will largely follow the same streamlines, but is also subject to diffusion, i.e., random collisions with neighboring molecules which cause it to deviate from the streamlines. Collisions are frequent but each cause only a small displacement, so the resulting path is erratic.

In absence of diffusion, we can find the trajectory of the particle by solving the ordinary differential equation

$$\frac{dX_t}{dt} = f(X_t).$$

Here, $X_t \in \mathbf{R}^2$ is the position in the plane of the particle at time t. The function $f(\cdot)$ is the flow field, so that $f(x)$ is a vector in the plane indicating the speed and direction of the water flow at position x. To obtain a unique solution, this equation needs an initial condition such as $X_0 = x_0$ where x_0 is the known position at time 0. The trajectory $\{X_t : t \in \mathbf{R}\}$ is exactly a streamline.

---

[1] The flow used here is irrotational, i.e., potential flow. Mathematically, this is convenient even if physically, it may not be the most meaningful choice.

To take the molecular diffusion into account, i.e., the seemingly random motion of a particle due to collisions with fluid molecules, we add "white noise" ξ_t to the equation

$$\frac{dX_t}{dt} = f(X_t) + g\,\xi_t. \tag{1.2}$$

Here, g is the (constant) noise intensity, and ξ_t is a two-vector. The trajectory in Figure 1.2 has been simulated as follows: We discretize time. At each time step, we first advect the particle in the direction of a streamline, then we shift it randomly by adding a perturbation which is sampled from a bivariate Gaussian where each component has mean 0 and variance $g^2 h$. Specifically,

$$X_{t+h} = X_t + f(X_t)\,h + g\,\xi_t^{(h)} \text{ where } \xi_t^{(h)} \sim N(0, hI).$$

Here, h is the time step and the superscript in $\xi_t^{(h)}$ indicates that the noise term depends on the time step, while I is a 2-by-2 identity matrix. If we let the particle start at a fixed position X_0, the resulting positions $\{X_h,\ X_{2h},\ X_{3h}, \ldots\}$ will each be a random variable, so together they constitute a stochastic process. This algorithm does not resolve the position between time steps; when we plot the trajectory, we interpolate linearly.

We hope that the position X_t at a given time t will not depend too much on the time step h as long as h is small enough. This turns out to be the case, if we choose the noise term in a specific way, viz.

$$\xi_t^{(h)} = B_{t+h} - B_t,$$

where $\{B_t\}$ is a particular stochastic process, namely *Brownian motion*. Thus, we should start by simulating the Brownian motion, and next compute the noise terms $\xi_t^{(h)}$ from the Brownian motion; we will detail exactly how to do this later. Brownian motion is key in the theory of stochastic differential equations for two reasons: First, it solves the simplest stochastic differential equation, $dX_t/dt = \xi_t$, and second, we use it to represent the noise term in *any* stochastic differential equation. With this choice of noise $\xi_t^{(h)}$, we can rewrite the recursion with the shorthand

$$\Delta X_t = f(X_t)\,\Delta t + g\,\Delta B_t \tag{1.3}$$

and since this turns out to converge as the time step $h = \Delta t$ goes to zero, we use the notation

$$dX_t = f(X_t)\,dt + g\,dB_t \tag{1.4}$$

for the limit. This is our preferred notation for a stochastic differential equation. In turn, (1.3) is an Euler-type numerical method for the differential equation (1.4), known as the Euler-Maruyama method.

If the particle starts at a given position $X_0 = x_0$, its position at a later time t will be random, and we would like to know the probability density of the position $\phi(x,t)$. It turns out that this probability density $\phi(x,t)$ is governed by a partial differential equation of advection-diffusion type, viz.

$$\frac{\partial \phi}{\partial t} = -\nabla \cdot (f\phi - \frac{1}{2}g^2\nabla\phi)$$

with appropriate boundary conditions. This is the same equation that governs the concentration of particles, if a large number of particles is released and move independently of each other. This equation is at the core of the theory of transport by advection and diffusion, and now also a key result in the theory of stochastic differential equations.

We can also ask, what is the probability that the particle hits the cylinder, depending on its initial position. This is governed by a related (specifically, adjoint) partial differential equation.

There is a deep and rich connection between stochastic differential equations and partial differential equations involving diffusion terms. This connection explains why we, in general, use the term *diffusion processes* for solutions to stochastic differential equations. From a practical point of view, we can analyze PDE's by simulating SDE's, or we can learn about the behavior of specific diffusion processes by solving associated PDE's analytically or numerically.

Population Growth and State-Dependent Noise

In the two previous examples, the wind turbine and the dispersing molecule, the noise intensity $g(x)$ in (1.1) was a constant function of the state x. If all models had that feature, this book would be considerably shorter; some intricacies of the theory only arise when the noise intensity is state-dependent. State-dependent noise intensities arise naturally in population biology, for example. A model for the growth of an isolated bacterial colony could be:

$$dX_t = rX_t(1 - X_t/K)\,dt + g(X_t)\,dB_t. \tag{1.5}$$

Here, $r > 0$ is the specific growth rate at low abundance, while $K > 0$ is the carrying capacity. Without noise, i.e., with $g(x) \equiv 0$, this is the *logistic growth model*; see Figure 1.3. Dynamics of biological systems is notoriously noisy, so we have included a noise term $g(X_t)\,dB_t$ and obtained a *stochastic logistic growth model*. Here, it is critical that the noise intensity $g(x)$ depends on the abundance x; otherwise, we can get negative abundances X_t! To avoid this, we must require that the noise intensity $g(x)$ vanishes at the origin, $g(0) = 0$, so that a dead colony stays dead. Figure 1.3 includes two realizations of the solution $\{X_t\}$ with the choice $g(x) = \sigma x$. For this situation, the theory allows us to answer the following questions:

1. How is the state X_t distributed, in particular as time $t \to \infty$? Again, we can pose partial differential equations of advection-diffusion type to answer this, and the steady-state distribution can be found in closed form. For example, we can determine the mean and variance.

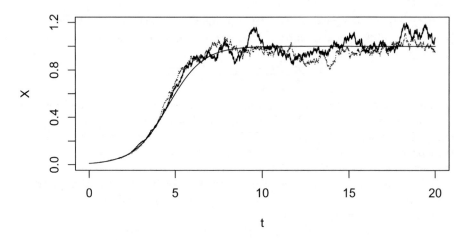

Figure 1.3 Two simulated sample paths (solid and dotted erratic curves) of the stochastic logistic growth model (1.5) with $g(x) = \sigma x$ of a bacterial population. A noise-free simulation is included (thin black line). Parameters are $r = K = 1$, $\sigma = 0.1$, $X_0 = 0.01$. The computational time step used is 0.001.

2. What is the temporal pattern of fluctuations in $\{X_t\}$? We shall see that $\{X_t\}$ is a *Markov process*, which allows us to characterize these fluctuations. We can assess their time scale through their stochastic Lyapunov exponent, which for this example leads to a time scale of $1/(r - \sigma^2/2)$, when the noise is weak and the process is in stochastic steady state.

3. Does a small colony risk extinction? For this particular model with these particular parameters, it turns out that the answer is "no". With other parameters, the answer is that the colony is doomed to extinction, and for other noise structures, the answer is that there is a certain probability of extinction, which depends on the initial size of the colony. These questions are answered by *stochastic stability theory* as well as by the theory of *boundary behavior and classification*.

However, before we can reach these conclusions, we must consider the equation (1.5) carefully. We call it a stochastic *differential* equation, but it should be clear from Figure 1.3 that the solutions are nowhere differentiable functions of time. This means that we should not take results from standard calculus for granted. Rather, we must develop a *stochastic calculus* which applies to diffusion processes. In doing so, we follow in the footsteps of Kiyosi Itô, who took as starting point an integral version of the equation (1.5), in order to circumvent the problem that stochastic differential equations have non-differentiable solutions. The resulting Itô calculus differs from standard calculus, most notably in its chain rule, which includes second order terms.

Overview of the Book

This book is in three parts. The core is Itô's stochastic calculus in part 2: It describes stochastic integrals, stochastic calculus, stochastic differential equations, and the Markov characterization of their solutions.

Before embarking on this construction, part 1 builds the basis. We first consider molecular diffusion as a transport processes (Chapter 2); this gives a physical reference for the mathematics. We then give a quick introduction to measure-theoretic probability (Chapter 3) after which we study Brownian motion as a stochastic process (Chapter 4). Chapter 5 concerns the very tractable special case of linear systems such as the wind turbine (Figure 1.1). At this point we are ready for the Itô calculus in part 2.

Finally, part 3 contains four chapters which each gives an introduction to an area of application. This concerns estimation and time series analysis (Chapter 10), quantifying expectations to the future (Chapter 11), stability theory (Chapter 12) and finally optimal control (Chapter 13).

Exercises, Solutions, and Software

In science and engineering, what justifies mathematical theory is that it lets us explain existing real-world systems and build new ones. The understanding of mathematical constructions must go hand in hand with problem-solving and computational skills. Therefore, this book contains exercises of different kinds: Some fill in gaps in the theory and rehearse the ability to argue mathematically. Others contain pen-and-paper exercises while yet others require numerical analysis on a computer. Solutions are provided at https://github.com/Uffe-H-Thygesen/SDEbook. There, also code which reproduces all figures is available. The computations use a toolbox, SDEtools for R, which is available at https://github.com/Uffe-H-Thygesen/SDEtools.

There is no doubt that stochastic differential equations are becoming more widely applied in many fields of science and engineering, and this by itself justifies their study. From a modeler's perspective, it is attractive that our understanding of processes and dynamics can be summarized in the drift term f in (1.1), while the noise term ξ_t (or B_t) manifests that our models are always incomplete descriptions of actual systems. The mathematical theory ties together several branches of mathematics – ordinary and partial differential equations, measure and probability, statistics, and optimization. As you develop an intuition for stochastic differential equations, you will establish interesting links between subjects that may at first seem unrelated, such as physical transport processes and propagation of noise. I have found it immensely rewarding to study these equations and their solutions. My hope is that you will, too.

I

Fundamentals

CHAPTER 2

Diffusive Transport and Random Walks

The theory of stochastic differential equations uses a fair amount of mathematical abstraction. If you are interested in applying the theory to science and technology, it may make the theory more accessible to first consider a physical phenomenon, which the theory aims to describe. One such phenomenon is molecular diffusion, which was historically a key motivation for the theory of stochastic differential equations.

Molecular diffusion is a transport process in fluids like air and water and even in solids. It is caused by the erratic and unpredictable motion of molecules which collide with other molecules. The phenomenon can be viewed at a *microscale*, where we follow the individual molecule, or at a *macroscale*, where it moves material from regions with high concentration to regions with low concentration (Figure 2.1).

The macroscale description of diffusion involves the concentration $C = C(x,t)$ of a substance and how it evolves in time. Here, x is the spatial position while t is the time; the concentration measures how many molecules there are in a given region. In this chapter, we derive and analyze the advection-diffusion equation which governs the concentration:

$$\frac{\partial C}{\partial t} = -\nabla \cdot (uC - D\nabla C).$$

In contrast, the microscale description of diffusion is that each single molecule moves according to a stochastic process, which is governed by a stochastic differential equation. It turns out that we can use the advection-diffusion equation to compute the statistics of this process. In turn, we can simulate the random motion of the molecule, which leads to Monte Carlo methods for analyzing the advection-diffusion equation. That is, there is a precise coupling between the microscale and the macroscale, between stochastic differential equations and partial differential equations.

DOI: 10.1201/9781003277569-2

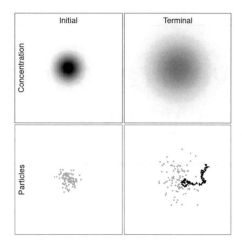

Figure 2.1 Diffusion in a container in two dimensions. *Top panels:* Concentration fields. *Bottom panels:* Position of 100 molecules. *Left panels:* Initially, the solute is concentrated at the center. *Right panels:* After some time, the solute is less concentrated. The bottom right panel also includes the trajectory of one molecule; notice its irregular appearance.

2.1 DIFFUSIVE TRANSPORT

In this section, we model how a substance spreads in space due to molecular diffusion. Think of smoke in still air, or dye in still water. The substance is distributed over a one-dimensional space **R**. Let $\mu_t([a, b])$ denote the amount of material present in the interval $[a, b]$ at time t. Mathematically, this μ_t is a *measure*. We may measure the substance in terms of number of molecules or moles, or in terms of mass, but we choose to let μ_t be dimensionless. We assume that μ_t admits a *density*, which is the concentration $C(\cdot, t)$ of the substance, so that the amount of material present in any interval $[a, b]$ can be found as the integral of the concentration over the interval:

$$\mu_t([a, b]) = \int_a^b C(x, t) \, dx.$$

The density has unit per length; if the underlying space had been two or three dimensional, then C would have unit per length squared or cubed, i.e., per area or per volume. The objective of this section is to pose a partial differential equation, the diffusion equation (2.3), which governs the time evolution of this concentration C.

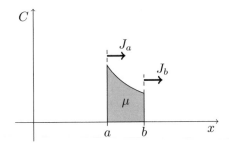

Figure 2.2 Conservation in one dimension. The total mass in the interval $[a, b]$ is $\mu_t([a, b]) = \int_a^b C(x, t)\, dx$, corresponding to the area of the shaded region. The net flow into the interval $[a, b]$ is $J(a) - J(b)$.

2.1.1 The Conservation Equation

We first establish the *conservation equation*

$$\frac{\partial C}{\partial t} + \frac{\partial J}{\partial x} = 0, \tag{2.1}$$

which expresses that mass is redistributed in space by continuous movements, but neither created, lost, nor teleported instantaneously between separated regions in space. To see that this equation holds, note that transport may then be quantified with a *flux* $J(x, t)$, which is the net amount of material that crosses the point x per unit time, from left to right. The flux has physical dimension "per time". Then, the amount of material in the interval $[a, b]$ is only changed by the net influx at the two endpoints, i.e.,

$$\frac{d}{dt}\mu_t([a, b]) = J(a, t) - J(b, t).$$

See Figure 2.2. Assume that the flux J is differentiable in x, then

$$J(a, t) - J(b, t) = -\int_a^b \frac{\partial J}{\partial x}(x, t)\, dx.$$

On the other hand, since μ_t is given as an integral, we can use the Leibniz integral rule to find the rate of change by differentiating under the integral sign:

$$\frac{d}{dt}\mu_t([a, b]) = \int_a^b \frac{\partial C}{\partial t}(x, t)\, dx.$$

Here, we assume that C is smooth so that the Leibniz integral rule applies. Combining these two expressions for the rate of change of material in $[a, b]$, we obtain:

$$\int_a^b \left[\frac{\partial C}{\partial t}(x, t) + \frac{\partial J}{\partial x}(x, t) \right] dx = 0.$$

Since the interval $[a, b]$ is arbitrary, we can conclude that the integrand is identically 0, or

$$\frac{\partial C}{\partial t}(x, t) + \frac{\partial J}{\partial x}(x, t) = 0.$$

This is known as the *conservation equation*. To obtain a more compact notation, we often omit the arguments (x, t), and we use a dot (as in $\dot{C}$) for time derivative and a prime (as in J') for spatial derivative. Thus, we can state the conservation equation compactly as

$$\dot{C} + J' = 0.$$

2.1.2 Fick's Laws

Fick's first law for diffusion states that the diffusive flux is proportional to the concentration gradient:

$$J(x, t) = -D\frac{\partial C}{\partial x}(x, t) \quad \text{or simply } J = -DC'. \tag{2.2}$$

This means that the diffusion will move matter from regions of high concentration to regions of low concentration. The constant of proportionality, D, is termed the diffusivity and has dimensions area per time (also when the underlying space has more than one dimension). The diffusivity depends on the diffusing substance, the background material it is diffusing in, and the temperature. See Table 2.1 for examples of diffusivities.

Fick's first law (2.2) is empirical but consistent with a microscopic model of molecule motion, as we will soon see. Combining Fick's first law with the conservation equation (2.1) gives Fick's second law, the diffusion equation:

$$\dot{C} = (DC')'. \tag{2.3}$$

This law predicts, for example, that the concentration will decrease at a peak, i.e., where $C' = 0$ and $C'' < 0$. In many physical situations, the diffusivity D is constant in space. In this case, we may write Fick's second law as

$$\dot{C} = DC'' \text{ when } D \text{ is constant in space,} \tag{2.4}$$

i.e., the rate of increase of concentration is proportional to the spatial curvature of the concentration. However, constant diffusivity is a special situation, and the general form of the diffusion equation is (2.3).

TABLE 2.1 Examples of Diffusivities

Process	Diffusivity $[\mathrm{m}^2/\mathrm{s}]$
Smoke particle in air at room temperature	2×10^{-5}
Salt ions in water at room temperature	1×10^{-9}
Carbon atoms in iron at 1250 K	2×10^{-11}

Biography: Adolph Eugen Fick (1829–1901)

A German pioneer in biophysics with a background in mathematics, physics, and medicine. Interested in transport in muscle tissue, he used the transport of salt in water as a convenient model system. In a sequence of papers around 1855, he reported on experiments as well as a theoretical model of transport, namely Fick's laws, which were derived as an analogy to the conduction of heat.

Exercise 2.1: For the situation in Figure 2.2, will the amount of material in the interval $[a, b]$ increase or decrease in time? Assume that (2.4) applies, i.e., the transport is diffusive and the diffusivity D is constant in space.

For the diffusion equation to admit a unique solution, we need an initial condition $C(x, 0)$ and spatial boundary conditions. Typical boundary conditions either fix the concentration C at the boundary, i.e., Dirichlet conditions, or the flux J. In the latter case, since the flux $J = uC - D\nabla C$ involves both the concentration C and its gradient ∇C, the resulting condition is of Robin type. In many situations, the domain is unbounded so that the boundary condition concerns the limit $|x| \to \infty$.

2.1.3 Diffusive Spread of a Point Source

We now turn to an important situation where the diffusion equation admits a simple solution in closed form: We take the spatial domain to be the entire real line **R**, we consider a diffusivity D which is constant in space and time, and we assume that the fluxes vanish in the limit $|x| \to \infty$. Consider the initial condition that one unit of material is located at position x_0, i.e.,

$$C(x, 0) = \delta(x - x_0),$$

where δ is the Dirac delta. The solution is then a Gaussian bell curve:

$$C(x, t) = \frac{1}{\sqrt{2Dt}} \phi\left(\frac{x - x_0}{\sqrt{2Dt}}\right). \tag{2.5}$$

Here, $\phi(\cdot)$ is the probability density function (p.d.f.) of a standard Gaussian variable,

$$\phi(x) = \frac{1}{\sqrt{2\pi}} \exp(-\frac{1}{2}x^2). \tag{2.6}$$

Thus, the substance is distributed according to a Gaussian distribution with mean x_0 and standard deviation $\sqrt{2Dt}$; see Figure 2.3. This standard deviation is a characteristic length scale of the concentration field which measures

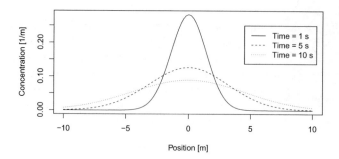

Figure 2.3 Diffusive spread. The concentration field at times $t = 1, 5, 10$ s with diffusivity $D = 1$ m^2/s and a unit amount of material, which initially is located at the point $x = 0$ m.

(half the) width of the plume; recall also that for a Gaussian, the standard deviation is the distance from the mean to inflection point. We see that length scales with the square root of time, or equivalently, time scales with length squared (Figure 2.4). This scaling implies that molecular diffusion is often to be considered a small-scale process: on longer time scales or larger spatial scales, other phenomena may take over and be more important. We will return to this point later, in Section 2.4.

Exercise 2.2: Insert solution (2.5) into the diffusion equation and verify that it satisfies the equation. In which sense does the solution also satisfy the initial condition?

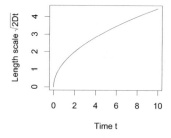

Figure 2.4 Square root relationship between time and diffusive length scale.

Exercise 2.3: Compute the diffusive length scale for smoke in air, and for salt in water, for various time scales between 1 second and 1 day.

Solution (2.5) is a *fundamental* solution (or Green's function) with which we may construct also the solution for general initial conditions. To see this, let $H(x, x_0, t)$ denote the solution $C(x, t)$ corresponding to the initial condition $C(x, 0) = \delta(x - x_0)$, i.e., (2.5). Since the diffusion equation is linear, a linear combination of initial conditions results in the same linear combination of solutions. In particular, we may write a general initial condition as a linear combination of Dirac deltas:

$$C(x, 0) = \int_{-\infty}^{+\infty} C(x_0, 0) \cdot \delta(x - x_0) \, dx_0.$$

We can then determine the response at time t from each of the deltas, and integrate the responses up:

$$C(x, t) = \int_{-\infty}^{+\infty} C(x_0, 0) \cdot H(x, x_0, t) \, dx_0. \tag{2.7}$$

Note that here we did not use the specific form of the fundamental solution; only linearity of the diffusion equation and existence of the fundamental solution. In fact, this technique works also when diffusivity varies in space and when advection is effective in addition to diffusion, as well as for a much larger class of problems. However, when the diffusivity is constant in space, we get a very explicit result, namely that the solution is the convolution of the initial condition with the fundamental solution:

$$C(x, t) = \int_{-\infty}^{+\infty} \frac{1}{(4\pi Dt)^{1/2}} \exp\left(-\frac{1}{2} \frac{|x - x_0|^2}{2Dt}\right) C(x_0, 0) \, dx_0.$$

2.1.4 Diffusive Attenuation of Waves

Another important situation which admits solutions in closed form is the diffusion equation (2.4) with the initial condition

$$C(x, 0) = \sin kx,$$

where k is a wave number, related to the wavelength L by the formula $kL = 2\pi$. In this case, the solution is

$$C(x, t) = \exp(-\lambda t) \sin kx \text{ with } \lambda = Dk^2. \tag{2.8}$$

Exercise 2.4: Verify this solution.

Thus, harmonic waves are eigenfunctions of the diffusion operator; that is, they are attenuated exponentially while preserving their shape. Note that the decay rate λ (i.e., minus the eigenvalue) is quadratic in the wave number

k. Another way of expressing the same scaling is that the half-time of the attenuation is

$$T_{1/2} = \frac{1}{\lambda} \log 2 = \frac{\log 2}{4\pi^2} \frac{L^2}{D},$$

i.e., the half time is quadratic in the wave length: Twice as long waves persist four times longer. See Figure 2.5. We recognize the square root/quadratic relationship between temporal and spatial scales from Figure 2.4.

Recall that we used the fundamental solution (2.5) to obtain the response of a general initial condition. We can do similarly with the harmonic solution (2.8), although we need to add the cosines or, more conveniently, use complex exponentials. Specifically, if the initial condition is square integrable, then it can be decomposed into harmonics as

$$C(x, 0) = \frac{1}{2\pi} \int_{-\infty}^{+\infty} \tilde{C}(k, 0) \, e^{ikx} \, dk,$$

where $\tilde{C}(k, 0)$ is the (spatial) Fourier transform

$$\tilde{C}(k, 0) = \int_{-\infty}^{+\infty} C(x, 0) \, e^{-ikx} \, dx.$$

Note that different authors use slightly different definitions of the Fourier transform. Now, each wave component $\exp(ikx)$ is attenuated to $\exp(-Dk^2t + ikx)$, so the Fourier transform of $C(x, t)$ is

$$\tilde{C}(k, t) = \tilde{C}(k, 0) \, e^{-Dk^2 t}.$$

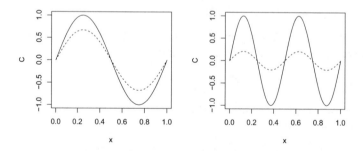

Figure 2.5 Attenuation of spatial waves by diffusion. Here the diffusivity is $D = 1$ and the terminal time is $T = 0.01$. *Solid lines:* At time 0, the amplitude is 1. *Dashed lines:* At time T, the wave is attenuated – how much depends on the wave number. *Left panel:* A long wave is attenuated slowly. *Right panel:* A shorter wave is attenuated more quickly.

We can now find the solution $C(x,t)$ by the inverse Fourier transform:

$$C(x,t) = \frac{1}{2\pi} \int_{-\infty}^{+\infty} \tilde{C}(k,0)\, e^{-Dk^2 t + ikx}\, dk.$$

One interpretation of this result is that short-wave fluctuations (large $|k|$) in the initial condition are smoothed out rapidly while long-wave fluctuations (small $|k|$) persist longer; the solution is increasingly dominated by longer and longer waves which decay slowly as the short waves disappear.

2.2   ADVECTIVE AND DIFFUSIVE TRANSPORT

In many physical situations, diffusion is not the sole transport mechanism: A particle with higher density than the surrounding fluid will have a movement with a downwards bias. If the fluid is flowing, then the particle will have a tendency to follow the flow (Figure 2.6). These situations both amount to a directional bias in the movement, so we focus on the latter.

Let the flow field be $u(x,t)$. If we use X_t to denote the position of a fluid element at time t, then X_t satisfies the differential equation

$$\frac{d}{dt}X_t = u(X_t, t).$$

Consider again a solute which is present in the fluid, and as before let $C(x,t)$ denote the concentration of the solute at position x and time t. If the material is a perfectly passive tracer (i.e., material is conserved and transported with the bulk motion of the fluid), then the flux of material is the advective flux:

$$J_A(x,t) = u(x,t)\, C(x,t).$$

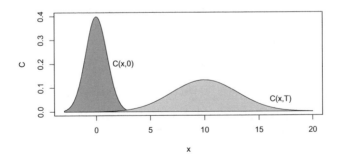

Figure 2.6   A fluid flow in one dimension which transports a substance. The plume is advected to the right with the flow; at the same time it diffuses out. The diffusivity is $D = 1$, the advection is $u = 2.5$, and the terminal time is $T = 4$.

If in addition molecular diffusion is in effect, then according to Fick's first law (2.2) this gives rise to a diffusive flux $J_D = -DC'$. We may assume that these two transport mechanisms operate independently, so that the total flux is the sum of the advective and diffusive fluxes:

$$J(x,t) = u(x,t)\ C(x,t) - D(x)\frac{\partial C}{\partial x}(x,t),$$

or simply $J = uC - DC'$. Inserting this into the conservation equation (2.1), we obtain the *advection-diffusion equation* for the concentration field:

$$\dot{C} = -(uC - DC')'. \tag{2.9}$$

A simple case is when u and D are constant, the initial condition is a Dirac delta, $C(x,0) = \delta(x - x_0)$, where x_0 is a parameter, and the flux vanishes as $|x| \to \infty$. Then the solution is:

$$C(x,t) = \frac{1}{\sqrt{4\pi Dt}}\exp\left(-\frac{(x - ut - x_0)^2}{4Dt}\right), \tag{2.10}$$

which is the probability density function of a Gaussian random variable with mean $x_0 + ut$ and variance $2Dt$. Advection shifts the mean with constant rate, as if there had been no diffusion, and diffusion gives rise to a linearly growing variance while preserving the Gaussian shape, as in the case of pure diffusion (i.e., diffusion without advection). This solution is important, but also a very special case: In general, when the flow is not constant, it will affect the variance, and the diffusion will affect the mean.

Exercise 2.5:

1. Verify the solution (2.10).

2. Solve the advection-diffusion equation (2.9) on the real line with constant u and D with the initial condition $C(x,0) = \sin(kx)$ or, if you prefer, $C(x,0) = \exp(ikx)$.

2.3 DIFFUSION IN MORE THAN ONE DIMENSION

Consider again the one-dimensional situation in Figure 2.2. In n dimensions, the interval $[a, b]$ is replaced by a region $V \subset \mathbf{R}^n$. Let $\mu_t(V)$ denote the amount of the substance present in this region. This measure can be written in terms of a volume integral of the density C:

$$\mu_t(V) = \int_V C(x,t)\ dx.$$

Here $x = (x_1, \dots, x_n) \in \mathbf{R}^n$ and dx is the volume of an infinitesimal volume element. The concentration C has physical dimension "per volume", i.e., SI unit m^{-n}, since $\mu_t(V)$ should still be dimensionless. The flux $J(x,t)$ is a vector field, i.e., a vector-valued function of space and time; in terms of coordinates, we have $J = (J_1, \ldots, J_n)$. The defining property of the flux J is that the net rate of exchange of matter through a surface ∂V is

$$\int_{\partial V} J(x,t) \cdot ds(x).$$

Here, ds is the surface element at $x \in \partial V$, a vector normal to the surface. The flux J has SI unit $m^{-n+1}s^{-1}$. Conservation of mass now means that the rate of change in the amount of matter present inside V is exactly balanced by the rate of transport over the boundary ∂V:

$$\int_V \dot{C}(x,t)\ dx + \int_{\partial V} J(x,t) \cdot ds(x) = 0, \tag{2.11}$$

where ds is directed outward. This balance equation compares a volume integral with a surface integral. We convert the surface integral to another volume integral, using the divergence theorem (the Gauss theorem), which equals the flow out of the control volume with the integrated divergence. Specifically,

$$\int_V \nabla \cdot J\ dx = \int_{\partial V} J \cdot ds.$$

In terms of coordinates, the divergence is $\nabla \cdot J = \partial J_1/\partial x_1 + \cdots + \partial J_n/\partial x_n$. Substituting the surface integral in (2.11) with a volume integral, we obtain

$$\int_V \left[\dot{C}(x,t) + \nabla \cdot J(x,t) \right]\ dx = 0.$$

Since the control volume V is arbitrary, we get

$$\dot{C} + \nabla \cdot J = 0, \tag{2.12}$$

which is the conservation equation in n dimensions, in differential form.

Fick's first law in n dimensions relates the diffusive flux to the gradient of the concentration field:

$$J = -D\nabla C,$$

where the gradient ∇C has coordinates $(\partial C/\partial x_1, \ldots, \partial C/\partial x_n)$. Often, the diffusivity D is a scalar material constant, so that the relationship between concentration gradient and diffusive flux is invariant under rotations. We then say that the diffusion is isotropic. However, in general D is a matrix (or a tensor, if we do not make explicit reference to the underlying coordinate system). Then, the diffusive flux is not necessarily parallel to the gradient, and its strength depends on the direction of the gradient. These situations can arise when the diffusion takes place in an anisotropic material, or when the diffusion is not molecular but caused by other mechanisms such as turbulence. Anisotropic diffusion is also the standard situation when the diffusion model does not describe transport in a physical space, but rather stochastic dynamics in a general state space of a dynamic system.

Fick's second law can now be written as

$$\dot{C} = \nabla \cdot (D\nabla C).$$

When the diffusivity is constant and isotropic, this reduces to $\dot{C} = D\nabla^2 C$. Here ∇^2 is the Laplacian $\nabla^2 = \frac{\partial^2}{\partial x_1^2} + \cdots + \frac{\partial^2}{\partial x_n^2}$, a measure of curvature.

To take advection into account, we assume a flow field u with coordinates $(u_1, \ldots, u_n)$. The advective flux is now uC and the advection-diffusion equation is

$$\dot{C} = -\nabla \cdot (uC - D\nabla C). \tag{2.13}$$

2.4 RELATIVE IMPORTANCE OF ADVECTION AND DIFFUSION

We have now introduced two transport processes, advection and diffusion, which may be in effect simultaneously. It is useful to assess the relative importance of the two.

Consider the solution (2.10) corresponding to constant advection u, constant diffusion D, and the initial condition $C(x,0) = \delta(x - x_o)$. At time t, the advection has moved the center of the plume a distance $|u|t$, while the diffusive length scale – the half width of the plume – is $\sqrt{2Dt}$. These length scales are shown in Figure 2.7 as functions of the time scale. Notice that initially, when time t is sufficiently small, the diffusive length scale is larger than the advective length scale, while for sufficiently large time t the advective length scale dominates. This justifies our earlier claim that diffusion is most powerful at small scales. The two length scales are equal when

$$t = \frac{2D}{u^2}.$$

In stead of fixing time and computing associated length scales, one may fix a certain length scale L and ask about the corresponding time scales associated with advective and diffusive transport: The advective time scale is L/u while

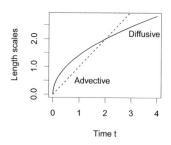

Figure 2.7 Advective (dashed) and diffusive (solid) length scale as function of time scale for the dimensionless model with $u = D = 1$.

the diffusive time scale is $L^2/2D$. We define the Péclet number as the ratio between the two:

$$\text{Pe} = 2\frac{\text{Diffusive time scale}}{\text{Advective time scale}} = \frac{Lu}{D}.$$

It is common to include factor 2 in order to obtain a simpler final expression, but note that different authors may include different factors. Regardless of the precise numerical value, a large Péclet number means that the diffusive time scale is larger than the advective time scale. In this situation, advection is a more effective transport mechanism than diffusion at the given length scale; that is, the transport is dominated by advection. Conversely, if the Péclet number is near 0, diffusion is more effective than advection at the given length scale. Such considerations may suggest to simplify the model by omitting the least significant term, and when done cautiously, this can be a good idea.

The analysis in this section has assumed that u and D were constant. When this is not the case, it is customary to use "typical" values of u and D to compute the Péclet number. This can be seen as a useful heuristic, but can also be justified by the non-dimensional versions of the transport equations, where the Péclet number enters. Of course, exactly which "typical values" are used for u and D can be a matter of debate, but this debate most often affects digits and not orders of magnitude. Even the order of magnitude of the Péclet number is a useful indicator if the transport phenomenon under study is dominated by diffusion or advection.

Example 2.4.1 *Consider the advection-diffusion equation (2.13) in two dimensions, where the flow $u(x)$ is around a cylinder. We non-dimensionalize space so that the cylinder is centered at the origin and has radius 1, and non-dimensionalize time so that the flow velocity far from the cylinder is 1. Then, the flow is, in polar coordinates (r, θ) with $x = r\cos\theta$, $y = r\sin\theta$,*

$$u_r(r, \theta) = (1 - r^{-2})\cos\theta, \quad u_\theta = -(1 + r^{-2})\sin\theta.$$

This is called irrotational flow *in fluid mechanics (Batchelor, 1967). A unit of material is released at time $t = 0$ at position $x = -3$, $y = -0.5$. We solve the advection-diffusion equation for $t \in [0, 2.5]$ for three values of the diffusivity D: $D = 1$, $D = 0.1$, and $D = 0.01$, leading to the three Péclet numbers 1, 10, and 100. Figure 2.8 shows the solution at time $t = 2.5$ for the three Péclet numbers. Notice how higher Péclet numbers (lower diffusivity D) imply a more narrow distribution of the material.*

2.5 THE MOTION OF A SINGLE MOLECULE

We can accept Fick's first equation as an empirical fact, but we would like to connect it to our microscopic understanding. In this section, we present a caricature of a microscopic mechanism which can explain Fickian diffusion: Each individual molecule moves in an erratic and unpredictable fashion, due

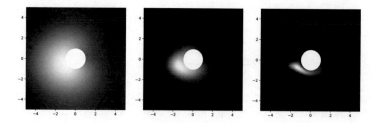

Figure 2.8 Concentrations $C(x, t = 2.5)$ from Example 2.4.1, for flow around a cylinder with diffusivities leading to different Péclet numbers. *Left panel:* Pe = 1. *Center panel:* Pe = 10. *Right panel:* Pe = 100. In each panel, the grayscale represents concentration relative to the maximum concentration.

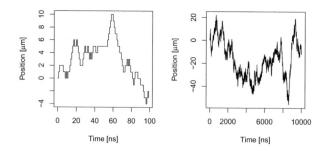

Figure 2.9 A random walk model of molecular motion. *Left:* A close-up where individual transitions are visible. *Right:* A zoom-out where the process is indistinguishable from Brownian motion.

to the exceedingly large number of collisions with other molecules, so that only a probabilistic description of its trajectory is feasible. This phenomenon is called Brownian motion.

Let $X_t \in \mathbf{R}$ denote the position at time t of a molecule, e.g., smoke in air, still considering one dimension only. At regularly spaced points of time, a time step h apart, the molecule is hit by an air molecule which causes a displacement $\pm k$. This happens independently of what has happened previously (Figure 2.9).[1]

---

[1] Physcially, collisions cause changes in velocity rather than position, but the simple picture is more useful at this point. We can argue that the velocity decays to 0 due to viscous friction and that the molecule drifts a certain distance during this decay. The simple model was used by Einstein (1905); extensions that take the velocity process into account were the Langevin (1908) equation and the Ornstein-Uhlenbeck process (Uhlenbeck and Ornstein, 1930) (Section 5.10).

Biography: Robert Brown (1773–1858)

A Scottish botanist who in 1827 studied pollen immersed in water under a microscope and observed an erratic motion of the grains. Unexplained at the time, we now attribute the motion to seemingly random collision between the pollen and water molecules. The physical phenomenon and its mathematical model are named Brownian motion, although many other experimentalists and theoreticians contributed to our modern understanding of the phenomenon.

In summary, the position $\{X_t : t \geq 0\}$ is a *random walk*:

$$X_{t+h} = \begin{cases} X_t + k & \text{w.p. (with probability) } p, \\ X_t & \text{w.p. } 1 - 2p, \\ X_t - k & \text{w.p. } p. \end{cases}$$

Here, $p \in (0, \frac{1}{2}]$ is a parameter. The displacement in one time step, $X_{t+h} - X_t$, has mean 0 and variance $2k^2p$. Displacements over different time steps are independent, so the central limit theorem applies. After many time steps, the probability distribution of the displacement X_{nh} will be well approximated by a Gaussian with mean 0 and variance $2k^2pn$:

$$X_{nh} \sim N(0, 2k^2pn). \tag{2.14}$$

That is, X_{nh} will (approximately) have the probability density function

$$\frac{1}{\sqrt{2k^2pn}} \cdot \phi(x/\sqrt{2k^2pn}),$$

where ϕ is still the p.d.f. of a standard Gaussian variable from (2.6). Next, assume that we release a large number N of molecules at the origin, and that they move independently. According to the law of large numbers, the number of molecules present between x_1 and x_2 at time nh will be approximately

$$N \int_{x_1}^{x_2} \frac{1}{\sqrt{2k^2pn}} \cdot \phi(x/\sqrt{2k^2pn}) \, dx.$$

Notice that this agrees with equation (2.5), assuming that we take $N = 1$, $D = k^2p/h$ and $t = nh$.

We see that our cartoon model of molecular motion is consistent with the results of the diffusion equation, if we assume that

Biography: Albert Einstein (1879–1955)

In his Annus Mirabilis, Einstein (1905) published not just on the photoelectric effect and on special relativity, but also on Brownian motion as the result of molecular collisions. His work connected pressure and temperature with the motion of molecules, gave credibility to statistical mechanics, and estimated the size of atoms. Soon confirmed in experiments by Jean Perrin, this work ended the debate whether atoms really exist.

1. molecules take many small jumps, i.e., h and k are small, so that the Central Limit Theorem is applicable at any length and time scale that our measuring devices can resolve, and

2. there is a large number of molecules present and they behave independently, so that we can ignore that their number will necessarily be integer, and ignore its variance.

Both of these assumptions are reasonable in many everyday situations where molecular systems are observed on a macroscale, for example, when breathing or making coffee.

To simulate the motion of a single molecule, if we only care about the displacements after many collisions, we may apply the approximation (2.14) recursively to get

$$X_t - X_s \sim N(0, 2D(t - s)), \text{ when } t > s.$$

Here we have assumed that the steps in time and space are consistent with the diffusivity, i.e., $D = k^2 p/h$. This process $\{X_t\}$ with independent and stationary Gaussian increments is called (mathematical) Brownian motion. Note that, physically, these properties should only hold when the time lag $t - s$ is large compared to the time h between molecular collisions, so mathematical Brownian motion is only an appropriate model of physical Brownian motion at coarse scale, i.e., for large time lags $t - s$. Mathematical Brownian motion is a fundamental process. It is simple enough that many questions regarding its properties can be given explicit and interesting answers, and we shall see several of these later, in Chapter 4.

In many situations, we choose to work with *standard* Brownian motion, where we take $2D = 1$ so that the displacement $B_{t+h} - B_t$ has variance equal to the time step h. When time has the physical unit of seconds s, notice that this means that B_t has the physical unit of $\sqrt{s}$!

2.6 MONTE CARLO SIMULATION OF PARTICLE MOTION

The previous section considered diffusion only. To add advection, the microscale picture is that each particle is advected with the fluid while subject to random collisions with other molecules which randomly perturb the particle. Thus, each particle performs a *biased* random walk. When the diffusivity D and the flow u are constant in space and time, the Gaussian solution (2.10) to the advection-diffusion equation applies. Then, a random walk model that is consistent with the advection-diffusion equation is to sample the increments $\Delta X = X_t - X_s$ from a Gaussian distribution

$$\Delta X \sim N(u \cdot (t - s), 2D(t - s)).$$

Now what if the flow $u = u(x, t)$ varies in space and time? To simulate the trajectory of a single particle, it seems to be a reasonable heuristic to divide the time interval $[0, T]$ into N subintervals

$$0 = t_0, t_1, \ldots, t_N = T.$$

We first sample X_0 from the initial distribution $C(\cdot, 0)$. Then, we sample the remaining trajectory recursively: At each sub-interval $[t_i, t_{i+1}]$, we approximate the flow field with a constant, namely $u(X_{t_i}, t_i)$. This gives us:

$$X_{t_{i+1}} \sim N(X_{t_i} + u(X_{t_i}) \cdot \Delta t_i, 2D \cdot \Delta t_i), \tag{2.15}$$

where $\Delta t_i = t_{i+1} - t_i$. It seems plausible that, as the time step in this recursion goes to 0, this approximation becomes more accurate so that the p.d.f. of X_t will approach the solution $C(\cdot, t)$ to the advection-diffusion equation (2.9). This turns out to be the case, although we are far from ready to prove it.

Example 2.6.1 (Flow past a cylinder revisited) *Example 2.4.1 and Figure 2.8 present solutions of the advection-diffusion equation for the case where the flow goes around a cylinder. In the introduction, Figure 1.2 (page 3) show a trajectory of a single molecule in the same flow. This trajectory is simulated with the recursion (2.15), using a Péclet number of 200.*

The Monte Carlo method, we have just described simulates the motion of single molecules, chosen randomly from the ensemble. Monte Carlo simulation can be used to compute properties of the concentration C, but can also answer questions that are not immediately formulated using partial differential equations; for example, concerning the time a molecule spends in a region. Monte Carlo methods are appealing in situations where analytical solutions to the partial differential equations are not available, and when numerical solutions are cumbersome, e.g., due to irregular geometries. Monte Carlo methods can also be useful when other processes than transport are active, for example, when chemical reactions take place on or between the dispersing particles.

2.7   CONCLUSION

Diffusion is a transport mechanism, the mathematical model of which involves the concentration field and the flux. Fick's laws tell us how to compute the flux for a given concentration field, which in turn specifies the temporal evolution of the concentration field. This is the *classical* approach to diffusion, in the sense of 19th century physics.

A microscopic model of diffusion involves exceedingly many molecules which each move erratically and unpredictably, due to collisions with other molecules. This statistical mechanical image of molecular chaos is consistent with the continuous fields of classical diffusion, but brings attention to the motion of a single molecule, which we model as a stochastic process, a so-called *diffusion process*. The probability density function associated with a single molecule is advected with the flow while diffusing out due to unpredictable collisions, in the same way the overall concentration of molecules spreads.

We can simulate the trajectory of a diffusing molecule with a stochastic recursion (Section 2.6). This provides a Monte Carlo particle tracking approach to solving the diffusion equation, which is useful in science and engineering: In each time step, the molecule is advected with the flow field but perturbed randomly from the streamline, modeling intermolecular collisions. This Monte Carlo method is particular useful in high-dimensional spaces or complex geometries where numerical solution of partial differential equations is difficult (and analytical solutions are unattainable).

Molecular diffusion is fascinating and relevant in its own right, but has even greater applicability because it serves as a reference and an analogy to other modes of dispersal; for example, of particles in turbulent flows, or of animals which move unpredictably (Okubo and Levin, 2001). At an even greater level of abstraction, a molecule moving randomly in physical space is a archetypal example of a dynamic system moving randomly in a general state space. When studying such general systems, the analogy to molecular diffusion provides not just physical intuition, but also special solutions, formulas and even software.

With the Monte Carlo approach to diffusion, the trajectory of the diffusing molecule is the focal point, while the classical focal points (concentrations, fluxes, and the advection-diffusion equation that connect them) become secondary, derived objects. This is the path that we follow from now on. In the chapters to come, we will depart from the physical notion of diffusion, in order to develop the mathematical theory of these random paths. While going through this construction, it is useful to have Figure 2.1 and the image of a diffusing molecule in mind. If a certain piece of mathematical machinery seems abstract, it may be enlightening to consider the question: How can this help describe the trajectory of a diffusing molecule?

Factbox: [The error function] The physics literature often prefers the "error function" to the standard Gaussian cumulative distribution function. The error function is defined as

$$\text{erf}(x) = \frac{2}{\sqrt{\pi}} \int_0^x e^{-s^2} \, ds,$$

and the complementary error function is

$$\text{erfc}(x) = 1 - \text{erf}(x) = \frac{2}{\sqrt{\pi}} \int_x^\infty e^{-s^2} \, ds.$$

These are related to the standard Gaussian distribution function $\Phi(x)$ by

$$\text{erfc}(x) = 2 - 2\Phi(\sqrt{2}x), \quad \Phi(x) = 1 - \frac{1}{2}\text{erfc}(x/\sqrt{2}).$$

$$\text{erf}(x) = 2\Phi(\sqrt{2}x) - 1, \quad \Phi(x) = \frac{1}{2} + \frac{1}{2}\text{erf}(x/\sqrt{2}).$$

2.8 EXERCISES

Exercise 2.6:

1. Solve the diffusion equation (2.4) on the real line with a "Heaviside step" initial condition

$$C(x,0) = \begin{cases} 0 & \text{when } x < 0, \\ 1 & \text{when } x > 0. \end{cases}$$

Use boundary conditions $\lim_{x \to +\infty} C(x,t) = 1$ and $\lim_{x \to -\infty} C(x,t) = 0$.

Hint: If you cannot guess the solution, use formula (2.7) and manipulate the integral into a form that resembles the definition of the cumulative distribution function.

2. Consider the diffusion equation (2.4) on the positive half-line $x \geq 0$ with initial condition $C(x,0) = 0$ and boundary conditions $C(0,t) = 1$, $C(\infty, t) = 0$.

Hint: Utilize the solution of the previous question, and the fact that in that question, $C(0,t) = 1/2$ for all $t > 0$.

Exercise 2.7: It is useful to have simple bounds on the tail probabilities in the Gaussian distribution, such as (Karatzas and Shreve, 1997):

$$\frac{x}{1+x^2}\phi(x) \leq 1 - \Phi(x) \leq \frac{1}{x}\phi(x).$$

which hold for $x \geq 0$. Here, as always, $\Phi(\cdot)$ is the c.d.f. of a standard Gaussian variable $X \sim N(0, 1)$, so that $1 - \Phi(x) = \mathbf{P}(X \geq x) = \int_x^\infty \phi(y) \, dy$ with $\phi(\cdot)$ being the density, $\phi(x) = \frac{1}{\sqrt{2\pi}} e^{-\frac{1}{2}x^2}$. A useful consequence is

$$1 - \Phi(x) = \phi(x) \cdot (x^{-1} + O(x^{-3})).$$

1. Plot the tail probability $1 - \Phi(x)$ for $0 \leq x \leq 6$. Include the upper and lower bound. Repeat, in a semi-logarithmic plot.

2. Show that the bounds hold. *Hint:* Show that the bounds hold as $x \to \infty$, and that the differential version of the inequality holds for $x \geq 0$ with reversed inequality signs.

Exercise 2.8: Consider pure diffusion in $n > 1$ dimensions with a scalar diffusion D, and a point initial condition $C(x, 0) = \delta(x - x_0)$, where $x \in \mathbf{R}^n$ and δ is the Dirac delta in n dimensions. Show that each coordinate can be treated separately, and thus, that the solution is a Gaussian in n dimensions corresponding to the n coordinates being independent, i.e.,

$$C(x, t) = \prod_{i=1}^n \frac{1}{\sqrt{2Dt}} \phi \left(\frac{e_i(x - x_0)}{\sqrt{2Dt}} \right) = \frac{1}{(4\pi Dt)^{n/2}} \exp \left(-\frac{1}{2} \frac{|x - x_0|^2}{2Dt} \right),$$

where e_i is the ith unit vector.

CHAPTER 3

Stochastic Experiments and Probability Spaces

To build the theory of stochastic differential equations, we need precise probabilistic arguments, and these require an axiomatic foundation of probability theory. This foundation is measure-theoretic and the topic of this chapter.

In science and engineering, probability is typically taught elementary, i.e., without measure theory. This is good enough for many applications, but it does not provide firm enough ground for more advanced topics like stochastic differential equations. One symptom of this is the existence of paradoxes in probability: Situations, or brain teasers, where different seemingly valid arguments give different results. Another symptom is that many elementary introductions to probability fail to give precise definitions, but in stead only offer synonyms such as "a probability is a likelihood".

The measure-theoretic approach to probability constructs a rigorous theory by considering stochastic experiments and giving precise mathematical definitions of the elements that make up such experiments: Sample spaces, realizations, events, probabilities, random variables, and information. To gain intuition, we consider simple experiments such as tossing coins or rolling dice, but soon we see that the theory covers also more complex experiments such as picking random functions.

At the end, we reach a construction which is consistent with the elementary approach, but covers more general settings. Therefore, there is no need to "unlearn" the elementary approach. The measure theoretic approach may seem abstract, but it gives precise mathematical meaning to concepts that not many people find intuitive. Once one has become familiar with the concepts, they can even seem natural, even if they are still technically challenging. A final argument in favor of the measure-theoretic approach is that the mathematical literature on stochastic differential equations is written in the language of measure theory, so the vocabulary is necessary for anyone working in this field, even if one's interest is applications rather than theory.

DOI: 10.1201/9781003277569-3

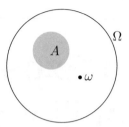

Figure 3.1 In a stochastic experiment, Chance picks an outcome ω from the sample space Ω. An event A is a subset of Ω and *occurs* or *is true* if $\omega \in A$. For the outcome ω in the figure, the event A did not occur.

3.1 STOCHASTIC EXPERIMENTS

The most fundamental concept in probability theory is the *stochastic experiment*. This is a mathematical model of real-world experiments such as rolling a die and observing the number of eyes.

Mathematically, a stochastic experiment involves a set containing all possible outcomes: The *sample space* Ω. We let ω denote an element in Ω; we call ω an outcome or a realization. For the die experiment, we set $\Omega = \{1, 2, 3, 4, 5, 6\}$. The stochastic experiment is that some mechanism, or the Goddess of Chance, selects one particular outcome ω from the sample space Ω (Figure 3.1).

A few other examples:

Statistical models: Statistical methods for data analysis are justified by postulating that the data has been produced by a stochastic experiment. For example, we may weigh an object n times, obtaining measurements $y_1, \ldots, y_n$, and postulating $y_i = \mu + e_i$. Here, μ is the (unknown) true weight of the object and e_i is the ith measurement error. We take the sample space to be $\Omega = \mathbf{R}^n$ and identify the outcome $\omega \in \Omega$ with the measurement errors, i.e., $\omega = (e_1, \ldots, e_n)$.

Monte Carlo simulation: In computer simulations of stochastic experiments, we rely on a random number generator which, ideally, produces an infinite series of independent and identically distributed numbers $\{Z_i \in \mathbf{R} : i \in \mathbf{N}\}$. Hence, the sample space is the set of sequences of real numbers, $\Omega = \mathbf{R}^{\mathbf{N}}$. If we take into account that computers only produce *pseudo*-random numbers, we may identify the realization ω with the *seed* of the random number generator; this allows to repeat the stochastic experiment, picking the *same* realization ω.

Diffusion and Brownian motion: In Chapter 2, we considered the experiment of releasing a large number of molecules; say, dye in water. A stochastic experiment is to pick one random molecule and record how its position changes in time. A mathematical model of this, in one dimension, is Brownian motion (Chapter 4), where the sample space is $\Omega = C(\bar{\mathbf{R}}_+, \mathbf{R})$, the set of continuous real-valued functions defined on $[0, \infty)$. We identify the randomly chosen continuous function $\omega \in \Omega$ with the trajectory of the molecule $\{B_t : t \geq 0\}$.

Once the sample space Ω is defined, we next need *events*. Events are statements about the outcome, such as "The die showed an even number" or "the molecule hits $x = 1$ before time $t = 1$". Once the experiment has been performed, the statement is either true or false. Mathematically, an event A is a subset of Ω containing those outcomes for which the statement is true. For example:

- For the die experiment, the statement "The die shows an even number" is the event $\{2, 4, 6\}$.

- For the linear regression model, an event is a subset of $\mathbf{R}^n$. One example is the event "all the measurement errors are positive", corresponding to $\mathbf{R}^n_+$.

This brings us to *probabilities*: The point of the stochastic model is to assign probabilities to each event. For the die-tossing experiment, if the die is fair, then $\mathbf{P}(\{2, 4, 6\}) = 1/2$, for example. This probability can be interpreted in different ways. The *frequentist* view is that if we toss the die repeatedly, we will eventually observe that the die has shown 2, 4 or 6 in half the tosses. The *subjective Bayesian* view is that we subjectively believe the event $\{2, 4, 6\}$ to be as probable as the alternative, $\{1, 3, 5\}$, and aims for consistency in such subjective beliefs. From an applied and pragmatic point of view, in some situations the frequentist view is justified, while in others the Bayesian view is more appropriate; even other interpretations exist. Fortunately, the mathematical construction in the following applies regardless of the interpretation.

Now the vocabulary is in place – sample space, outcomes, events, probabilities – we need to specify the mathematical properties of these objects. First, which events do we consider? For the example of the die, it is simple: Any subset of Ω is allowed, including the empty set and Ω itself. Moreover, if the die is fair, the probability of an event depends only on the number of elements in it, $\mathbf{P}(A) = |A|/|\Omega|$.

For the statistical model, we could start by trying to make an event out of each and every subset A of $\mathbf{R}^n$. In an elementary course on probability and statistics, we would maybe postulate a probability density function $f(e)$ for each measurement error and claim independence such that

$$\mathbf{P}(A) = \int_A f(e_1) \cdots f(e_n) \, de_1 \cdots de_n.$$

Unfortunately, it turns out that there are subsets A of $\mathbf{R}^n$ that are so pathological that this integral is not defined – even for $n = 1$ and regardless of how regular f is. You can accept this as a curious mathematical fact, or you can look up the "Vitali set" in (Billingsley, 1995) or on Wikipedia. We need to exclude such "non-measurable" subsets of $\mathbf{R}^n$: They do not correspond to events.

In elementary probability and statistics, we tend to ignore non-measurable subsets of $\mathbf{R}^n$, which have little relevance in applications. However, for more complicated sample spaces Ω, which appear in the study of stochastic processes and in particular stochastic differential equations, this difficulty cannot be ignored: Not all subsets of the sample space can be events, and we often need careful analysis to determined which ones are.

When not every set $A \subset \Omega$ can be an event, which ones should be? Some events are required for the theory to be useful. For example, in the scalar case $\Omega = \mathbf{R}$, we want intervals to correspond to events, so that the statement $\omega \in [a, b]$ is an event for any a and b. Moreover, it is imperative that our machinery of logic works: If A is an event, then the complementary set A^c must also be an event, so that the statement "not A" is valid. Next, if also B is an event, then the intersection $A \cap B$ must also be an event, so that the statement "A and B" is valid. More generally, in stochastic processes we often consider infinite sequences of events, for instance when analyzing convergence. So if $\{A_i : i \in \mathbf{N}\}$ is a sequence of events, then the statement "for each integer i, the statement A_i holds" should be valid. In terms of subsets of sample space Ω, this means that $A_1 \cap A_2 \cap \ldots$ must be an event.

Let $\mathcal{F}$ denote the collection of events which we consider. Mathematically speaking, the requirements on $\mathcal{F}$ that we have just argued for, means that $\mathcal{F}$ is a σ-algebra:

Definition 3.1.1 (σ-algebra of events) *Given a sample space Ω, a σ-algebra $\mathcal{F}$ of events is a family of subsets of Ω for which:*

1. *The certain event is included, $\Omega \in \mathcal{F}$.*

2. *For each event $A \in \mathcal{F}$, the complementary set A^c is also an event, $A^c \in \mathcal{F}$.*

3. *Given a sequence of events $\{A_i \in \mathcal{F} : i \in \mathbf{N}\}$, it is an event that all A_i occur, i.e., $\cap_i A_i \in \mathcal{F}$.*

Given a sequence of events $\{A_i \in \mathcal{F} : i \in \mathbf{N}\}$, also the union $\cup_i A_i$ is an event. *Exercise: Verify this!*. We often say, for short, that σ-algebras are characterized by being closed under countable operations of union and intersection.

Example 3.1.1 (The Borel algebra) *A specific σ-algebra which we will encounter frequently in this book, is related to the case $\Omega = \mathbf{R}$. We previously argued that the intervals $[a, b]$ should be events for the theory to be useful in many situations. The smallest σ-algebra which contains the intervals is called the Borel-algebra and denoted $\mathcal{B}(\mathbf{R})$ or simply $\mathcal{B}$.*

The Borel algebra contains all open sets and all closed sets, as well as many others. This collection of sets is large enough to contain the sets one encounters in practice, and the fact that the Vitali set and other non-Borel sets exist is more an excitement to mathematicians than a nuisance to practitioners.

Biography: Félix Édouard Justin _Émile_ Borel (1871–1956)
The French mathematician Borel was one of the leading figures in the development of measure theory. He applied measure theory to real functions of real variables, as well as to the foundations of probability. He was also politically active, argued for a united Europe, was Minister of the Navy, and was imprisoned during World War II for assisting the Resistance.

In the case $\Omega = \mathbf{R}^n$, we require $\mathcal{F}$ to include (hyper)rectangles of the form $(a_1, b_1) \times (a_2, b_2) \times \ldots \times (a_n, b_n)$, for $a_i, b_i \in \mathbf{R}$ and also use the name Borel-algebra, $\mathcal{B}(\mathbf{R}^n)$, for the smallest σ-algebra that contains these hyper-rectangles. More generally, if Ω is a topological space (i.e., we have defined a system of open subsets of Ω), then the Borel algebra on Ω is the smallest σ-algebra of subsets of Ω that contain all the open sets.

It is important to notice that for a given sample space Ω, there can be several systems of events $\mathcal{F}$. In many applications there is an obvious choice, but in general, to specify a stochastic experiment, we must state not only what can happen (i.e., Ω) but also which questions we can ask (i.e., $\mathcal{F}$).

Having outcomes and events in place, we need to assign a probability $\mathbf{P}(A)$ to each event $A \in \mathcal{F}$. The way we do this must be consistent:

Definition 3.1.2 (Probability measure) _A measure $\mathbf{P}$ is a map $\mathcal{F} \mapsto [0, \infty]$ which is countably additive, i.e.: $\mathbf{P}(\cup_i A_i) = \sum_i \mathbf{P}(A_i)$ whenever $A_1, A_2, \ldots$ are mutually exclusive events ($A_i \cap A_j = \emptyset$ for $i \neq j$). A probability measure is a measure for which $\mathbf{P}(\Omega) = 1$._

This definition explains why we call the sets in $\mathcal{F}$ _measurable_: $\mathcal{F}$ consists exactly of those sets, for which the measure $\mathbf{P}(A)$ is defined. Notice that probability is additive only for _countable_ collections of sets, just as the σ-algebra $\mathcal{F}$ must only be closed under _countable_ unions of sets.

An event with probability 0 is called a _null event_. Conversely, if an event has probability 1, then we say that this event occurs _almost surely_ (a.s.) or _with probability 1_ (w.p. 1).

Example 3.1.2 _Consider the uniform distribution on $[0, 1)$; i.e., $\Omega = [0, 1)$, $\mathcal{F}$ is the Borel algebra on Ω, and probability corresponds to length, $\mathbf{P}([a, b]) = b - a$ for $0 \leq a \leq b < 1$. Now, the rational numbers form a countable set $\mathbf{Q}$, so the set $A = \Omega \cap \mathbf{Q}$ is measurable and has measure $P(A) = 0$: When sampling a real number uniformly from $[0, 1)$, the probability of getting a rational number is 0. At the same time, the rationals are dense in $[0, 1)$ (every real number can_

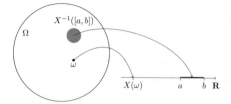

Figure 3.2 A real-valued random variable is a map $X : \Omega \mapsto \mathbf{R}$ such that the preimage $X^{-1}([a, b])$ of any interval $[a, b]$ is an event, i.e., an element in $\mathcal{F}$.

be approximated with a rational number with arbitrary accuracy). So almost no real numbers are rational, but every real number is almost rational.

To summarize: Our mathematical model of a stochastic experiment involves a sample space Ω, a family of events $\mathcal{F}$, and a probability measure $\mathbf{P}$, which all satisfy the assumptions in the previous. Together, the triple $(\Omega, \mathcal{F}, \mathbf{P})$ is called a *probability space* and constitutes the mathematical model of a stochastic experiment.

3.2 RANDOM VARIABLES

A random variable is a quantity which depends on the outcome of the stochastic experiment; mathematically, it is a function defined on Ω. In the real-valued case, we have $X : \Omega \mapsto \mathbf{R}$.

It is a great idea to define random variables as functions on sample space: We are very familiar with functions and have a large toolbox for their analysis, and now we can apply this machinery to random variables.

Just like there may be subsets of Ω which are not valid events, there may be functions $\Omega \mapsto \mathbf{R}$ which are not valid random variables. For example, let A be a non-measurable subset of Ω, i.e., $A \notin \mathcal{F}$, and take X to be the indicator function of A:

$$X(\omega) = \mathbf{1}_A(\omega) = \mathbf{1}(\omega \in A) = \begin{cases} 1 \text{ if } \omega \in A, \\ 0 \text{ else.} \end{cases}$$

Then the statement $X = 1$ (which is shorthand for $\{\omega \in \Omega : X(\omega) = 1\}$) is no event; we cannot assign a probability to it, zero or non-zero, in a meaningful way. This X is a real-valued function on sample space, but does not qualify to be a random variable.

To avoid such degenerate cases, we require that the statement "$X \in [a, b]$" corresponds to an event, for any a and b. Generalizing to the multidimensional case:

Definition 3.2.1 (Random variable) *A $\mathbf{R}^d$-valued random variable is a mapping $X : \Omega \mapsto \mathbf{R}^d$ such that*

$$\{\omega \in \Omega : X(\omega) \in B\} \in \mathcal{F}$$

for any Borel set $B \in \mathcal{B}(\mathbf{R}^d)$.

We say that X is a *measurable* function from $(\Omega, \mathcal{F})$ to $(\mathbf{R}^d, \mathcal{B}(\mathbf{R}^d))$. Notice that this definition concerns not just the domain Ω and the codomain $\mathbf{R}^d$, but also the σ-algebras on these sets. When discussing measurable functions, it is convenient to introduce the *preimage*:

Definition 3.2.2 (Preimage) *Given a function $X : \Omega \mapsto \mathbf{R}^d$, and a set $B \subset \mathbf{R}^d$, the preimage is*

$$X^{-1}(B) = \{\omega \in \Omega : X(\omega) \in B\}.$$

Note that the preimage of a function is different from the inverse of a function, although we use the same notation: The preimage maps subsets of $\mathbf{R}^d$ to subsets of Ω. In most cases in this book, X maps a high-dimensional sample space Ω to a low-dimensional space such as $\mathbf{R}$, so the function X will not be invertible. However, should the function X happen to be invertible, then the preimage of a singleton $\{x\}$ is a singleton $\{\omega\}$.

With the notion of preimage, we can say that a function X from $(\Omega, \mathcal{F})$ to $(\mathbf{R}^d, \mathcal{B}(\mathbf{R}^d))$ is measurable if $X^{-1}(B) \in \mathcal{F}$ for any $B \in \mathcal{B}(\mathbf{R}^d)$.

We can now define objects which are familiar from elementary probability. The cumulative distribution function (c.d.f.) of a real-valued random variable is

$$F_X(x) = \mathbf{P}\{X \leq x\} = \mathbf{P}\{\omega \in \Omega : X(\omega) \leq x\} = \mathbf{P}\{X^{-1}((-\infty, x])\}.$$

Note again the notation, where we often omit the ω argument; $\{X \leq x\}$ is a shorthand for $\{\omega \in \Omega : X(\omega) \leq x\}$. If F_X is differentiable and its derivative is continuous, then we define the probability density function $f_X(x)$ as this derivative:[1]

$$f_X(x) = \frac{dF_X}{dx}(x).$$

Once we have defined one random variable X, we can derive others from it. For example, X^2 is a random variable. In general, if $g : \mathbf{R} \mapsto \mathbf{R}$ is *Borel measurable*, then $g(X)$ is a random variable. Practically all the functions $g : \mathbf{R} \mapsto \mathbf{R}$ we encounter in applications are Borel measurable; for example, the piecewise continuous functions.

---

[1]F_X does not have to be C^1; absolute continuity suffices.

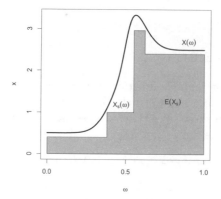

Figure 3.3 Expectation as integrals over sample space Ω. Here, $\Omega = [0, 1]$. A non-negative random variable X is bounded below by a simple random variable X_s. If the probability measure is uniform, then expectation corresponds to area under the curve; e.g., $\mathbf{E}X_s$ corresponds to the gray area.

3.3 EXPECTATION IS INTEGRATION

Recall that in the elementary (non-measure theoretic) approach to probability, we define expectation $\mathbf{E}X$ of a continuous random variable as an integral $\int x\, f_X(x)\, dx$ where f_X is the probability density; in the case of a discrete random variable, the integral is replaced with a sum. We now present the measure-theoretic definition of expectation, which is consistent with the elementary notion in both cases.

First, consider the case of a "simple" random variable X_s, i.e., one that attains a finite number of possible values $x_1, \ldots, x_n$. Then the elementary definition of the expectation is

$$\mathbf{E}X_s = \sum_{i=1}^{n} x_i \mathbf{P}(X_s = x_i)$$

and this definition is applicable in the measure-theoretic construction as well. If $\Omega \subset \mathbf{R}$ is an interval as in Figure 3.3, then the right hand side corresponds to an area which can be seen as the integral over Ω of a piecewise constant function $X_s : \Omega \mapsto \mathbf{R}$, so we can write

$$\mathbf{E}X_s = \int_{\Omega} X_s(\omega)\, d\mathbf{P}(\omega).$$

Importantly, we interpret this as an integral also when Ω is more complex than $\mathbf{R}^n$; e.g., when ω is a function or a sequence. Next, consider an arbitrary non-negative random variable X. Then we may construct a simple random variable X_s which is a lower bound on X, i.e., such that $0 \le X_s(\omega) \le X(\omega)$. We require that $\mathbf{E}X$ satisfies $\mathbf{E}X \ge \mathbf{E}X_s$. Also, we require that if X_s is a "good" approximation of X, then $\mathbf{E}X_s$ is "near" $\mathbf{E}X$. This leads us to define

$$\mathbf{E}X = \sup\{\mathbf{E}X_s : X_s \text{ simple}, 0 \le X_s \le X\}.$$

Recall that the supremum (sup) is the smallest upper bound. Note that the expectation $\mathbf{E}X$ is always defined and non-negative, but may equal $+\infty$.

Exercise 3.1: Show that the set $\{\mathbf{E}X_s : X_s \text{ simple}, 0 \le X_s \le X\}$ can be written as either $[0, c)$ or $[0, c]$ for some $0 \le c \le \infty$.

This procedure, where we approximate a non-negative function from below with simple functions, is also used in integration theory. The result is the *Lebesgue* integral of X over the sample space Ω with respect to the probability measure, so we can write:

$$\mathbf{E}X = \int_\Omega X(\omega) \, d\mathbf{P}(\omega).$$

Finally, for a random variable X which attains both positive and negative values, we define the positive part $X^+ = X \vee 0$ and the negative part $X^- = (-X) \vee 0$ (here, $\vee$ is the maximum operator: $a \vee b = \max(a, b)$). Note that X^+ and X^- are non-negative random variables and that $X = X^+ - X^-$. We now define the expectation

$$\mathbf{E}X = \mathbf{E}X^+ - \mathbf{E}X^- \quad \text{if } \mathbf{E}X^+ < \infty, \quad \mathbf{E}X^- < \infty.$$

We may state the condition that both positive and negative part have finite expectation more compactly: We require that $\mathbf{E}|X| < \infty$.

Notice that this construction of the integral of a function X over Ω does not rely on partitioning the domain Ω into ever finer subdomains, as e.g. the Riemann integral would do. This is crucial when outcomes $\omega \in \Omega$ are functions or sequences.

This definition of expectation has the nice properties we are used to from elementary probability, and which we expect from integrals:

Theorem 3.3.1

1. *(Linearity) Let $a, b \in \mathbf{R}$ and let X, Y be random variables with $\mathbf{E}|X| < \infty$, $\mathbf{E}|Y| < \infty$. Then $\mathbf{E}|aX + bY| < \infty$ and $\mathbf{E}(aX + bY) = a\mathbf{E}X + b\mathbf{E}Y$.*

2. *(Markov's inequality) Let X be a non-negative random variable and let $c \ge 0$. Then $\mathbf{E}X \ge c \cdot \mathbf{P}(X \ge c)$.*

3. *(Jensen's inequality) Let X be a random variable with $\mathbf{E}|X| < \infty$ and let $g : \mathbf{R} \mapsto [0, \infty)$ be convex. Then $\mathbf{E}g(X) \ge g(\mathbf{E}X)$.*

4. *(Fatou's lemma) Let $\{X_n : n \in \mathbf{N}\}$ be a sequence of non-negative random variables. Then $\mathbf{E} \liminf_{n \to \infty} X_n \leq \liminf_{n \to \infty} \mathbf{E} X_n$.*

Defining expectation as an integral has the convenience that it covers both the discrete case where $X(\Omega)$ is a finite or countable set, and the continuous case where e.g. $X(\Omega) = \mathbf{R}$, so there is no need to state every result in both a continuous version and a discrete version. The definition is also consistent with a Lebesgue-Stieltjes integral

$$\mathbf{E} X = \int_{-\infty}^{+\infty} x \, dF_X(x)$$

where we integrate, not over sample space Ω, but over the possible values of X, i.e., the real axis. Also this definition covers both the continuous and discrete case in one formula. A much more in-depth discussion of expectations and integrals over sample space can be found in e.g. (Williams, 1991) or (Billingsley, 1995).

Exercise 3.2: Define $\Omega = (0, 1]$, $\mathcal{F} = \mathcal{B}(\Omega)$, and let $\mathbf{P}$ be the uniform distribution on Ω. Let $G : [0, \infty) \mapsto [0, 1]$ be a continuous strictly decreasing function with $G(0) = 1$ and $G(x) \to 0$ as $x \to \infty$. Let $X(\omega) = G^{-1}(\omega)$.

1. Show that G is the complementary distribution function of X, i.e., $\mathbf{P}(X > x) = G(x)$. This result is useful for stochastic simulation: If we can simulate a uniform random variable, and invert the complementary distribution function, then we have a recipe for simulating a random variable with that distribution.

2. Show geometrically that

$$\mathbf{E} X = \int_\Omega X(\omega) \, \mathbf{P}(d\omega) = \int_0^\infty G(x) \, dx$$

by showing that the two integrals describe the area of the same set in the $(x, G(x))$ plane (or in the (X, ω) plane). This is a convenient way of computing expectations in some situations.

3. Extend the result to the case where G is merely nonincreasing right continuous and $X(\omega) = \sup\{x \in [0, \infty) : G(x) \geq \omega\})$.

3.4 INFORMATION IS A σ-ALGEBRA

Consider an observer who has *partial* knowledge about the outcome of a stochastic experiment. This is a very important situation in statistics as well as in stochastic processes. We say that the observer can *resolve* a given event $A \in \mathcal{F}$, if the observation always lets her know if the event has occurred or not.

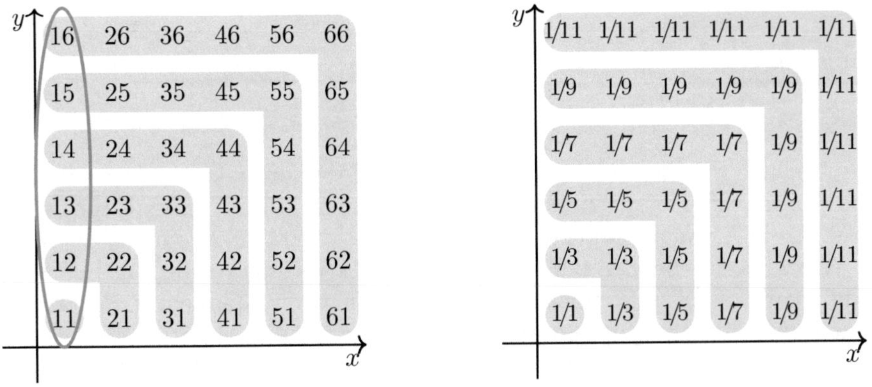

Figure 3.4 Information and conditional expectation when tossing two dice. *Left panel:* Numbers indicate the outcome ω. Light gray regions illustrate the information σ-algebra $\mathcal{H}$ generated by observing the maximum of the two dice: The open dark gray ellipsis contains the event "The first die shows one', which is not contained in $\mathcal{H}$. *Right panel:* Numbers indicate the conditional probability that the first die shows 1, conditional on $\mathcal{H}$.

For example, when tossing two dice, the sample space is $\Omega = \{1, \ldots, 6\}^2$; we will denote the outcomes $11, 12, \ldots, 16, 21, \ldots, 66$. See Figure 3.4, left panel. Now assume an observer does not see the dice, but is told the maximum of the two. Which events can she resolve? She will certainly know if the event $\{11\}$ is true; this will be the case iff (if and only if) the maximum is 1. Similar, the maximum being 2 corresponds to the event $\{12, 21, 22\}$. Generally, she can resolve the event $\{1z, z1, 2z, z2, \ldots, zz\}$ for any $z \in \{1, \ldots, 6\}$. There are certainly events which she cannot resolve, for example, the event that the first die shows 1: She will not generally know if this event is true; only if she is told that the maximum is 1.

Using the symbol $\mathcal{H}$ for *all* the events that the observer can resolve, we note that $\mathcal{H}$ will be a σ-algebra. For example, if she knows whether each of the events A and B occurred, then she also knows if $A \cap B$ occurred. The σ-algebra $\mathcal{H}$ will be contained in the original system of events $\mathcal{F}$. In summary:

> The information available to an observer is described by a set of events $\mathcal{H}$, which is a sub-σ-algebra to $\mathcal{F}$.

In most situations, the information $\mathcal{H}$ stems from observing a random variable:

Definition 3.4.1 (Information generated by a random variable) *Let* $X : \Omega \mapsto \mathbf{R}^d$ *be a random variable on a probability space* $(\Omega, \mathcal{F}, \mathbf{P})$ *and let*

$\mathcal{B}(\mathbf{R}^d)$ be the Borel algebra on $\mathbf{R}^d$. The information generated by X is the σ-algebra

$$\sigma(X) = \{A \in \mathcal{F} : A = X^{-1}(B) \text{ for some } B \in \mathcal{B}(\mathbf{R}^d)\}.$$

A related question is if the observer who holds information $\mathcal{H}$ always knows the realized value of a random variable X. This will be the case iff the event $X \in B$ is in $\mathcal{H}$, for any Borel set B. In that case we say that X is $\mathcal{H}$-measurable.

We will often be in situations where two observers have different information about a stochastic experiment. Then, we have two different σ-algebras $\mathcal{G}$ and $\mathcal{H}$, which are both sub-σ-algebras to $\mathcal{F}$. An extreme situation is that one observer's information is contained in the others. Let's say that one observer, Gretel, has measured X and therefore holds information $\mathcal{G} = \sigma(X)$ while another observer, Hans, has measured Y and holds information $\mathcal{H} = \sigma(Y)$. In which situations does Gretel know also Hans' observation Y? That will be the case if Y is $\sigma(X)$-measurable, in which case $\sigma(Y) \subset \sigma(X)$. A lemma due to Doob and Dynkin (see e.g. (Williams, 1991)) states that the first observer will know the realized value of Y, if and only if it is possible to compute Y from X. To be precise:

Lemma 3.4.1 (Doob-Dynkin) *Let* $X : \Omega \mapsto \mathbf{R}^m$ *and* $Y : \Omega \mapsto \mathbf{R}^n$ *be random variables on a probability space* $(\Omega, \mathcal{F}, \mathbf{P})$. *Then* Y *is measurable w.r.t.* $\sigma(X)$ *if and only if there exists a (Borel measurable) function* $g : \mathbf{R}^m \mapsto \mathbf{R}^n$ *such that* $Y(\omega) = g(X(\omega))$ *for all* $\omega \in \Omega$.

Maybe you think that this is unnecessary formalism; that a statement such as *"The observer has observed $Y = y$"* is sufficient. In this case, consider exercise 3.18, which is a slightly modified version of *Borel's paradox*.

3.5 CONDITIONAL EXPECTATIONS

What does an observer want to do with the obtained information? The basic use of information is to compute *conditional expectations* of random variables. We now aim to define conditional expectations, such as

$$\mathbf{E}\{X|\mathcal{H}\}$$

where X is a random variable on $(\Omega, \mathcal{F}, \mathbf{P})$ and $\mathcal{H}$ is a sub-σ-algebra to $\mathcal{F}$, describing the information.

Figure 3.4 (right panel) illustrates the situation for the case of tossing two dice and observing the maximum. In the figure, X is the indicator variable which takes the value 1 when the first die shows 1, and 0 otherwise. Then, $\mathbf{E}\{X|\mathcal{H}\}$ is the conditional probability that the first die shows one.

First, note that the conditional expectation $\mathbf{E}\{X|\mathcal{H}\}$ is a random variable: If we repeat the experiment, then the observer makes different observations and will therefore have a different expectation. In the figure, this corresponds to $\mathbf{E}\{X|\mathcal{H}\}$ being assigned a value for each outcome $\omega \in \Omega$.

Second, $\mathbf{E}\{X|\mathcal{H}\}$ must depend only on the information available to the observer. For example, the outcomes 12, 21, 22 all lead to an observed maximum of 2, so these three outcomes must lead to the same realized value of $\mathbf{E}\{X|\mathcal{H}\}$. In general, $\mathbf{E}\{X|\mathcal{H}\}$ must be measurable w.r.t. $\mathcal{H}$. When the information stems from a measurement of Y, i.e., $\mathcal{H} = \sigma(Y)$, then the Doob-Dynkin lemma tells us that there must exist some (measurable) function g such that $\mathbf{E}\{X|Y\} = g(Y)$: We must be able to compute the conditional expectation from the available data. Note that we allow the shorthand $\mathbf{E}\{X|Y\}$ for $\mathbf{E}\{X|\sigma(Y)\}$.

Third, we must specify the value of the random variable $\mathbf{E}\{X|\mathcal{H}\}$ for each ω. In Figure 3.4, we have used elementary probability. For example, if we observe a maximum of 2, then there are three outcomes consistent with that observation. In only one of these, the first die shows 1, and since the distribution is uniform, we get a conditional probability of $1/3$. In general, when $\mathcal{H} = \sigma(Y)$ and Y is discrete, we get

$$\mathbf{E}\{X|Y = y_i\} = \frac{\mathbf{E}\{X \cdot \mathbf{1}(Y = y_i)\}}{\mathbf{P}\{Y = y_i\}}.$$

The right hand side can be seen as averaging X only over that part of the sample space which is consistent with the observation $Y = y_i$. This expression defines the random variable $\mathbf{E}\{X|Y\}$ on the entire sample space Ω; it is constant on each event $Y^{-1}(y_i)$. The expression only makes sense because $\mathbf{P}\{Y = y_i\} > 0$, but if we multiply both sides with $\mathbf{P}\{Y = y_i\}$, we obtain an "integral" version which holds trivially also when $\mathbf{P}\{Y = y_i\} = 0$. Defining $g(y_i)$ as the right hand side, we use the identity $g(y_i) \cdot \mathbf{P}\{Y = y_i\} = \mathbf{E}\{g(Y) \cdot \mathbf{1}(Y = y_i)\}$ to obtain a more appealing form:

$$\mathbf{E}\{g(Y) \cdot \mathbf{1}(Y = y_i)\} = \mathbf{E}\{X \cdot \mathbf{1}(Y = y_i)\}.$$

This serves as our definition of conditional expectation with respect to *any* information σ-algebra $\mathcal{H}$:

Definition 3.5.1 *If X is a random variable on $(\Omega, \mathcal{F}, \mathbf{P})$ such that $\mathbf{E}|X| < \infty$, and $\mathcal{H} \subset \mathcal{F}$ is an information sub-σ-algebra, then the conditional expectation of X w.r.t. $\mathcal{H}$ is the random variable $Z = \mathbf{E}\{X|\mathcal{H}\}$ which is measurable w.r.t. $\mathcal{H}$, and for which*

$$\mathbf{E}\{Z \cdot \mathbf{1}_H\} = \mathbf{E}\{X \cdot \mathbf{1}_H\} \tag{3.1}$$

holds for any $H \in \mathcal{H}$.

It follows that $\mathbf{E}\{ZY\} = \mathbf{E}\{XY\}$ for any $\mathcal{H}$-measurable random variable Y such that $\mathbf{E}|XY| < \infty$. Figure 3.5 illustrates the situation in the continuous case $\Omega = \mathbf{R}^2$ where $\omega = (x, y)$, $X(\omega) = x$, $Y(\omega) = y$, and when the information $\mathcal{H}$ is obtained by observing Y: The conditional expectation $Z = \mathbf{E}\{X|Y\}$ is a random variable, i.e., a function defined on the plane. It is Y-measurable,

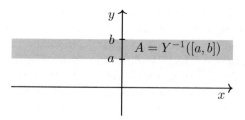

Figure 3.5 Conditional expectation w.r.t. $\sigma(Y)$: Horizontal strips $A = Y^{-1}([a,b]) = \{\omega = (x,y) : a \leq y \leq b\}$ generate $\sigma(Y)$ and are the typical elements in $\sigma(Y)$. The conditional expectation of X w.r.t. Y corresponds to averaging X w.r.t. **P** over such a thin horizontal strip.

i.e., constant along any horizontal line. Finally, the equation (3.1) says the integrals of Z and X over the horizontal strip in Figure 3.5 must coincide. Thus, Z is X averaged over horizontal lines.

Definition 3.5.1 hides an implicit theorem, namely that the conditional expectation is well defined in the sense that it exists and is unique. See (Billingsley, 1995) or (Williams, 1991). The conditional expectation is only "almost surely unique" since it is defined in terms of expectations, and therefore can be modified on a $\mathcal{H}$-measurable set of **P**-measure 0 and still satisfy the definition. So whenever we write an equation involving realizations of the conditional expectation, we should really add the qualification "almost surely". We do not do this. If the information stems from measurements of a continuous random variable Y such that $\mathcal{H} = \sigma(Y)$, then there may exist a *continuous* g such that $\mathbf{E}\{X|\mathcal{H}\} = g(Y)$; in this case, g is unique. This is reassuring, since from a modeller's perspective it would be worrying if conclusions depend discontinuously on an observed random variable, or are not uniquely defined. We will assume that g is chosen to be continuous whenever possible. This allows us to use the notation

$$\mathbf{E}\{X|Y = y\}$$

meaning "$g(y)$ where $g(Y) = \mathbf{E}\{X|Y\}$ (w.p. 1) and g is taken to be continuous".

Exercise 3.3: Consider again the stochastic experiment of tossing two dice and observing the maximum (Figure 3.4). Let Z be the conditional expectation of the first die given $\mathcal{H}$. Compute $Z(\omega)$ for each ω and display the results in a two dimensional table similar to Figure 3.4.

3.5.1 Properties of the Conditional Expectation

Some useful properties of conditional expectations are summarized in the following theorem.

Theorem 3.5.1 *Given a probability space* $(\Omega, \mathcal{F}, \mathbf{P})$, *a sub-$\sigma$-algebra $\mathcal{H}$ of $\mathcal{F}$, and random variables X and Y such that $\mathbf{E}|X| < \infty$ and $\mathbf{E}|Y| < \infty$. Then:*

1. $\mathbf{E}\{aX + bY | \mathcal{H}\} = a\mathbf{E}\{X | \mathcal{H}\} + b\mathbf{E}\{Y | \mathcal{H}\}$ *for* $a, b \in \mathbf{R}$ *(Linearity of conditional expectation)*

2. $\mathbf{E}\mathbf{E}\{X | \mathcal{H}\} = \mathbf{E}X$ *(The Law of Total Expectation).*

3. *Let $\mathcal{G}$ be a σ-algebra on Ω such that $\mathcal{F} \supset \mathcal{G} \supset \mathcal{H}$. Then* $\mathbf{E}[\mathbf{E}\{X | \mathcal{G}\} | \mathcal{H}] = \mathbf{E}\{X | \mathcal{H}\}$. *This "tower" property generalizes the law of total expectation.*

4. $\mathbf{E}\{X | \mathcal{H}\} = X$ *if and only if X is $\mathcal{H}$-measurable.*

5. $\mathbf{E}\{XY | \mathcal{H}\} = X\mathbf{E}\{Y | \mathcal{H}\}$ *whenever X is $\mathcal{H}$-measurable. ("Taking out what is known")*

The tower property deserves an explanation. Assume that Fred conducts a stochastic experiment and knows the outcome ω; his information is $\mathcal{F}$. He gives some information to both Hansel and Gretel, but also some information to Gretel only. So Gretel's information $\mathcal{G}$ contains Hansel's information $\mathcal{H}$, $\mathcal{F} \supset \mathcal{G} \supset \mathcal{H}$. Fred asks the two siblings to write down their expectations of the random variable X; they write $\mathbf{E}\{X | \mathcal{H}\}$ and $\mathbf{E}\{X | \mathcal{G}\}$, respectively. Fred then asks Hansel what he expects that Gretel wrote. According to the tower property, Hansel expects Gretel's result to be the same as his own; $\mathbf{E}[\mathbf{E}\{X | \mathcal{G}\} | \mathcal{H}] = \mathbf{E}\{X | \mathcal{H}\}$.

To show the tower property, define $Z = \mathbf{E}\{X | \mathcal{G}\}$. We claim that $\mathbf{E}\{Z | \mathcal{H}\} = \mathbf{E}\{X | \mathcal{H}\}$. The only thing to show is that

$$\mathbf{E}\{Z\mathbf{1}_H\} = \mathbf{E}\{X\mathbf{1}_H\}$$

for $H \in \mathcal{H}$. But since $\mathcal{G} \supset \mathcal{H}$, this H is also in $\mathcal{G}$: Any question that Hansel can answer, Gretel can also answer. So this equation follows from the definition of $\mathbf{E}\{X | \mathcal{G}\}$.

The proofs of the other claims are fairly straightforward and a good exercise.

3.5.2   Conditional Distributions and Variances

From the conditional expectation of a random variable we can define other conditional statistics, such as

1. The conditional probability of an event:

$$\mathbf{P}(A | \mathcal{H}) = \mathbf{E}\{\mathbf{1}_A | \mathcal{H}\}.$$

2. The conditional distribution function of a random variable

$$F_{X | \mathcal{H}}(x) = \mathbf{P}(X \le x | \mathcal{H}).$$

3. The conditional density of a random variable

$$f_{X|\mathcal{H}}(x) = \frac{d}{dx} F_{X|\mathcal{H}}(x),$$

wherever it exists.

4. The conditional variance of a random variable X such that $\mathbf{E}|X|^2 < \infty$:

$$\mathbf{V}\{X|\mathcal{H}\} = \mathbf{E}\{(X - \mathbf{E}[X|\mathcal{H}])^2|\mathcal{H}\} = \mathbf{E}\{X^2|\mathcal{H}\} - (\mathbf{E}\{X|\mathcal{H}\})^2.$$

These conditional statistics are all $\mathcal{H}$-measurable random variables. When $\mathcal{H}$ is generated by a random variable Y, each of these statistics will be functions of Y, in which case we can write e.g. $f_{X|Y}(x, y)$ for the conditional density of X at x given $Y = y$. When the involved distributions admit densities, we have the important relationship

$$f_{X,Y}(x, y) = f_Y(y)f_{X|Y}(x, y)$$

between the joint density $f_{X,Y}$, the marginal density f_Y, and the conditional density $f_{X|Y}$.

Exercise 3.4: Show that if X is $\mathcal{H}$-measurable and $\mathbf{E}|X|^2 < \infty$, then $\mathbf{V}\{X|\mathcal{H}\} = 0$.

It is useful to be able to manipulate conditional variances. Two fundamental formulas are the following: Let X and Y be random variables such that $\mathbf{E}|X|^2$, $\mathbf{E}|Y|^2$ and $\mathbf{E}|XY|^2$ all are finite. If furthermore Y is $\mathcal{H}$-measurable, then we can "take out what is known":

$$\mathbf{V}\{XY|\mathcal{H}\} = Y^2\mathbf{V}\{X|\mathcal{H}\}$$

and

$$\mathbf{V}\{X + Y|\mathcal{H}\} = \mathbf{V}\{X|\mathcal{H}\}.$$

These formulas generalize the well known formulas for $\mathbf{V}(aX) = a^2\mathbf{V}X$, $V(a + X) = \mathbf{V}X$ where a is a real constant. They can be understood in the way that given the information in $\mathcal{H}$, Y is known and can hence be treated as if deterministic.

We have a very useful decomposition formula for the variance, which is also known as the Law of Total Variance:

$$\mathbf{V}X = \mathbf{E}\mathbf{V}\{X|\mathcal{H}\} + \mathbf{V}\mathbf{E}\{X|\mathcal{H}\}. \tag{3.2}$$

In the next section, we shall see that this decomposition can be interpreted in terms of estimators and estimation errors. A generalization of this result is derived in Exercise 3.17.

Exercise 3.5: Verify the variance decomposition formula.

3.6 INDEPENDENCE AND CONDITIONAL INDEPENDENCE

When random variables are independent, it is a great simplification.

Definition 3.6.1 (Independence) *We say that events $A, B \in \mathcal{F}$ are independent if $\mathbf{P}(A \cap B) = \mathbf{P}(A)\,\mathbf{P}(B)$. We say that two σ-algebras $\mathcal{G}, \mathcal{H} \subset \mathcal{F}$ are independent if events G and H are independent whenever $G \in \mathcal{G}$ and $H \in \mathcal{H}$. We say that two random variables X, Y are independent if $\sigma(X)$ and $\sigma(Y)$ are independent.*

Thus, X and Y are independent iff $P\{X \in A, Y \in B\} = \mathbf{P}\{X \in A\}\mathbf{P}\{Y \in B\}$ for any Borel sets A, B. We use the symbol $\perp\!\!\!\perp$ for independence: $A \perp\!\!\!\perp B$, $\mathcal{G} \perp\!\!\!\perp \mathcal{H}$, $X \perp\!\!\!\perp Y$.

Theorem 3.6.1 *Let two random variables X and Y be independent and such that $\mathbf{E}|X| < \infty$, $\mathbf{E}|Y| < \infty$. Then $\mathbf{E}XY = \mathbf{E}X\,\mathbf{E}Y$. If also $\mathbf{E}X^2 < \infty$, $\mathbf{E}Y^2 < \infty$, then $\mathbf{V}(X + Y) = \mathbf{V}X + \mathbf{V}Y$.*

Proof: We leave this as an exercise. To show the result for the mean, start by assuming that X and Y are simple. ■

It is often possible to reach results swiftly using independence, in concert with the rules of expectation and variance that we have established. The following exercise is illustrative.

Exercise 3.6: Let $\{X_i : i \in \mathbf{N}\}$ be a collection of independent random variables, each Gaussian distributed with mean $\mu = 1$ and variance $\sigma^2 = 2$. Let N be a random variable, independent of all X_i, and Poisson distributed with mean $\lambda = 5$. Finally, define $Y = \sum_{i=1}^{N} X_i$. Determine the mean and variance of Y.

Independence is often too much to ask for; stochastic processes is all about dependence between random variables. Then, we can use the notion of *conditional independence*.

Definition 3.6.2 (Conditional independence) *We say that two events $A, B \in \mathcal{F}$ are conditionally independent given a σ-algebra $\mathcal{G} \subset \mathcal{F}$, if*

$$\mathbf{P}\{A \cap B | \mathcal{G}\} = \mathbf{P}\{A|\mathcal{G}\} \cdot \mathbf{P}\{B|\mathcal{G}\}$$

(almost surely). We say that two σ-algebras $\mathcal{H}, \mathcal{I} \subset \mathcal{F}$ are conditionally independent given $\mathcal{G}$ if any events $A \in \mathcal{H}$, $B \in \mathcal{I}$ are conditionally independent given $\mathcal{G}$. We say that two random variables X and Y are conditionally independent given $\mathcal{G}$, if $\sigma(X)$ and $\sigma(Y)$ are conditionally independent given $\mathcal{G}$.

It is convenient to depict dependence structures between random variables in a probabilistic graphical model (Figure 3.6); among other benefits, this can help establishing an overview of the random variables in a model. In such a graph, each node represents a random variable, while different conventions exist for the precise meaning of an edge. (Figure 3.6c). The graph may be

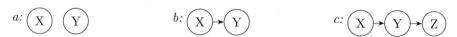

Figure 3.6 Probabilistic graphical models. *a:* Two random variables X and Y are independent. *b:* X and Y are not independent; the model specifies the marginal distribution of X and the conditional distribution of Y given X. *c:* X and Z are conditionally independent given Y.

directed (Figure 3.6*b*) to indicate for which random variables the marginal distribution is given, and for which random variables the conditional distribution is given. A directed acyclic graph (also called a *Bayesian network*) corresponds directly to computer code that simulates the random variables: In Figure 3.6*b*, Y is the result of a calculation that involves X (and a random number generator, if Y is not $\sigma(X)$-measurable). For such an acyclic graph, the joint distribution of all variables can be computed readily. For example, in Figure 3.6*c* we have the joint density

$$f_{X,Y,Z}(x,y,z) = f_X(x)\, f_{Y|X}(x,y)\, f_{Z|Y}(y,z) \tag{3.3}$$

whenever the distributions admit densities. Bayes' rule can be used to revert the information flow in the graph, e.g.,

$$f_{X|Y}(x,y) = \frac{f_{X,Y}(x,y)}{f_Y(y)} = \frac{f_{Y|X}(x,y)\, f_X(x)}{f_Y(y)}. \tag{3.4}$$

Thus, the joint density in Figure 3.6*c* can alternatively be written as

$$f_{X,Y,Z}(x,y,z) = f_Y(y)\, f_{X|Y}(x,y)\, f_{Z|Y}(y,z). \tag{3.5}$$

Mathematically, the two forms of $f_{X,Y,Z}$, (3.3) and (3.5), are equivalent, but they correspond to different directed graphs and thus different simulation algorithms, differing in whether X or Y is considered the *root node* which is simulated first.

Conditional independence simplifies estimation: In Figure 3.6c, say that we aim to estimate X based on Y and Z. Then we get the same result if we base the estimation only on Y. That is, it holds that $\mathbf{E}\{X|Y,Z\} = \mathbf{E}\{X|Y\}$. A slightly more general version of this statement is the following:

Theorem 3.6.2 *Let $\mathcal{G}$, $\mathcal{H}$, and $\mathcal{I}$ be sub-σ-algebras of $\mathcal{F}$ such that $\mathcal{G}$ and $\mathcal{I}$ are conditionally independent given $\mathcal{H}$. Let X be a $\mathcal{G}$-measurable random variable such that $\mathbf{E}|X| < \infty$. Then $\mathbf{E}\{X|\mathcal{H}\} = \mathbf{E}\{X|\mathcal{H},\mathcal{I}\}$.*

Proof: Define $Z = \mathbf{E}\{X|\mathcal{H}\}$; we aim to show that $Z = \mathbf{E}\{X|\mathcal{H},\mathcal{I}\}$. Clearly Z is measurable w.r.t $\sigma(\mathcal{H},\mathcal{I})$. To see that $\mathbf{E}\{X\mathbf{1}_K\} = \mathbf{E}\{Z\mathbf{1}_K\}$ holds for any $K \in \sigma(\mathcal{H},\mathcal{I})$, note that it suffices to show this for a K of the form $K = H \cap I$

where $H \in \mathcal{H}$ and $I \in \mathcal{I}$. We get

$$
\begin{aligned}
\mathbf{E}\{X\mathbf{1}_H\mathbf{1}_I\} &= \mathbf{E}\mathbf{E}\{X\mathbf{1}_H\mathbf{1}_I\}|\mathcal{H}\} \\
&= \mathbf{E}\left(\mathbf{E}\{X|\mathcal{H}\}\,\mathbf{1}_H\,\mathbf{P}\{I|\mathcal{H}\}\right) \\
&= \mathbf{E}\left(Z\mathbf{1}_H\mathbf{1}_I\right).
\end{aligned}
$$

Here, we have used first the law of total expecatation, then that X and I are conditionally independent given $\mathcal{H}$ and that $\mathbf{1}_H$ is $\mathcal{H}$-measurable, and the final equality comes from the definition of $\mathbf{P}\{I|\mathcal{H}\}$, since $Z\mathbf{1}_H$ is $\mathcal{H}$-measurable. ■

3.7   LINEAR SPACES OF RANDOM VARIABLES

Since random variables are functions defined on sample space, many standard techniques and results from analysis of functions apply to random variables.

Given a sample space Ω and a σ-algebra $\mathcal{F}$, all the random variables on $(\Omega, \mathcal{F})$ form a linear (vector) space. That is: If X_1 and X_2 are random variables defined on $(\Omega, \mathcal{F})$ and c_1 and c_2 are real numbers, then also $X : \Omega \mapsto \mathbf{R}$ given by

$$
X(\omega) = c_1 X_1(\omega) + c_2 X_2(\omega) \text{ for } \omega \in \Omega
$$

is a random variable. (See Exercise 3.10 for the measurability).

This linear space can be equipped with a norm. We focus on the $\mathcal{L}_p$ norms for $p > 0$:

$$
\|X\|_p = (\mathbf{E}|X^p|)^{1/p} = \left(\int_\Omega |X(\omega)|^p\,\mathbf{P}(d\omega)\right)^{1/p}
$$

which each define $\mathcal{L}_p(\Omega, \mathcal{F}, \mathbf{P})$, i.e., a linear space of those real-valued random variables for which the norm is finite. Letting $p \to \infty$, we obtain the $\mathcal{L}_\infty$-norm

$$
\|X\|_\infty = \operatorname*{ess\,sup}_{\omega \in \Omega} |X(\omega)|.
$$

The "ess" stands for "essential" and indicates that $X(\omega)$ is allowed to exceed $\|X\|_\infty$ on an event of probability 0.

Exercise 3.7: Show that if $p > q > 0$ and $X \in \mathcal{L}_p$, then $X \in \mathcal{L}_q$. Moreover, $\|X\|_q \leq \|X\|_p$. Give an example where $\|X\|_q = \|X\|_p$, and another example where $\|X\|_q \ll \|X\|_p$.

Of particular interest is the $\mathcal{L}_1$-norm; i.e., mean abs:

$$
\|X\|_1 = \mathbf{E}|X| = \int_\Omega |X(\omega)|\,\mathbf{P}(d\omega)
$$

and the $\mathcal{L}_2$-norm, i.e., root mean square:

$$
\|X\|_2 = \sqrt{\mathbf{E}X^2} = \sqrt{\int_\Omega X^2(\omega)\mathbf{P}(d\omega)}.
$$

For $X \in \mathcal{L}_2(\Omega, \mathcal{F}, \mathbf{P})$, both the mean and the mean square is finite, so $\mathcal{L}_2(\Omega, \mathcal{F}, \mathbf{P})$ consists of those variables which have finite variance, since

$$\mathbf{V}X = \mathbf{E}X^2 - (\mathbf{E}X)^2.$$

In many applications, the space $\mathcal{L}_2$ is large enough to contain all variables of interest, yet the space has many nice properties. Most importantly, the norm can be written in terms of an inner product, i.e., $\|X\|_2 = \sqrt{\langle X, X \rangle}$ where $\langle X, Y \rangle = \mathbf{E}XY$. This means that many results from standard Euclidean geometry applies, which is extremely powerful. For example, the *Schwarz inequality* applies to random variables X, Y such that $\mathbf{E}|X|^2 < \infty$, $\mathbf{E}|Y|^2 < \infty$:

$$|\mathbf{E}XY| \leq \mathbf{E}|XY| \leq \sqrt{\mathbf{E}X^2 \cdot \mathbf{E}Y^2}$$

or, in $\mathcal{L}_2$ terminology, $|\langle X, Y \rangle| \leq \langle |X|, |Y| \rangle \leq \|X\|_2 \cdot \|Y\|_2$. The Schwarz inequality implies that the covariance of two $\mathcal{L}_2$-variables is finite:

$$\mathbf{Cov}(X, Y) = \mathbf{E}XY - (\mathbf{E}X)(\mathbf{E}Y) = \mathbf{E}[(X - \mathbf{E}X)(Y - \mathbf{E}Y)]$$

and that the correlation coefficient $\mathbf{Cov}(X, Y)/\sqrt{\mathbf{V}X \, \mathbf{V}Y}$ is always in the interval $[-1, 1]$. An impressive use of this geometric view is in the following result, which relates mean-square estimation, conditional expectation, and orthogonal projection in $\mathcal{L}_2$:

Theorem 3.7.1 *Let X be a random variable in $\mathcal{L}_2(\Omega, \mathcal{F}, \mathbf{P})$ and let $\mathcal{H}$ be a sub-σ-algebra of $\mathcal{F}$. Use the conditional expectation as an estimator of X, i.e., define $\hat{X} = \mathbf{E}(X|\mathcal{H})$, and let $\tilde{X} = X - \hat{X}$ be the corresponding estimation error. Then:*

1. *The estimator is unbiased, i.e., $\mathbf{E}\tilde{X} = 0$ or equivalently $\mathbf{E}\hat{X} = \mathbf{E}X$.*

2. *The estimator and the estimation error are uncorrelated:*

$$\mathbf{E}\tilde{X}\hat{X} = 0.$$

3. *The mean square (or variance) of X can be decomposed in a term explained by $\mathcal{H}$, and a term unexplained by $\mathcal{H}$:*

$$\mathbf{E}|X|^2 = \mathbf{E}|\hat{X}|^2 + \mathbf{E}|\tilde{X}|^2, \qquad \mathbf{V}X = \mathbf{V}\hat{X} + \mathbf{V}\tilde{X}.$$

4. *$\hat{X}$ is the least squares estimator of X, i.e., if Z is any $\mathcal{L}_2$ random variable which is measurable w.r.t. $\mathcal{H}$, then $\mathbf{E}(Z - X)^2 \geq \mathbf{E}(\hat{X} - X)^2$.*

Proof: That the estimator is unbiased follows directly from the tower property:
$$\mathbf{E}\hat{X} = \mathbf{E}\mathbf{E}\{X|\mathcal{H}\} = \mathbf{E}X.$$

To show that estimator and estimation error are uncorrelated:

$$\mathbf{E}\tilde{X}\hat{X} = \mathbf{E}(X - \hat{X})\hat{X}$$
$$= \mathbf{E}\mathbf{E}\{(X - \hat{X})\hat{X}|\mathcal{H}\}$$
$$= \mathbf{E}\left[\hat{X}\mathbf{E}\{X - \hat{X}|\mathcal{H}\}\right]$$
$$= \mathbf{E}\left[\hat{X} \cdot 0\right]$$
$$= 0$$

since $\mathbf{E}\{\hat{X} - X|\mathcal{H}\} = \hat{X} - \mathbf{E}\{X|\mathcal{H}\} = 0$. The decomposition of 2-norms (or variance) follows directly from this orthogonality, and is essentially the variance decomposition we established in the previous section. Finally, let Z be $\mathcal{H}$-measurable. Then

$$\mathbf{E}(Z - X)^2 = \mathbf{E}((Z - \hat{X}) + (\hat{X} - X))^2$$
$$= \mathbf{E}(Z - \hat{X})^2 + \mathbf{E}(\hat{X} - X)^2 + 2\mathbf{E}[(Z - \hat{X})(\hat{X} - X)].$$

Adding and subtracting the candidate solution $\hat{X}$ is a standard trick, which appears on several occasions when working in $\mathcal{L}_2$, also in filtering and optimal control. Now, for the last term we have

$$\mathbf{E}[(Z - \hat{X})(\hat{X} - X)] = \mathbf{E}\mathbf{E}\{(Z - \hat{X})(\hat{X} - X)|\mathcal{H}\}$$
$$= \mathbf{E}\left[(Z - \hat{X})\mathbf{E}\{\hat{X} - X|\mathcal{H}\}\right]$$
$$= \mathbf{E}\left[(Z - \hat{X}) \cdot 0\right]$$
$$= 0,$$

since Z and $\hat{X}$ are $\mathcal{H}$-measurable. It follows that

$$\mathbf{E}(Z - X)^2 = \mathbf{E}(Z - \hat{X})^2 + \mathbf{E}(\hat{X} - X)^2 \geq \mathbf{E}(\hat{X} - X)^2$$

and we see that equality holds if and only if $Z = \hat{X}$ w.p. 1. ■

This is a projection result in the following sense: The information $\mathcal{H}$ defines a linear sub-space of $\mathcal{L}_2$, namely those random variables which are $\mathcal{H}$-measurable. The random variable X can now be decomposed into two orthogonal terms: The one, $\hat{X}$, resides in this linear sub-space, while the other, $\tilde{X}$, resides in the orthogonal complement. So $\mathbf{E}\{X|\mathcal{H}\}$ is the projection of X on the linear space of $\mathcal{H}$-measurable random variables.

The $\mathcal{L}_2$ theory of random variables is extremely powerful, and we will see that is central in the construction of stochastic differential equations. It is particularly operational in the case of multivariate Gaussian variables, where all computations reduce to linear algebra (Exercise 3.20). It is a great simplification that in this case, zero correlation implies independence. A word of caution is that some students forget that this hinges critically on the assumption of joint Gaussianity. Therefore, the following exercise is worthwhile:

Exercise 3.8:

1. Give an example of two random variables X and Y which are each standard Gaussian, such that X and Y are uncorrelated but not independent.

2. Show that X and Y are uncorrelated whenever $\mathbf{E}\{Y|X\}$ is constant and equal to $\mathbf{E}Y$. Next, show this condition is not necessary: Give an example of two random variables X and Y which are uncorrelated, but such that $E\{Y|X\}$ is not constant and equal to $\mathbf{E}Y$.

3.8 CONCLUSION

In the teaching of probability in science and engineering, it is an on-going debate when students should be introduced to measure theory, if at all. The elementary approach, without measure theory, is sufficient for many applications in statistics and in stochastic processes. The measure theoretic language, and train of thought, takes time to get used to and even more time to master! For most students in science and engineering, the time is better spent with issues that relate more directly to applications.

However, for continuous-time continuous-space processes, and in particular stochastic differential equations, the elementary approach is not firm enough. Here, we need the axioms and rigor of measure theory. Even students who focus on applications will one day need to read a journal article which uses the measure-theoretic language.

When you first encounter measure-theoretic probability, the property of measurability is often a source of confusion. In mathematical analysis, non-measurable sets and functions are esoteric phenomena; few people outside the mathematics departments know about them, and only few people inside mathematics departments actually study them. Subsets of **R** which are not Borel do not appear in applications. However, when *partial* information exists in the form of a sub-σ-algebra $\mathcal{H}$, then non-measurable events or random variables *w.r.t.* $\mathcal{H}$ are everyday phenomena. So, requiring a random variable to be $\mathcal{F}$-measurable is a technical condition which does not limit the applicability of the theory. On the other hand, if a given random variable X is $\mathcal{H}$-measurable, then this states that the observer who has access to the information $\mathcal{H}$ is also able to determine the realized value of X.

Another confusion arises when random variables have infinite variance, or the expectation is undefined because the mean-abs is infinite. From an applied point of view, the typical situation is that these moments are well defined and finite. There are notable exceptions; phenomena such as heavy tails and long-range dependence give rise to infinite moments and appear in turbulence, finance, and social sciences, just to name a few. It is useful to build intuition about distributions where the moments diverge, such as the Cauchy distribution and the Pareto distribution. However, it is not central to this book, where the typical situation is that the random variables we encounter

TABLE 3.1 A Summary of Technical Terms and Their Interpretation

Technical term	Interpretation
Basic σ-algebra $\mathcal{F}$	All events, i.e., all statements about the outcome of the stochastic experiment that we consider.
Information σ-algebra $\mathcal{H} \subset \mathcal{F}$	The information available to an observer; i.e., the "yes/no" questions that the observe can answer.
X is $\mathcal{H}$-measurable	The information in $\mathcal{H}$ is enough to determine the realized value of X.
$\mathcal{H} = \sigma(X)$	The information in $\mathcal{H}$ is (or could be) obtained by observing X.
$\mathcal{G} \subset \mathcal{H}$	Any question that can be answered with $\mathcal{G}$ can also be answered with $\mathcal{H}$.
$X \in \mathcal{L}_1$	X has finite mean.
$X \in \mathcal{L}_2$	X has finite mean square (and thus finite variance)
$X \perp Y$	X and Y are ($\mathcal{L}_2$ and) uncorrelated.
$X \perp\!\!\!\perp Y$	X and Y are independent.

have well defined mean and variance, even if it would be too restrictive to require it throughout. In contrast, the machinery of the space $\mathcal{L}_2(\Omega, \mathcal{F}, \mathbf{P})$ *is* central. For example, we will often first assume that all variances are well defined, carry through an argument using the $\mathcal{L}_2$ machinery, and finally expand the applicability so that it covers also the rare situations where variances are infinite.

In summary, for anyone working with stochastic differential equations, or advanced applications of probability, the measure-theoretic approach is fundemental. In this chapter, we have not constructed this foundation brick by brick; that would have required the better part of a book and the better part of a course. But at least we have introduced the language and outlined the principles, and this allows us to develop the theory of stochastic differential equations using the standard terminology (Table 3.1), which is measure-theoretic.

3.9 NOTES AND REFERENCES

A seminal account of the material in this chapter is (Kolmogorov, 1933); more modern and in-depth treatments are (Billingsley, 1995; Williams, 1991).

3.9.1 The Risk-Free Measure

In science and engineering, the frequentist and the Bayesian interpretations of probabilities are the most common. In finance, an additional notion is related to pricing: Markets are notoriously unpredictable and affected by e.g. harvests, production of wind energy, and the success of research and development projects. So it is natural to model markets as stochastic experiments with a sample space Ω and a family $\mathcal{F}$ of events. For any event $A \in \mathcal{F}$, imagine a contract where the issuer after the experiment pays the holder a fixed amount, say v, if event A occurs, and nothing else. What is the fair price of such a contract before the experiment? Importantly, ideal markets are *arbitrage free*, i.e., it is not possible to make a profit without taking a risk. The implies that the price of such a contract should be $v\mathbf{Q}(A)$ where $\mathbf{Q}$ is a probability measure on $(\Omega, \mathcal{F})$ – for example, if $\mathbf{Q}(A \cup B)$ were greater than $\mathbf{Q}(A) + \mathbf{Q}(B)$ for disjoint events A, B, then one could make a risk-free profit by selling a $A \cup B$-contract and buying an A-contract and a B-contract. Now, for a contract where the issuer pays the holder $X(\omega)$ after the experiment, the fair price is $\mathbf{E}X$ where expectation is with respect to $\mathbf{Q}$. Since pricing involves only expectation and not e.g. the variance, we say that the measure is risk-free. This measure $\mathbf{Q}$ is not identical to the "real-world" measure $\mathbf{P}$ a frequentist or a Bayesian would consider: Market prices are not expectations w.r.t. $\mathbf{P}$, but typically reward taking risks.

3.9.2 Convergence for Sequences of Events

This section and the next concern convergence, which is of paramount importance in the theory of stochastic differential equations. Questions of convergence can be technically challenging, in particular when one is not yet comfortable with the measure-theoretic apparatus. Here, we provide a collection of results which will be useful in the following chapters.

Let $A_1 \subset A_2 \subset \cdots$ be an increasing sequence of events in $\mathcal{F}$. Its limit is

$$\lim_{n\to\infty} A_n = \bigcup_{n\in\mathbf{N}} A_n.$$

For a decreasing sequence of events $A_1 \supset A_2 \supset \cdots$, we define the limit as $\cap_{n\in\mathbf{N}}A_n$. These limits are both measurable since $\mathcal{F}$ is a σ-algebra.

Lemma 3.9.1 *If $\{A_n : n \in \mathbf{N}\}$ is an increasing, or decreasing, sequence of events, then*

$$\mathbf{P}(\lim_{n\to\infty} A_n) = \lim_{n\to\infty} \mathbf{P}(A_n).$$

For any sequence of events $\{A_n : n \in \mathbf{N}\}$, we can establish an increasing sequence $\{I_n : n \in \mathbf{N}\}$

$$I_n = \bigcap_{i\geq n} A_i.$$

We define the limit of $\{I_n\}$ as the *limit inferior* of $\{A_n : n \in \mathbf{N}\}$:

$$\liminf_{n \to \infty} A_n = \bigcup_{n \in \mathbf{N}} I_n = \bigcup_{n \in \mathbf{N}} \bigcap_{i \geq n} A_i.$$

For a given outcome $\omega \in \Omega$, we say that the events $\{A_n\}$ occur *eventually* if $\omega \in \liminf_{n \to \infty} A_n$. This will be the case if there exists an m such that $\omega \in A_n$ for all $n \geq m$.

Similarly, from $\{A_n : n \in \mathbf{N}\}$ we can define a decreasing sequence of events

$$U_n = \bigcup_{i \geq n} A_i$$

and from this define the limit superior of $\{A_n\}$:

$$\limsup_{n \to \infty} A_n = \bigcap_{n \in \mathbf{N}} U_n = \bigcap_{n \in \mathbf{N}} \bigcup_{i \geq n} A_i.$$

For an outcome ω, we say that the events $\{A_n\}$ occur *infinitely often* if $\omega \in \limsup_{n \to \infty} A_n$. This will hold iff, for any $n \in \mathbf{N}$ exists an $i \geq n$ such that $\omega \in A_i$.

Is it easy to see that $\liminf_{n \to \infty} A_n \subset \limsup_{n \to \infty} A_n$ holds: Any sequence of events that occurs eventually, also occurs infinitely often.

If it holds that $\liminf_{n \to \infty} A_n = \limsup_{n \to \infty} A_n$, then we say that the sequence of events converges, and we define the limit

$$\lim_{n \to \infty} A_n = \liminf_{n \to \infty} A_n = \limsup_{n \to \infty} A_n.$$

Often, the event A_n is a "bad" event and we want to ensure that there is probability 0 that the events A_n occur infinitely often. Equivalently, the event A_n^c is "good" and we want to make sure that with probability 1 the events $\{A_n^c\}$ occur eventually. A first result is:

Lemma 3.9.2 *If* $\liminf_{n \to \infty} \mathbf{P}(A_n) = 0$, *then* $\mathbf{P}(\liminf_{n \to \infty} A_n) = 0$.

Proof: Let $B_m = \bigcap_{n \geq m} A_n$; then $\mathbf{P}(B_m) \leq \mathbf{P}(A_n)$ for all $n \geq m$. Since $\liminf \mathbf{P}(A_n) = 0$ it follows that $\mathbf{P}(B_m) = 0$ for all m. Hence also $\mathbf{P}(\bigcup_{m \in \mathbf{N}} B_m) = 0$ from which the conclusion follows. ■

However, the conclusion here is that A_n occurs *eventually* with probability 0; the condition $\mathbf{P}(A_n) \to 0$ does not rule out that the events A_n may occur infinitely often, i.e., that $\mathbf{P}(\limsup_{n \to \infty} A_n) > 0$. A standard example is:

Exercise 3.9: Let $\Omega = [0, 1)$, let $\mathcal{F}$ be the usual Borel algebra, and let the measure $\mathbf{P}$ be Lebesgue measure, i.e., the length. Then consider the sequence $\{A_n : n \in \mathbf{N}\}$ given by

$A_1 = [0/1, 1/1)$,

$A_2 = [0/2, 1/2)$, $A_3 = [1/2, 2/2)$,

$A_4 = [0/4, 1/4)$, $A_5 = [1/4, 2/4)$, $A_6 = [3/4, 4/4)$, $A_7 = [3/4, 4/4)$,

$\dots$

Show that for this sequence $\mathbf{P}(A_n) \to 0$ while $\mathbf{P}(\limsup A_n) = 1$.

In stead, a useful result is the (first) Borel-Cantelli lemma:

Lemma 3.9.3 (Borel-Cantelli I) *If* $\sum_{n=1}^{\infty} \mathbf{P}(A_n) < \infty$, *then* $\mathbf{P}(\limsup_{n\to\infty} A_n) = 0$.

Given this lemma, it is reasonable to ask what can be concluded if the sum $\sum_{n=1}^{\infty} \mathbf{P}(A_n)$ diverges. Without further assumptions, not much can be said. *Exercise: Construct an example of a sequence such that* $\sum \mathbf{P}(A_n)$ *diverges and where* $\liminf_{n\to\infty} A_n = \Omega$, *and one where* $\limsup_{n\to\infty} A_n = \emptyset$. However, if we add the requirement that the events are independent, then a much stronger conclusion can be drawn:

Lemma 3.9.4 (Borel-Cantelli II) *Let* A_n *be a sequence of* independent *events such that* $\sum_{n=1}^{\infty} \mathbf{P}(A_n) = \infty$. *Then* $\mathbf{P}(\limsup_{n\to\infty} A_n) = 1$.

3.9.3   Convergence for Random Variables

Throughout this book, we will often consider sequences of random variables and ask questions about their convergence. Since random variables are functions on Ω, there are several modes of convergence for random variables. The most common modes of convergence are given in the following:

Definition 3.9.1 *Let random variables* X *and* $\{X_i : i \in \mathbf{N}\}$ *be defined on a probability triple* $(\Omega, \mathcal{F}, \mathbf{P})$ *and take values in* $\mathbf{R}^n$. *We say that, as* $i \to \infty$,

1. $X_i \to X$ *almost surely (a.s.) if* $\mathbf{P}\{\omega : X_i(\omega) \to X(\omega)\} = 1$. *We also say* $X_i \to X$ *with probability 1 (w.p. 1).*

2. $X_i \to X$ *in* $\mathcal{L}_p$ *if* $\|X_i - X\|_p \to 0$. *This is equivalent to* $\mathbf{E}|X_i - X|^p \to 0$, *provided that* $p < \infty$. *When* $p = 1$ *we use the term* convergence in mean *and when* $p = 2$ *we say* convergence in mean square. *The cases* $p = 1$, $p = 2$ *and* $p = \infty$ *are the most common.*

3. $X_i \to X$ *in* probability *if, for any* $\epsilon > 0$:
$$\mathbf{P}(|X_i - X| > \epsilon) \to 0 \text{ as } i \to \infty.$$

4. $X_i \to X$ *in* distribution *(or in law or weakly) if*
$$\mathbf{P}(X_i \in B) \to \mathbf{P}(X \in B)$$
for any Borel set B *such that* $\mathbf{P}(X \in \partial B) = 0$.

Note that several of the definitions use the norm $|\cdot|$ in $\mathbf{R}^n$; recall that it does not matter which norm in $\mathbf{R}^n$ we use, since all norms on $\mathbf{R}^n$ are equivalent. So in a given situation, we may choose the most convenient norm. In most

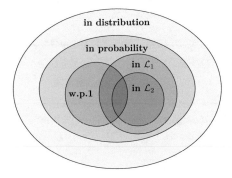

Figure 3.7 Modes of convergence for random variables, illustrated by sets of sequences. For example, a sequence which converges in probability also converges in distribution.

situations, this will either be the Euclidean 2-norm $|x|_2 = |x_1^2 + \cdots + x_n^2|^{1/2}$, the max norm $|x|_\infty = \max\{|x_1|, \ldots, |x_n|\}$, or the sum norm $|x|_1 = |x_1| + \cdots + |x_n|$.

Almost sure convergence corresponds to pointwise convergence, except possibly on a set of measure 0. Note that there may be realizations ω for which the convergence does not happen; there typically are.

Regarding convergence in distribution, the requirement that $\mathbf{P}(X \in \partial B) = 0$ cannot be disregarded: Consider for example the sequence $\{X_i\}$ with $X_i \sim N(0, i^{-2})$. Then $\mathbf{P}(X_i = 0) = 0$ for all i but the weak limit X has $\mathbf{P}(X = 0) = 1$. For scalar random variables, the requirement is that the distribution functions $F_i(x) = \mathbf{P}(X_i \leq x)$ converge pointwise to $F(x) = \mathbf{P}(X \leq x)$ at any point x where F is continuous.

As the following theorem states, the different modes of convergence are not completely independent (see also Figure 3.7).

Theorem 3.9.5 *Given a sequence $\{X_i : i \in \mathbf{N}\}$ of random variables, and a candidate limit X, on a probability space $\{\Omega, \mathcal{F}, \mathbf{P}\}$ and taking values in $\mathbf{R}^n$.*

1. *If $X_i \to X$ in $\mathcal{L}_p$ and $p \geq q > 0$, then $X_i \to X$ in $\mathcal{L}_q$. In particular, mean square ($\mathcal{L}_2$) convergence implies convergence in mean (in $\mathcal{L}_1$).*

2. *If $X_i \to X$ in $\mathcal{L}_p$ for $p > 0$, then $X_i \to X$ in probability.*

3. *If $X_i \to X$ almost surely, then $X_i \to X$ in probability.*

4. *If $X_i \to X$ in probability, then $X_i \to X$ in distribution.*

It is useful to think through situations where variables converge in one sense but not in another, because it illuminates the difference between the modes of convergence. For example:

Example 3.9.1

1. If X_i and X are all i.i.d, then $X_i \to X$ in distribution but in no other sense – convergence in distribution concerns the distributions only and not the random variables themselves.

2. Almost sure convergence does not in general imply convergence of moments: Consider a uniform distribution on $[0, 1)$, i.e., $\Omega = [0, 1)$, $\mathcal{F}$ the usual Borel algebra on $[0, 1)$, and $\mathbf{P}(B) = |B|$ for $B \in \mathcal{F}$. Now let $p > 0$ and

$$X_i(\omega) = i^{1/p} \cdot \mathbf{1}(\omega \in [0, 1/i)) = \begin{cases} i & \text{when } 0 \leq \omega \leq \frac{1}{i} \\ 0 & \text{else.} \end{cases}$$

Then $X_i \to 0$ w.p. 1, but $\|X_i\|_p = 1$ for all i.

Another example, which concerns Brownian motion and is related to stability theory, is the subject of Exercise 4.15.

3. In turn, convergence in moments does not in general imply almost sure convergence. A standard counter-example involves the same probability space and considers a sequence of random variables constructed as indicator variables of the sets in Exercise 3.9:

$$X_1 = \mathbf{1}_{[0,\frac{1}{1})},$$
$$X_2 = \mathbf{1}_{[0,\frac{1}{2})}, \quad X_3 = \mathbf{1}_{[\frac{1}{2},\frac{2}{2})},$$
$$X_4 = \mathbf{1}_{[0,\frac{1}{4}]}, \quad X_5 = \mathbf{1}_{[\frac{1}{4},\frac{2}{4})}, \quad X_6 = \mathbf{1}_{[\frac{2}{4},\frac{3}{4})}, \quad X_7 = \mathbf{1}_{[\frac{3}{4},\frac{4}{4})},$$
$$\ldots$$

Then $X_i \to 0$ in $\mathcal{L}_p$ for $1 \leq p < \infty$ and hence also in probability, but not with probability 1: For every ω and every $n \in \mathbf{N}$, there exists an $i > n$ such that $X_i(\omega) = 1$.

We will see another example of convergence in $\mathcal{L}_2$ but not w.p. 1 in the following, when discussing the Law of the Iterated Logarithm for Brownian motion (Theorem 4.3.4; page 74).

With extra assumptions, however, the converse statements hold:

- If X_i converges weakly to a deterministic limit, i.e., $X_i \to X$ where X is a constant function on Ω, then the convergence is also in probability.

- Monotone convergence: If $\{X_i(\omega) : i \in \mathbf{N}\}$, for each ω, is a non-negative non-decreasing sequence which converges to $X(\omega)$, then either $\mathbf{E}X_i \to \infty$ and $\mathbf{E}X = \infty$, or $X \in \mathcal{L}_1$ and $X_i \to X$ in $\mathcal{L}_1$.

- Dominated convergence: If there is a bound $Y \in \mathcal{L}_1(\Omega, \mathcal{F}, \mathbf{P})$, and random variables such that $|X_i(\omega)| \leq Y(\omega)$ for each ω and each i, and $X_i \to X$ almost surely, then $X \in \mathcal{L}_1$ and the convergence is in $\mathcal{L}_1$. If Y is constant, then this theorem is called the bounded convergence theorem.

- *Fast convergence:* If $X_i \to X$ in probability "fast enough", then the convergence is also almost sure. Specifically, if for all $\epsilon > 0$

$$\sum_{i=1}^{\infty} \mathbb{P}(|X_i - X| > \epsilon) < \infty,$$

then $X_i \to X$ almost surely. This follows from the first Borel-Cantelli lemma. This also implies that if $X_i \to X$ in $\mathcal{L}_p$, $1 \le p < \infty$, fast enough so that $\sum_{i=1}^{n} \mathbb{E}|X_i - X|^p < \infty$, then $X_i \to X$ almost surely.

- *Convergence of a subsequence:* If $X_i \to X$ in probability, then there exists an increasing subsequence $\{n_i : i \in \mathbf{N}\}$ such that $X_{n_i} \to X$ converges fast and hence also almost surely. For example, for the sequence in Example 3.9.1, item 3, the sequence $\{X_{2^i} : i \in \mathbf{N}\}$ converges to 0 almost surely.

Regarding convergence in $\mathcal{L}_p$, a situation that appears frequently is that we are faced with a sequence of random variables $\{X_n : n \in \mathbf{N}\}$ and aim to show that it converges to some limit X which is unknown to us. A useful property of the $\mathcal{L}_p$ spaces is that they are *complete*: If the sequence X_n has the Cauchy property that the increments tend to zero, i.e.

$$\sup_{m,n>N} \|X_m - X_n\|_p \to 0 \text{ as } N \to \infty$$

then there exists a limit $X \in \mathcal{L}_p$ such that $X_n \to X$ as $n \to \infty$. Recall (or prove!) that a convergent series necessarily is Cauchy; the word "complete" indicates that the spaces $\mathcal{L}_p$ include the limits of Cauchy sequences, so that a sequence is Cauchy if and only if it converges.

Some classical theorems concern convergence for averages:

Theorem 3.9.6 (Central limit theorem of Lindeberg-Lévy) *Let $\{X_i : i \in \mathbf{N}\}$ be a sequence of independent and identically distributed random variables with mean μ and variance $0 < \sigma^2 < \infty$. Then*

$$\frac{1}{\sigma\sqrt{n}} \sum_{i=1}^{n} (X_i - \mu) \to N(0,1) \text{ in distribution, as } n \to \infty.$$

Theorem 3.9.7 (Weak law of large numbers) *Let $\{X_i : i \in \mathbf{N}\}$ be a sequence of independent and identically distributed random variables with mean μ. Then*

$$\frac{1}{n} \sum_{i=1}^{n} X_i \to \mu \text{ in probability, as } n \to \infty.$$

Theorem 3.9.8 (Strong law of large numbers) *Let $\{X_i : i \in \mathbf{N}\}$ be a sequence of independent and identically distributed random variables with mean μ and variance $\sigma^2 < \infty$. Then*

$$\frac{1}{n} \sum_{i=1}^{n} X_i \to \mu \text{ almost surely and in } \mathcal{L}_2 \text{ as } n \to \infty.$$

3.10 EXERCISES

Fundamental Probability

Exercise 3.10: Consider the plane $\mathbf{R}^2$ with its Borel algebra $\mathcal{B}(\mathbf{R}^2)$. Show that the set $A = \{(x, y) \in \mathbf{R}^2 : x + y \leq c\}$ is Borel, for any $c \in \mathbf{R}$. *Hint:* The Borel algebra is built from rectangular sets, so show that A can be constructed from such rectangular sets using countably many operations.

Exercise 3.11 Independence vs. Pairwise Independence: Let X and Y be independent and identically distributed Bernoulli variables taking values on $\{-1, 1\}$ and with probability parameter $p = 1/2$, i.e., $\mathbf{P}(X = -1) = \mathbf{P}(X = 1) = 1/2$. Let $Z = XY$. Show that X, Y and Z are pairwise independent, but not all independent.

The Gaussian Distribution

Exercise 3.12 Simulation of Gaussian Variables: This is a clean and easy way to simulate from the Gaussian distribution. Let $\Omega = (0, 1] \times [0, 1)$, let $\mathcal{F}$ be the usual Borel algebra on Ω, and let $\mathbf{P}$ be the uniform measure, i.e., area. For $\omega = (\omega_1, \omega_2)$, define $\Theta(\omega) = 2\pi\omega_2$ and $S(\omega) = -2 \log \omega_1$.

1. Show that S is exponentially distributed with mean 2.

2. Define $X = \sqrt{S} \cos \Theta$, $Y = \sqrt{S} \sin \Theta$. Show that X and Y are independent and each are standard Gaussian. *Hint:* Brute force works; write up the p.d.f. of (S, Θ) and derive from that the p.d.f. of (X, Y). Alternatively, show first that there is a one-to-one mapping between the p.d.f. of (X, Y) and that of (S, Θ). Then show that if (X, Y) is i.i.d. and standard Gaussian, then (S, Θ) distributed as in this construction.

Note: There are (slightly) more computationally efficient ways to simulate Gaussians.

Exercise 3.13 Moments in the Gaussian Distribution: Consider a standard Gaussian variable, $X \sim N(0, 1)$. Show that the moments of X are given by the following formula:

$$\mathbf{E}|X|^p = \sqrt{2^p/\pi}\, \Gamma(\frac{p}{2} + \frac{1}{2})$$

Hint: Write up the integral defining the moments, use symmetry, and substitute $u = \frac{1}{2}x^2$. Recall the definition of the gamma function $\Gamma(x) = \int_0^\infty t^{x-1}e^{-t}\, dt$.

In particular, show that $\mathbf{E}|X| = \sqrt{2/\pi} \approx 0.798$, $\mathbf{E}|X|^2 = 1$, $\mathbf{E}|X|^3 = \sqrt{8/\pi}$, $\mathbf{E}|X|^4 = 3$, so that $\mathbf{V}(X^2) = 2$. Double-check these results by Monte Carlo simulation.

Exercise 3.14 Probability That Two Gaussians Have the Same Sign:
Let X, Y be two scalar random variables, jointly Gaussian distributed with
mean 0, variances σ_X^2, σ_Y^2 and covariance σ_{XY}. Show that the probability that
the two variables have the same sign is

$$\mathbf{P}\{XY > 0\} = \frac{1}{2} + \frac{1}{\pi} \arcsin \frac{\sigma_{XY}}{\sqrt{\sigma_X^2 \sigma_Y^2}}.$$

Hint: Write the vector (X, Y) as a linear combination of two independent
standard Gaussian variables (U, V), identify the set in the (u, v) plane for
which the condition is met, and compute the probability of this set using
rotational invariance.

Conditioning

Exercise 3.15 Conditional Expectation, Graphically: Consider the
probability space $(\Omega, \mathcal{F}, \mathbf{P})$ with $\Omega = [0, 1]^2$, $\mathcal{F}$ the usual Borel-algebra on
Ω, and $\mathbf{P}$ the Lebesgue measure, i.e., area.
   For $\omega = (x, y) \in \Omega$, define $X(\omega) = x$, $Y(\omega) = y$, and $Z(\omega) = x + y$.

1. Sketch level sets (contour lines) for X, Y, Z, $\mathbf{E}\{Z|X\}$ and $\mathbf{E}\{X|Z\}$.

2. Define and sketch (continuous) g and h such that $\mathbf{E}\{Z|X\} = g(X)$ and
   $\mathbf{E}\{X|Z\} = h(Z)$.

Exercise 3.16 The Tower Property: Fred rolls a die and observes the
outcome. He tells Gretel and Hansel if the number of eyes is odd or even.
He also tells Gretel if the number of eyes is greater or smaller than 3.5. He
then asks Gretel and Hansel to estimate the number of eyes (using conditional
expectation). For each outcome, what is Gretel's estimate? What is Hansel's
estimate? What is Hansel's estimate of Gretel's estimate? What is Gretel's
estimate of Hansel's estimate?

**Exercise 3.17 A Variance Decomposition; More Information Means
Less Variance, on Average:** Let X be a random variable on $(\Omega, \mathcal{F}, \mathbf{P})$
such that $\mathbf{V}X < \infty$, and let $\mathcal{H} \subset \mathcal{G}$ be sub-σ-algebras of $\mathcal{F}$.

1. Show that $\mathbf{V}\{X|\mathcal{H}\} = \mathbf{E}[\mathbf{V}\{X|\mathcal{G}\}|\mathcal{H}] + \mathbf{V}[\mathbf{E}\{X|\mathcal{G}\}|\mathcal{H}]$.

2. Show that $\mathbf{E}[\mathbf{V}\{X|\mathcal{G}\}|\mathcal{H}] \leq \mathbf{V}\{X|\mathcal{H}\}$.

3. Construct an example for which $\mathbf{V}\{X|\mathcal{G}\} > \mathbf{V}\{X|\mathcal{H}\}$ is an event with
   positive probability.

Exercise 3.18 Borel's Paradox: Continuing Exercise 3.12, show that

$$\mathbf{E}\{S|\Theta\} = 2 \text{ and } \mathbf{E}\{S|Y\} = Y^2 + 1.$$

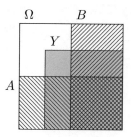

Figure 3.8 Venn diagram illustrating conditional independence. Probability corresponds to surface area. A and B are conditionally independent given Y, but not conditionally independent given $\neg Y$.

Conclude that $\mathbf{E}\{S|\Theta \in \{0, \pi\}\} = 2$ while $\mathbf{E}\{S|Y = 0\} = 1$, even if the event $\Theta \in \{0, \pi\}$ is the same as the event $Y = 0$. (Ignore the null event $S = 0$.) Discuss: What is the conditional expectation of S given that the point (X, Y) is on the x-axis?

Exercise 3.19 Conditional Independence: First, consider the Venn diagram in Figure 3.8. Show that A and B are conditionally independent given Y. Then show that A and B are not conditionally independent given $\neg Y$. Next, construct an example involving two events A and B which are conditionally independent given a σ-algebra $\mathcal{G}$, and where there is an event $G \in \mathcal{G}$ such that A and B are not conditionally independent given G. *Hint:* You may want to choose an example where $G = \Omega$; i.e., conditional independence does not imply unconditional independence.

Exercise 3.20: Show the following result: Let X and Y be jointly Gaussian distributed stochastic variables taking values in $\mathbf{R}^m$ and $\mathbf{R}^n$ with mean μ_X and μ_Y, respectively, and with

$$\mathbf{V}X = \Sigma_{xx}, \quad \mathbf{Cov}(X, Y) = \Sigma_{xy}, \quad \mathbf{V}Y = \Sigma_{yy}.$$

Assume $\Sigma_{yy} > 0$. Then

$$\mathbf{E}\{X|Y\} = \mu_X + \Sigma_{xy}\Sigma_{yy}^{-1}(Y - \mu_Y)$$

and

$$\mathbf{V}\{X|Y\} = \Sigma_{xx} - \Sigma_{xy}\Sigma_{yy}^{-1}\Sigma_{yx}.$$

Finally, the conditional distribution of X given Y is Gaussian.

Exercise 3.21 Conditional Expectations under Independence: Let X be a random variable on a probability space $(\Omega, \mathcal{F}, \mathbf{P})$ and let $\mathcal{H}$ be a sub-σ-algebra of $\mathcal{F}$ such that $\sigma(X)$ and $\mathcal{H}$ are independent. Show that $\mathbf{E}\{X|\mathcal{H}\} = \mathbf{E}X$ (*Hint:* Assume first that X is simple). Next, give an example where $\mathbf{E}\{X|\mathcal{H}\} = \mathbf{E}X$, but $\sigma(X)$ and $\mathcal{H}$ are not independent.

CHAPTER 4

Brownian Motion

In Chapter 2, we considered the physics of Brownian motion and its relationship to diffusion. In this chapter, we consider Brownian motion as a stochastic process, using the notions of measure theory of Chapter 3, and we describe its mathematical properties.

Brownian motion is a key process in the study of stochastic differential equations: First, it is the solution of the simplest stochastic differential equation, and therefore serves as the main illustrative example. Second, we will later use Brownian motion to generate noise that perturbs a general ordinary differential equation; this combination characterizes exactly a stochastic differential equation. These are two good reasons to study Brownian motion in detail.

Brownian motion as a stochastic process has several remarkable properties. Most importantly, it has independent increments, which implies that the variance of Brownian motion grows linearly with time. This humble statement has profound consequences; for example, the sample paths of Brownian motion are continuous everywhere but not differentiable at any point. The detailed study of Brownian motion in this chapter will later allow us to develop the stochastic calculus that concerns non-differentiable functions and governs stochastic differential equations.

Even if Brownian motion is a rich phenomenon, it has many simplifying properties. It is a Gaussian process, which means that many statistics can be computed analytically in closed form. It is also a self-similar process, which makes it is easier to analyze it. Perhaps even more importantly, Brownian motion is the prime example of two central classes of processes: The *martingales*, which formalize unbiased random walks, and the *Markov processes*, which are connected to the state space paradigm of dynamic systems. All these properties become useful in the construction of stochastic differential equations.

DOI: 10.1201/9781003277569-4

4.1 STOCHASTIC PROCESSES AND RANDOM FUNCTIONS

A stochastic process is an indexed collection of random variables

$$\{X_t : t \in \mathbf{T}\}$$

where t represents time and $\mathbf{T}$ is the time domain. We focus on the continuous-time case where $\mathbf{T}$ is the set of reals $\mathbf{R}$ or an interval of reals, and where the stochastic process takes values in $\mathbf{R}^n$. Thus, for each time $t \in \mathbf{T} \subset \mathbf{R}$, we have a random variable $X_t : \Omega \mapsto \mathbf{R}^n$.

Formulated differently, a stochastic process is a function which takes values in $\mathbf{R}^n$ and has two arguments: The realization $\omega \in \Omega$, and the time $t \in \mathbf{T}$:

$$X : \Omega \times \mathbf{T} \mapsto \mathbf{R}^n.$$

Thus, for fixed time t, we have the function $\omega \mapsto X_t(\omega)$, which by definition is a random variable, i.e. measurable. Conversely, for fixed realization $\omega \in \Omega$, we have the sample path $t \mapsto X_t(\omega)$. So a stochastic process specifies a way to sample randomly a function $\mathbf{T} \mapsto \mathbf{R}^n$, including a σ-algebra and a probability measure on the space of functions $\mathbf{T} \mapsto \mathbf{R}^n$.

Continuous-time stochastic processes involve an infinite number of random variables and therefore require a more complicated sample space Ω than do e.g. statistical models with only a finite number of variables. The sample space Ω may, for example, be a space of functions $\omega : \mathbf{T} \mapsto \mathbf{R}^n$. It is the complexity of these sample spaces that require and justify the rigor and precision of the measure-theoretic approach to probability.

4.2 DEFINITION OF BROWNIAN MOTION

Recalling Brownian motion as described in Section 2.5, we define:

Definition 4.2.1 *[Brownian motion] Let $\{B_t : t \geq 0\}$ be a stochastic process on some probability space $(\Omega, \mathcal{F}, \mathbf{P})$. We say that $\{B_t\}$ is Brownian motion, if it satisfies the following properties:*

1. *The process starts at $B_0 = 0$.*

2. *The increments of B_t are independent. That is to say, let time points $0 \leq t_0 < t_1 < t_2 < \ldots < t_n$ be given and let the corresponding increments be $\Delta B_i = B_{t_i} - B_{t_{i-1}}$ where $i = 1, \ldots, n$. Then these increments $\Delta B_1, \ldots, \Delta B_n$ are independent.*

3. *The increments are Gaussian with mean 0 and variance equal to the time lag:*

$$B_t - B_s \sim N(0, t - s)$$

   whenever $0 \leq s \leq t$.

4. *For almost all realizations ω, the sample path $t \mapsto B_t(\omega)$ is continuous.*

Biography: Norbert Wiener (1894–1964)
An American wonder kid who obtained his Ph.D. degree at the age of 19. His work on Brownian motion (1923) explains why this process is often referred to as the "Wiener process". During World War II, he worked on automation of anti-aircraft guns; this work lead to what is now known as the Wiener filter for noise removal. He fathered the theory of "cybernetics" which formalized the notion of feed-back, and stimulated work on artificial intelligence. Photo ©ShutterStock.

Sometimes we also use the word Brownian motion to describe the shifted-and-scaled process $\alpha B_t + \beta$ for $\alpha, \beta \in \mathbf{R}$. In that case we call the case $\alpha = 1$, $\beta = 0$ *standard* Brownian motion. Similarly, if $B_0 = \beta \neq 0$, then we speak of Brownian motion starting at β.

Although we now know the defining properties of Brownian motion, it is not yet clear if there actually exists a process with these properties. Fortunately, we have the following theorem:

Theorem 4.2.1 *Brownian motion exists. That is, there exists a probability triple $(\Omega, \mathcal{F}, \mathbf{P})$ and a stochastic process $\{B_t : t \geq 0\}$ which together satisfy the conditions in Definition 4.2.1.*

In many situations, we do not need to know what the probability triple is, but it can be illuminating. The standard choice is to take Ω to be the space $C([0, \infty), \mathbf{R})$ of continuous real-valued functions $\bar{\mathbf{R}}_+ \mapsto \mathbf{R}$, and to identify the realization ω with the sample path of the Brownian motion, i.e., $B_t(\omega) = \omega(t)$. The σ-algebra $\mathcal{F}$ is the smallest σ-algebra which makes B_t measurable for each $t \geq 0$. In other words, the smallest σ-algebra such that $\{\omega : a \leq \omega(t) \leq b\}$ is an event, for any choice of a, b and $t \geq 0$. The probability measure $\mathbf{P}$ is fixed by the statistics of Brownian motion. This construction is called *canonical* Brownian motion. The probability measure $\mathbf{P}$ on $C([0, \infty))$ is called Wiener measure, after Norbert Wiener.

This construction agrees with the interpretation we have offered earlier: Imagine an infinite collection of Brownian particles released at the origin at time 0. Each particle moves along a continuous trajectory; for each possible continuous trajectory there is a particle which follows that trajectory. Now pick one random of these particles. The statistical properties of Brownian motion specify what we mean with a "random" particle.

There are also other constructions, where we can generate the initial part of the sample path $\{B_t(\omega) : 0 \leq t \leq 1\}$ from a sequence of standard Gaussian variables $\{\xi_i(\omega) : i = 1, 2, \ldots\}$; for example, the *Brownian bridge* (next section and Exercise 4.11) and the *Wiener expansion* (Exercise 4.13).

4.3 PROPERTIES OF BROWNIAN MOTION

Physical Dimension and Unit of Brownian Motion

In the mathematical literature, we typically consider all variables dimensionless. However, in science and engineering it is tremendously important to keep track of dimensions and units, and sometimes a simple look at dimensions can supply quick answers to complicated questions. Moreover, countless errors in computations, designs and constructions can be traced back to wrong dimensions and units.

From our definition, the variance of standard Brownian motion equals time. This means that the dimension of Brownian motion must equal the square root of time. If we do computations in SI units, that means that the unit of B_t is $\sqrt{s}$, the square root of a second. Of course, in some applications we may prefer to measure time, e.g., in years, in which case the unit of B_t is the square root of a year.

If we construct a model that involves a particular process $\{X_t\}$, and wish to model this process as Brownian motion, then X_t typically comes with a dimension. For example, X_t may be a distance. In that case, we can write $X_t = \alpha B_t$ where $\{B_t\}$ is standard Brownian motion and the scale α has dimension length per square root of time; in fact we saw in Chapter 2 that $\alpha = \sqrt{2D}$, where D is the diffusivity measured in length squared per time.

Finite-Dimensional Distributions

A stochastic process involves an infinite number of random variables. To characterize its distribution, one must restrict attention to a finite number of variables. The so-called finite-dimensional distributions do exactly this. Take an arbitrary natural number n and then n time points $0 \le t_1 < t_2 < \cdots < t_n$, then we must specify the joint distribution of the vector random variable

$$\bar{B} = (B_{t_1}, B_{t_2}, \ldots, B_{t_n})$$

For Brownian motion, the distribution of this vector is Gaussian with mean

$$\mathbf{E}\bar{B} = (0, 0, \ldots, 0)$$

and covariance

$$\mathbf{E}\bar{B}^\top\bar{B} = \begin{bmatrix} t_1 & t_1 & \cdots & t_1 \\ t_1 & t_2 & \cdots & t_2 \\ \vdots & \ddots & \ddots & \vdots \\ t_1 & t_2 & \cdots & t_n \end{bmatrix}. \tag{4.1}$$

The expression for the covariance can be summarized with the statement that $\mathbf{Cov}(B_s, B_t) = s$ whenever $0 \le s \le t$; alternatively, $\mathbf{E}B_s B_t = s \wedge t = \min(s, t)$.

Exercise 4.1: Prove that this is the joint distribution of $\bar{B}$.

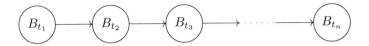

Figure 4.1 Probabilistic graphical model of Brownian motion evaluated at a set of time points $0 \leq t_1 < \cdots < t_n$.

Hint: Use the properties of the increments; show first that $\bar{B}$ is Gaussian, then find its mean, then the variance of each element, and finally the covariance of two elements.

A probabilistic graphical model of Brownian motion, evaluated at time points $0 < t_1 < t_2 < \cdots < t_n$, is seen in Figure 4.1. Note that B_{t_i} is, given its neighbors $B_{t_{i-1}}$ and $B_{t_{i+1}}$, conditionally independent of the rest, i.e. of $\{B_{t_j} : j = 1, \ldots, i-2, i+2, \ldots, n\}$. We will exploit this conditional independence throughout.

When all finite-dimensional distributions are Gaussian, we say that the process itself is Gaussian, so Brownian motion is a Gaussian process. Note that this requires much more than just that B_t is Gaussian for each t.

The finite-dimensional distributions contain much information about a process, even if not all (Exercise 4.22). A famous result is Kolmogorov's extension theorem, which says that if you prescribe the finite-dimensional distributions, then it is possible to construct a stochastic process which has exactly these finite-dimensional distributions.

Simulation

Simulation is a key tool in the study of stochastic processes. The following R-code from SDEtools defines a function rBM which takes as input a vector of time points, and returns a single sample path of Brownian motion, evaluated at those time points.

```
rBM <- function(t)
    cumsum(rnorm(n=length(t),
                 sd=sqrt(diff(c(0,t)))))
```

Exercise 4.2: Test the function by simulating sufficiently many replicates of $(B_0, B_{1/2}, B_{3/2}, B_2)$ to verify the covariance of this vector, and the distribution of B_2.

This way of simulating sample paths is sufficient for most applications, but sometimes it is useful to be able to refine the grid of time points. This can be done with the *Brownian bridge*, where we consider the law of Brownian motion conditional on the end points. First, recall the formula for conditioning in the Gaussian distribution (Exercise 3.20). This shows that, for $0 \leq s \leq t$,

$$\mathbf{E}\{B_s|B_t\} = B_t \, s/t, \quad \mathbf{V}\{B_s|B_t\} = s(1 - s/t).$$

That is, the conditional mean interpolates the two end points linearly, while the conditional variance is a quadratic function of time with an absolute slope of 1 at the two end points. This allows us to insert an additional time point in the grid – note that independence of increments implies that we only need to condition on the neighboring time points (Figure 4.1). Next, we can repeat this recursively until we have obtained the desired resolution in time. See also Exercise 4.11.

This construction has also theoretical implications: Since it allows us to simulate Brownian motion with arbitrary accuracy, it also provides a probability space on which we can define Brownian motion, as an alternative to the canonical construction involving the Wiener measure. To see this, recall (Chapter 3) that any Monte Carlo simulation model corresponds to a stochastic experiment, in which the sample space Ω is the space of sequences $\omega = (\omega_1, \omega_2, \ldots)$ of real numbers. Indeed, when simulating the Brownian bridge on $t \in [0, 1]$, with a random end point $B_1 \sim N(0, 1)$ and with increasing resolution, one obtains a sequence of random processes which converges in $\mathcal{L}_2$; the limit is Brownian motion.

Self-Similarity

Brownian motion is a self-similar process: if we rescale time, we can also rescale the motion so that we recover the original process. Specifically, if B_t is Brownian motion, then so is $\alpha^{-1} B_{\alpha^2 t}$, for any $\alpha > 0$. *Exercise: Verify this claim.* This means that Brownian motion itself possesses no characteristic time scales which makes it an attractive component in models. Notice that the rescaling is linear in space and quadratic in time, in agreement with the scaling properties of diffusion (Section 2.1.3 and, in particular, Figure 2.4).

A graphical signature of self-similarity is seen in Figure 4.2. The sample paths themselves are not self-similar, i.e., they each appear differently under the three magnifications. However, they are statistically indistinguishable. If the axis scales had not been provided in the figure, it would not be possible to infer the zoom factor from the panels.

A useful consequence of self-similarity is that the moments of Brownian motion also scale with time:

$$\mathbf{E}|B_t|^p = \mathbf{E}|\sqrt{t}B_1|^p = t^{p/2} \cdot \mathbf{E}|B_1|^p.$$

Numerical values can be found for these moments $\mathbf{E}|B_1|^p$ (Exercise 3.13), but in many situations the scaling relationships $\mathbf{E}|B_t|^p \sim t^{p/2}$ are all that is needed. To remember these scaling relationships, keep in mind that the physical dimension of Brownian motion is the square root of time, then the scalings follow from dimensional analysis.

The Total and Quadratic Variation

The increments $B_{t+h} - B_t$ are stationary (i.e., follow the same distribution regardless of $t \geq 0$ for fixed $h \geq 0$), have mean 0, and are independent for

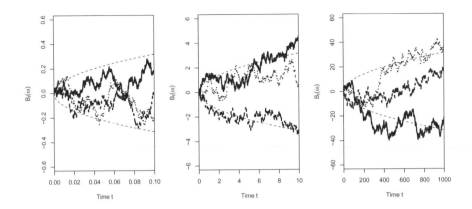

Figure 4.2 Self-similarity of Brownian motion. Three realizations of Brownian motion shown at three different magnifications. The standard deviation of B_t is included (thin dashed curves).

non-overlapping intervals. These properties are key in the analysis of Brownian motion. They also imply that the sample paths of Brownian motion are, although continuous, very erratic. This is evident from Figure 4.2. One mathematical expression of this feature is that Brownian motion has unbounded *total variation*. To explain this, consider Brownian motion on the interval $[0, 1]$, and consider a *partition* of this interval:

Definition 4.3.1 (Partition of an Interval) *Given an interval $[S, T]$, we define a partition Δ as an increasing sequence $S = t_0 < t_1 < \cdots < t_n = T$. For a partition Δ, let $\#\Delta$ be the number of sub-intervals, i.e., $\#\Delta = n$, and let the* mesh *of the partition be the length of the largest sub-interval, $|\Delta| = \max\{t_i - t_{i-1} : i = 1, \ldots, n\}$.*

Define the sum

$$V_\Delta = \sum_{i=1}^{\#\Delta} |B_{t_i} - B_{t_{i-1}}|. \tag{4.2}$$

We call this a discretized total variation associated with the partition Δ. We define the total variation of Brownian motion on the interval $[0, 1]$ as the limit in probability $V = \lim\sup_{|\Delta| \to 0} V_n$ as the partition becomes finer so that its mesh vanishes, whenever the limit exists. Then it can be shown (Exercise 4.20) that $V = \infty$, w.p. 1, which agrees with the discrete time simulation in Figure 4.3.

One consequence of the unbounded total variation is that the length of the Brownian path is infinite. That is, a particle that performs Brownian motion (in 1, 2, or more dimensions) will travel an infinite distance in finite time, almost surely. A physicist would be concerned about this property: It implies

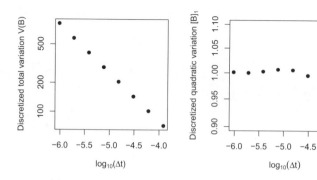

Figure 4.3 Estimating the variation (left panel) and the quadratic variation (right panel) of one sample path of Brownian motion on $[0, 1]$ with discretization. Notice the logarithmic axis in the total variation. As the time discretization becomes finer, the total variation diverges to infinity, while the quadratic variation appears to converge to 1.

that a Brownian particle has infinite speed and infinite kinetic energy. The explanation is that the path of a physical particle differs from mathematical Brownian motion on the very fine scale. The difference may be insignificant in a specific application such as finding out where the particle is going, so that Brownian motion is a useful model, but the difference explains that physical particles have finite speeds while Brownian particles do not.

In turn, Brownian motion has finite quadratic variation:

Definition 4.3.2 (Quadratic Variation) *The quadratic variation of a process $\{X_s : 0 \le s \le t\}$ is the limit*

$$[X]_t = \lim_{|\Delta| \to 0} \sum_{i=1}^{\#\Delta} |X_{t_i} - X_{t_{i-1}}|^2 \text{ (limit in probability)}$$

whenever it exists. Here, Δ is a partition of the interval $[0, t]$.

Theorem 4.3.1 *The quadratic variation of Brownian motion equals the time, $[B]_t = t$, for all $t \ge 0$.*

Proof: For each partition $\Delta = \{0 = t_0, t_1, \ldots, t_n = t\}$ of the time interval $[0, t]$, define S_Δ as

$$S_\Delta = \sum_{i=1}^{\#\Delta} |\Delta B_i|^2 \text{ where } \Delta B_i = B_{t_i} - B_{t_{i-1}}.$$

Since ΔB_i is Gaussian with mean 0 and variance $\Delta t_i = t_i - t_{i-1}$, and since the increments are independent, we have

$$\mathbf{E}S_\Delta = \sum_{i=1}^{\#\Delta} \mathbf{E}|\Delta B_i|^2 = \sum_{i=1}^{\#\Delta} \Delta t_i = t$$

and

$$\mathbf{V}S_\Delta = \sum_{i=1}^{\#\Delta} \mathbf{V}|\Delta B_i|^2 = \sum_{i=1}^{\#\Delta} 2(\Delta t_i)^2 \le 2|\Delta| \sum_{i=1}^{\#\Delta} \Delta t_i = 2|\Delta|t.$$

Thus, as $|\Delta| \to 0$, S_Δ converges to the deterministic limit t in $\mathcal{L}_2$ and therefore also in probability. ■

To appreciate these results, notice that for a continuously differentiable function $f : \mathbf{R} \mapsto \mathbf{R}$, the total variation over the interval $[0, 1]$ is finite, and in fact equals $\int_0^1 |f'(t)| \, dt$, while the quadratic variation is 0. We can therefore conclude that the sample paths of Brownian are not differentiable (w.p. 1). In fact, the sample paths of Brownian motion are *nowhere* differentiable, w.p. 1.

Exercise 4.3: Show that Brownian motion is continuous in mean square, but not differentiable in mean square, at any given point. That is, show that the increments $B_{t+h} - B_t$ converge to 0 in mean square as $h \to 0$, for any $t \ge 0$, but that the difference quotients

$$\frac{1}{h}(B_{t+h} - B_t)$$

do not have a mean square limit. *Note:* See also Exercise 4.6.

It is remarkable that the quadratic variation of Brownian motion is deterministic. One consequence of this concerns statistical estimation: Let $\{X_t\}$ be scaled Brownian motion, $X_t = \alpha B_t$ where $\{B_t\}$ is standard Brownian motion. Then the quadratic variation of $\{X_t\}$ is $[X]_t = \alpha^2 t$. Now, consider the situation where the scale α is unknown but we have observed a segment of a sample path $\{X_t : 0 \le t \le T\}$. Then we can compute $[X]_T$ and from that compute $\alpha = \sqrt{[X]_T/T}$, reaching the correct value, regardless of how short the interval $[0, T]$ is. Thus, each sample path of $\{X_t\}$ holds infinite information about α in any finite time interval. In practice, of course, finite sampling frequencies and measurement errors will introduce errors on the estimate. The quadratic variation of Brownian motion is central in stochastic calculus, notably in Itô's lemma (Section 7.3).

The Maximum over a Finite Interval

How far to the right of the origin does Brownian motion move in a given finite time interval $[0, t]$? Define the maximum

$$S_t = \max\{B_s : 0 \le s \le t\}.$$

The following theorem shows a surprising connection between the distribution of S_t and the distribution of B_t:

Theorem 4.3.2 (Distribution of the maximum of Brownian motion)
For any $t, x > 0$, we have

$$\mathbf{P}(S_t \geq x) = 2\mathbf{P}(B_t \geq x) = 2\Phi(-x/\sqrt{t}) \tag{4.3}$$

where, as always, Φ is the cumulative distribution function of a standard Gaussian variable.

Note that the maximum process $\{S_t : t \geq 0\}$ is also self-similar; for example, $\alpha^{-1}S_{\alpha^2 t}$ has the same distribution as S_t whenever $\alpha > 0$.
Proof: (Sketch) First, since the path of the Brownian motion is continuous and the interval $[0, t]$ is closed and bounded, the maximum is actually attained, so $S_t(\omega)$ is well defined for each ω. Thanks to the monotone convergence theorem, S_t is measurable. Next, notice that

$$\mathbf{P}(S_t \geq x) = \mathbf{P}(S_t \geq x, B_t \leq x) + \mathbf{P}(S_t \geq x, B_t \geq x) - \mathbf{P}(S_t \geq x, B_t = x)$$

and for the last term we find

$$\mathbf{P}(S_t \geq x, B_t = x) \leq \mathbf{P}(B_t = x) = 0.$$

We now aim to show that $\mathbf{P}(S_t \geq x, B_t \leq x) = \mathbf{P}(S_t \geq x, B_t \geq x) = \mathbf{P}(B_t \geq x)$. Consider a realization ω for which $B_t(\omega) \geq x$. Let $\tau = \tau(\omega)$ be the "hitting time", a random variable defined by

$$\tau(\omega) = \inf\{s : B_s(\omega) = x\}.$$

This is the first, but definitely not the last, time we encounter such a hitting time; see Section 4.5. Note that $\tau(\omega) \leq t$ since we assumed that $B_t(\omega) \geq x > 0 = B_0$, and since the sample path is continuous. Define the reflected trajectory (see Figure 4.4)

$$B_s^{(r)}(\omega) = \begin{cases} B_s(\omega) & \text{for } 0 \leq s \leq \tau(\omega), \\ 2x - B_s(\omega) & \text{for } \tau(\omega) \leq s \leq t. \end{cases}$$

We see that each sample path with $S_t \geq x, B_t \geq x$ corresponds in this way to exactly one sample path with $S_t \geq x, B_t \leq x$. Moreover, the reflection operation does not change the absolute values of the increments, and therefore the original and the reflected sample path are equally likely realizations of Brownian motion. This is the argument that works straightforwardly in the case of a discrete-time random walk on $\mathbf{Z}$ (Grimmett and Stirzaker, 1992), but for Brownian motion some care is needed to make the statement and the argument precise. The key is to partition the time interval into ever finer grids; see (Rogers and Williams, 1994a) or (Karatzas and Shreve, 1997). Omitting the details of this step, we reach the conclusion

$$\mathbf{P}(S_t \geq x, B_t > x) = \mathbf{P}(S_t \geq x, B_t \leq x)$$

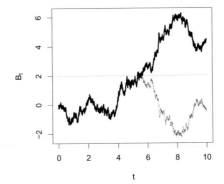

Figure 4.4 The reflection argument used to derive the distribution of the maximum S_t. When B_t (thick solid line) first hits the level $x = 2$, we cut the trajectory and reflect the rest of the trajectory (thin dotted line).

and therefore
$$\mathbf{P}(S_t \geq x) = 2\mathbf{P}(S_t \geq x, B_t \geq x).$$

Now if the end point exceeds x, then obviously the process must have hit x, so $B_t \geq x \Rightarrow S_t \geq x$. Hence

$$\mathbf{P}(S_t \geq x, B_t \geq x) = \mathbf{P}(B_t \geq x)$$

and therefore
$$\mathbf{P}(S_t \geq x) = 2\mathbf{P}(B_t \geq x).$$

■

In many situations involving stochastic differential equations we need to bound the effect of random permutations. It is therefore quite useful that the maximum value of the Brownian motion follows a known distribution with finite moments. However, our main motivation for including this result is that it leads to hitting time distributions for Brownian motion, as the next section shows.

Brownian Motion Is Null-Recurrent

Brownian motion in one dimension always hits any given point on the real line, and always returns to the origin again, but the expected time until it does so is infinite.

To state this property precisely, let $x \neq 0$ be an arbitrary point and define again the hitting time $\tau(\omega)$ to be the first time the sample path hits x, i.e., $\tau = \inf\{t > 0 : B_t = x\}$. By convention the infimum over an empty set is infinity, so $\tau = \infty$ means that the sample path never hits x.

Theorem 4.3.3 *The distribution of the hitting time* $\tau = \inf\{t > 0 : B_t = x\}$ *is given by*

$$\mathbf{P}\{\tau \le t\} = 2\Phi(-|x|/\sqrt{t})$$

so τ *has the probability density function*

$$f_\tau(t) = \frac{d\mathbf{P}(\tau \le t)}{dt} = xt^{-3/2}\phi(x/\sqrt{t}).$$

In particular, $\mathbf{P}(\tau < \infty) = 1$ *and* $\mathbf{E}\tau = \infty$.

The p.d.f. of τ is plotted in Figure 4.5 for $x = 1$. Notice the heavy power-law tail with a slope of $-3/2$.

Proof: Assume that $x > 0$; the case $x < 0$ follows using symmetry. Then, recall from our discussion of the maximum $S_t = \max\{B_s : 0 \le s \le t\}$, that $\tau \le t \Leftrightarrow S_t \ge x$ and in particular

$$\mathbf{P}(\tau \le t) = \mathbf{P}(S_t \ge x) = 2\Phi(-x/\sqrt{t}).$$

Now it is clear that $\mathbf{P}(\tau \le t) \to 1$ as $t \to \infty$, so τ is finite w.p. 1. The p.d.f. of τ is found by differentiating. The power law tail implies a divergent expectation:

$$\mathbf{E}\tau = \int_0^\infty t f_\tau(t)\, dt = \infty.$$

To see this, note that $f_\tau(t) \ge \frac{x}{5}t^{-3/2}$ whenever $t \ge x^2$. Hence

$$\mathbf{E}\tau = \int_0^\infty t f_\tau(t)\, dt \ge \int_{x^2}^\infty \frac{x}{5}t^{-1/2}\, dt = \infty.$$

■

Asymptotics and the Law of the Iterated Logarithm

We know that Brownian motion B_t scales with the square root of time in the sense that $B_t/\sqrt{t}$ is identically distributed for all $t > 0$; in fact follows a standard Gaussian distribution, $B_t/\sqrt{t} \sim N(0,1)$. We are now concerned with the behavior of the sample path of $B_t/\sqrt{t}$ in the limit $t \to \infty$.

Theorem 4.3.4 (The Law of the Iterated Logarithm)

$$\limsup_{t \to \infty} \frac{B_t}{2\sqrt{t \log \log t}} = 1,$$

with probability one.

This result states quite precisely how far from the origin the Brownian motion will deviate, in the long run, and this can be used to derive asymptotic properties and bounds on more general diffusion processes.

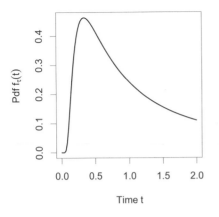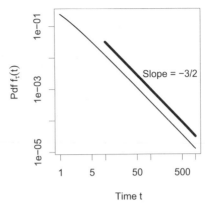

Figure 4.5 The p.d.f. of the hitting time $\tau = \inf\{t : B_t = 1\}$. *Left panel:* The initial part of the curve. *Right panel:* The tail of the curve. Notice the log-scales. Included is also a straight line corresponding to a power law decay $\sim t^{-3/2}$.

Since Brownian motion is symmetric, if follows immediately that almost surely

$$\liminf_{t \to \infty} \frac{B_t}{2\sqrt{t \log \log t}} = -1.$$

Now, since the path of the Brownian motion is continuous and, loosely said, makes never-ending excursions to $\pm 2\sqrt{t \log \log t}$, it also follows that the sample path almost always re-visits the origin: Almost surely there exists a sequence $t_n(\omega)$ such that $t_n \to \infty$ and $B_{t_n} = 0$.

Although the law of the iterated logarithm is simple to state and use, it is a quite remarkable result. The scaled process $B_t/(2\sqrt{t \log \log t})$ converges (slowly!) to 0 in $\mathcal{L}_2$ as $t \to \infty$ (*Exercise: Verify this!*), but the sample path will continue to make excursions away from 0, and the ultimate size of these excursions are equal to 1, no more, no less. Stated in a different way, when we normalize the Brownian motion, and view it in logarithmic time, we reach the process $\{X_s : s \geq 0\}$ given by $X_s = B_t/\sqrt{t}$ with $t = \exp(s)$ (we will consider such time changes further in Section 7.7). This process $\{X_s\}$ is Gaussian stationary (compare Figure 4.6).[1] Hence, it will eventually break any bound, i.e., $\limsup_{s \to \infty} X_s = \infty$. But if the bounds grow slowly with logarithmic time s as well, i.e., $\pm 2\sqrt{\log s}$ (compare Figure 4.6), then the process will ultimately just touch the bound.

---

[1]You may want to verify the stationarity, i.e., that $\mathbf{E}X_s$, $\mathbf{V}X_s$ and $\mathbf{E}X_s X_{s+h}$ do not depend on time s. Here $h \geq 0$ is a time lag. We will discuss stationary processes further in Chapter 5.

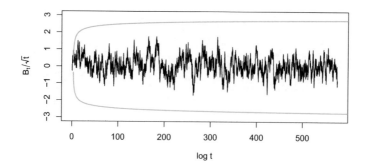

Figure 4.6 Rescaled Brownian motion in logarithmic time. Included is also the growing bounds from the Law of the Iterated Logarithm.

The proof of the law of the iterated logarithm is out of our scope; see (Williams, 1991).

Exercise 4.4: Using the law of the iterated logarithm, show that (almost surely) $\limsup_{t\to\infty} t^{-p} B_t = 0$ for $p > 1/2$, while $\limsup_{t\to\infty} t^{-p} B_t = \infty$ for $0 \le p \le 1/2$.

Invariant under Time-Inversion

If $\{B_t : t \ge 0\}$ is Brownian motion, then also the process $\{W_t : t \ge 0\}$ given by

$$W_0 = 0 , \quad W_t = tB_{1/t} \text{ for } t > 0,$$

is Brownian motion.

Exercise 4.5: Show that this $\{W_t\}$ satisfies the conditions in the definition of Brownian motion. *Hint:* To establish continuity of W_t at $t = 0$, use the law of the iterated logarithm, in particular the results established in Exercise 4.4.

This result is particularly useful, because it can be used to connect properties in the limit $t \to \infty$ with properties in the limit $t \to 0$. For example, from the discussion of the law of the iterated logarithm we learned that Brownian motion almost always revisits the origin in the long run. By time inversion, it then follows that Brownian motion almost always revisits the origin immediately after time 0. To be precise, with probability 1 there exists a sequence t_n such that $t_n \to 0$ and $B_{t_n} = 0$.

Exercise 4.6: Following up on Exercise 4.3, show that the sample paths of Brownian motion are not differentiable at the origin, almost surely. Specifically, show that

$$\limsup_{t \searrow 0} \frac{1}{t} B_t = +\infty, \qquad \liminf_{t \searrow 0} \frac{1}{t} B_t = -\infty$$

almost surely. *Hint:* Use time inversion and the law of the iterated logarithm, in particular Exercise 4.4.

4.4 FILTRATIONS AND ACCUMULATION OF INFORMATION

Recall that we used a σ-algebra of events to model information. This is a static concept, i.e., the information does not change with time. For stochastic processes, we need to consider also the situation that information changes with time, so that we obtain a *family* of σ-algebras, parametrized by time. Our interest is accumulation of information obtained by new observations, and not, for example, loss of information due to limited memory. We therefore define a *filtration* to be a family of σ-algebras, i.e., $\{\mathcal{F}_t : t \in \mathbf{R}\}$, which is *increasing* in the sense that

$$\mathcal{F}_s \subset \mathcal{F}_t \text{ whenever } s < t.$$

We can think of a filtration as the information available to an observer who monitors an evolving stochastic experiment – as time progresses, this observer is able to answer more and more questions about the stochastic experiment. In our context, the information $\mathcal{F}_t$ almost invariably comes from observation of some stochastic process X_t, so that

$$\mathcal{F}_t = \sigma(X_s : 0 \leq s \leq t).$$

In this case, we say that the filtration $\{\mathcal{F}_t : t \geq 0\}$ is *generated* by the process $\{X_t : t \geq 0\}$. A related situation is that the information $\mathcal{F}_t$ is sufficient to determine X_t, for any $t \geq 0$. If X_t is $\mathcal{F}_t$-measurable, for any $t \geq 0$, then we say that the process $\{X_t : t \geq 0\}$ is *adapted* to the filtration $\{\mathcal{F}_t : t \geq 0\}$. Since a filtration is an *increasing* family of σ-algebras, it follows that also earlier values of the stochastic process are measurable, i.e., X_s is measurable with respect to $\mathcal{F}_t$ whenever $0 \leq s \leq t$. Of course, a process is adapted to its own filtration, i.e., the filtration it generates.

The underlying probability space $(\Omega, \mathcal{F}, \mathbf{P})$ and a filtration $\{\mathcal{F}_t : t \geq 0\}$ together constitute a *filtered probability space*. We will make frequent use of filtrations throughout this book. Our first application of filtrations is to describe a class of processes known as *martingales*.

4.5 THE MARTINGALE PROPERTY

Brownian motion is a prime of example of a class of stochastic processes called *martingales*,[2] which can be seen as *unbiased* random walks: The expected value

[2]The term *martingale* has an original meaning which is fairly far from its usage in stochastic processes: A martingale can be a part of a horse's harness, a piece of rigging on a tall ship, or even a half belt on a coat; such martingales provide control and hold things down. Gamblers in 18th century France used the term for a betting strategy where one doubles the stake after a loss; if the name should indicate that this controls the losses, then it is quite misleading. In turn, the accumulated winnings (or losses) in a fair game is a canonical example of a stochastic process with the martingale property.

Biography: Joseph Leo Doob (1910–2004)

An American mathematician and Harvard graduate, he spent the majority of his career at the University of Illinois. Doob was central in the development of the theory of martingales (Doob, 1953). Inspired by the seminal work of Kakutani, he connected potential theory to the theory of stochastic processes (Doob, 2001). Photo credit: CC BY-SA 2.0 DE.

of future increments is always 0. It turns out that this property is tremendously important, for (at least) two reasons: First, from a modeller's perspective, the mean value of a stochastic process is obviously important, and therefore it is attractive to model the contribution from the noise itself as a martingale. Second, a surprising number of conclusions can be drawn from the martingale property, and therefore it is useful for the analysis of a model to identify martingales that appear in connection to the model.

Definition 4.5.1 *Given a probability space* $(\Omega, \mathcal{F}, \mathbf{P})$ *with filtration* $\{\mathcal{F}_t : t \geq 0\}$, *a stochastic process* $\{M_t : t \geq 0\}$ *is a martingale (w.r.t.* $\mathcal{F}_t$ *and* $\mathbf{P}$*) if*

1. *The process* $\{M_t\}$ *is adapted to the filtration* $\{\mathcal{F}_t\}$.

2. *For all times* $t \geq 0$, $\mathbf{E}|M_t| < \infty$.

3. $\mathbf{E}\{M_t | \mathcal{F}_s\} = M_s$ *whenever* $t \geq s \geq 0$.

The first condition states that the σ-algebra $\mathcal{F}_t$ contains enough information to determine M_t, for each $t \geq 0$. The second condition ensures that the expectation in the third condition exists. The third condition, which is referred to as "the martingale property", expresses that $\{M_t\}$ is an *unbiased* random walk: At time s, the conditional expectation of the future increment $M_t - M_s$ is 0.

The time argument t can be discrete ($t \in \mathbf{N}$) or continuous ($t \in \mathbf{R}_+$); our main interest is the continuous-time case. If we just say that $\{M_t\}$ is a martingale, and it is obvious from the context which probability measure should be used to compute the expectations, then it is understood that the filtration $\{\mathcal{F}_t :\geq 0\}$ is the one generated by the process itself, i.e., $\mathcal{F}_t = \sigma(M_s : s \leq t)$.

Exercise 4.7: Show that Brownian motion is a martingale w.r.t. its own filtration.

In greater generality, let there be be given a filtered probability space $(\Omega, \mathcal{F}, \{\mathcal{F}_t\}, \mathbf{P})$ and a process $\{B_t : t \geq 0\}$ with continuous sample paths.

We say that $\{B_t\}$ is Brownian motion on $(\Omega, \mathcal{F}, \{\mathcal{F}_t\}, \mathbf{P})$, if $B_0 = 0$, $\{B_t\}$ is adapted to $\{\mathcal{F}_t\}$, and the increment $B_t - B_s$ is distributed as $N(0, t - s)$ and independent of $\mathcal{F}_s$, for all $0 \leq s \leq t$. We then also say that $\{B_t\}$ is Brownian motion w.r.t. $\{\mathcal{F}_t\}$. It follows that $\{B_t\}$ is then also a martingale w.r.t. $\{\mathcal{F}_t\}$. The filtration $\{\mathcal{F}_t\}$ could be the one generated by $\{B_t\}$, but it can also include other information about other random variables or stochastic processes; the definition requires just that the filtration does not contain information about the future increments.

Exercise 4.8: Let $\{B_t\}$ be Brownian motion w.r.t. a filtration $\{\mathcal{F}_t\}$. Show that the process $\{B_t^2 - t : t \geq 0\}$ is a martingale w.r.t. $\{\mathcal{F}_t\}$. *Note:* The so-called Lévy characterization of Brownian motion says that a converse result also holds: If $\{M_t\}$ is a continuous martingale such that also $\{M_t^2 - t\}$ is a martingale, then $\{M_t\}$ is Brownian motion.

Martingales play an important role in mathematical finance: Let M_t be the (discounted) market price of an asset (e.g., a stock) at time t, and assume that everybody in the market has access to the information $\mathcal{F}_s$ at time s. Then the fair price of the asset at time $s \in [0, t]$ must be $M_s = \mathbf{E}\{M_t | \mathcal{F}_s\}$ where expectation is w.r.t. the risk-free measure $\mathbf{Q}$. I.e., the (discounted) price $\{M_t\}$ is a martingale w.r.t. $\mathbf{Q}$ and $\mathcal{F}_t$.

In the context of gambling, martingales are often said to characterize *fair* games: if M_t is the accumulated winnings of a gambler at time t, and $\{M_t : t \geq 0\}$ is a martingale, then the game can be said to be fair. In this context, an important result is that it is impossible to beat the house on average, i.e., obtain a positive expected gain, by quitting the game early. To present this, we must first define the admitted strategies for quitting. Let τ be a random time, i.e., a random variable taking values in $[0, \infty]$, specifying when some phenomenon happens. We allow the value $\tau = \infty$ to indicate that the phenomenon never occurs. An important distinction is if we always know whether the phenomenon has occurred, or if we may be unaware that it has occurred. We use the term *Markov time*, or *stopping time*, to describe the first situation:

Definition 4.5.2 (Markov time, stopping time) *A random variable τ taking values in $[0, \infty]$ is denoted a Markov time (or a stopping time) (w.r.t. $\mathcal{F}_t$) if the event*

$$\{\omega : \tau(\omega) \leq t\}$$

is contained in $\mathcal{F}_t$ for any $t \geq 0$.

Probably the most important example of stopping times are *hitting times*, which we have already encountered: If $\{X_t : t \geq 0\}$ is a stochastic process taking values in $\mathbf{R}^n$, $\{\mathcal{F}_t\}$ is its filtration, and $B \subset \mathbf{R}^n$ is a Borel set, then the time of first entrance

$$\tau = \inf\{t \geq 0 : X_t \in B\}$$

is a stopping time (with respect to $\mathcal{F}_t$). Recall that by convention we take the infimum of an empty set to be ∞. On the other hand, the time of last exit

$$\sup\{t \geq 0 : X_t \in B\}$$

is *not* in general a stopping time, since we need to know the future in order to tell if the process will ever enter B again.

In the context of gambling, a stopping time represents a strategy for quitting the game. The following result states that you can't expect to win if you quit a fair game early.

Lemma 4.5.1 *On a probability space $(\Omega, \mathcal{F}, \mathbf{P})$, let the process $\{M_t : t \geq 0\}$ be a continuous martingale with respect to a filtration $\{\mathcal{F}_t : t \geq 0\}$. Let τ be a stopping time. Then the stopped process*

$$\{M_{t\wedge\tau} : t \geq 0\}$$

is a martingale with respect to $\{\mathcal{F}_t : t \geq 0\}$. In particular, $\mathbf{E}(M_{t\wedge\tau}|\mathcal{F}_0) = M_0$. Here, $\wedge$ is the "min" symbol: $a \wedge b = \min(a,b)$.

Proof:(Sketch) To see that $M_{t\wedge\tau}$ is $\mathcal{F}_t$-measurable, note that an observer with acess to the information $\mathcal{F}_t$ knows the value of $t \wedge \tau$ and therefore also $M_{t\wedge\tau}$. For a rigorous proof, which uses the continuity of the sample paths of $\{M_t\}$, see (Williams, 1991).

To verify that $\mathbf{E}|M_{t\wedge\tau}| < \infty$, consider first the case where τ is discrete, i.e., takes value in an increasing deterministic sequence $0 = t_0 < t_1 < t_2 < \cdots$. Assume that $\mathbf{E}|M_{t_i\wedge\tau}| < \infty$, then

$$\begin{aligned}\mathbf{E}|M_{t_{i+1}\wedge\tau}| &= \mathbf{E}|\mathbf{1}(\tau \leq t_i)M_\tau + \mathbf{1}(\tau > t_i)M_{t_{i+1}}|\\ &\leq \mathbf{E}|\mathbf{1}(\tau \leq t_i)M_{\tau\wedge t_i}| + \mathbf{E}|\mathbf{1}(\tau > t_i)M_{t_{i+1}}|\\ &\leq \mathbf{E}|M_{\tau\wedge t_i}| + \mathbf{E}|M_{t_{i+1}}|\\ &< \infty.\end{aligned}$$

It follows by iteration that $\mathbf{E}|M_{t_i\wedge\tau}| < \infty$ for all i. Similarly,

$$\begin{aligned}\mathbf{E}\{M_{t_{i+1}\wedge\tau}|\mathcal{F}_{t_i}\} &= \mathbf{E}\{\mathbf{1}(\tau \leq t_i)M_{t_{i+1}\wedge\tau} + \mathbf{1}(\tau > t_i)M_{t_{i+1}\wedge\tau}|\mathcal{F}_{t_i}\}\\ &= \mathbf{E}\{\mathbf{1}(\tau \leq t_i)M_\tau + \mathbf{1}(\tau > t_i)M_{t_{i+1}}|\mathcal{F}_{t_i}\}\\ &= \mathbf{1}(\tau \leq t_i)M_\tau + \mathbf{1}(\tau > t_i)M_{t_i}\\ &= M_{\tau\wedge t_i}\end{aligned}$$

Again, by iteration, $\{M_{t_i\wedge\tau} : i = 0,1,2,\ldots\}$ is a martingale.

We outline the argument for the general case where τ is not necessarily discrete: We approximate τ with discrete stopping times $\{\tau_n : n \in \mathbf{N}\}$ which converge monotonically to τ, for each ω. For each approximation τ_n, the stopped process $M_{t\wedge\tau_n}$ is a martingale, and in the limit $n \to \infty$ we find, with the monotone convergence theorem,

$$\mathbf{E}\{M_{t\wedge\tau}|\mathcal{F}_s\} = \lim_{n\to\infty} \mathbf{E}\{M_{t\wedge\tau_n}|\mathcal{F}_s\} = M_s$$

which shows that $\{M_{t\wedge\tau} : t \geq 0\}$ is also a martingale. ∎

Of course, it is crucial that τ is a stopping time, and not just any random time: It is not allowed to sneak peek a future loss and stop before it occurs. It is also important that the option is to quit the game *early*, i.e., the game always ends no later than a fixed time t. Consider, for example, the stopping time $\tau = \inf\{t : B_t \geq 1\}$ where B_t is Brownian motion. Then τ is finite almost surely (since Brownian motion is recurrent, Section 4.3), and of course $B_\tau = 1$ so that, in particular, $\mathbf{E}B_\tau \neq B_0$. But this stopping time τ is not bounded so stopping at τ is not a strategy to quit *early*. In contrast, if we stop the Brownian motion whenever it hits 1, or at a fixed terminal time $t > 0$, whichever happens first, we get the stopped process $B_{t\wedge\tau}$ for which we now know that $\mathbf{E}B_{t\wedge\tau} = 0$.

Although this result should seem fairly obvious – except perhaps to die-hard gamblers – it has a somewhat surprising corollary: It bounds (in probability) the maximum value of the sample path of a *non-negative* martingale:

Theorem 4.5.2 *[The martingale inequality] Let $\{M_t : t \geq 0\}$ be a non-negative continuous martingale. Then*

$$\mathbf{P}(\max_{s\in[0,t]} M_s \geq c) \leq \frac{\mathbf{E}M_0}{c}.$$

This inequality is key in stochastic stability theory (Chapter 12) where it is used to obtain bounds on the solutions of stochastic differential equations. **Proof:** Define the stopping time $\tau = \inf\{t : M_t \geq c\}$ and consider the stopped process $\{M_{t\wedge\tau} : t \geq 0\}$. This is a martingale, and therefore

$$\mathbf{E}M_0 = \mathbf{E}M_{t\wedge\tau}.$$

By Markov's inequality,

$$\mathbf{E}M_{t\wedge\tau} \geq c\, \mathbf{P}(M_{t\wedge\tau} \geq c).$$

Combining, we obtain $\mathbf{E}M_0 \geq c\, \mathbf{P}(M_{t\wedge\tau} \geq c)$. Noting that $M_{t\wedge\tau} \geq c$ if and only if $\max\{M_s : 0 \leq s \leq t\} \geq c$, the conclusion follows. ■

Exercise 4.9: A gambler plays repeated rounds of a fair game. At each round, he decides the stakes. He can never bet more than his current fortune, and he can never lose more than he bets. His initial fortune is 1. Show that the probability that he ever reaches a fortune of 100, is no greater than 1%.

The definition of a martingale concerns only expectations, so a martingale does not necessarily have finite variance. However, many things are simpler if the variances do in fact exist, i.e., if $\mathbf{E}|M_t|^2 < \infty$ for all $t \geq 0$. In this case, we say that $\{M_t\}$ is an $\mathcal{L}_2$ martingale. Brownian motion, for example, is an $\mathcal{L}_2$ martingale. In the remainder of this section, $\{M_t\}$ is an $\mathcal{L}_2$ martingale.

Exercise 4.10: Show that if $\{M_t : t \geq 0\}$ is a martingale such that $\mathbf{E}|M_t|^2 < \infty$ for all t, then the increments

$$M_t - M_s \text{ and } M_v - M_u,$$

are uncorrelated, whenever $0 \leq s \leq t \leq u \leq v$. (*Hint:* When computing the covariance $\mathbf{E}(M_v - M_u)(M_t - M_s)$, condition on $\mathcal{F}_u$.) Next, show that the variance of increments is additive:

$$\mathbf{V}(M_u - M_s) = \mathbf{V}(M_u - M_t) + \mathbf{V}(M_t - M_s).$$

Finally, show that the variance is increasing, i.e.,

$$\mathbf{V}M_s \leq \mathbf{V}M_t, \text{ whenever } 0 \leq s \leq t.$$

This property of uncorrelated increments generalizes the independent increments of Brownian motion.

Since the variance is increasing, an important characteristic of an $\mathcal{L}_2$ martingale is how fast and how far the variance increases. To illustrate this, we may ask what happens as $t \to \infty$. Clearly the variance must either diverge to infinity, or converge to a limit, $\lim_{t\to\infty} \mathbf{V}M_t < \infty$. A very useful result is that if the variance converges, then also the process itself converges. Specifically:

Theorem 4.5.3 (Martingale Convergence, $\mathcal{L}_2$ Version) *Let $\{M_t : t \geq 0\}$ be a continuous martingale such that the variance $\{\mathbf{V}M_t : t \geq 0\}$ is bounded. Then there exists a random variable M_∞ such that $M_t \to M_\infty$ in $\mathcal{L}_2$ and w.p. 1.*

Proof: Let $0 = t_1 < t_2 < \cdots$ be an increasing divergent sequence of time points, i.e., $t_i \to \infty$ as $i \to \infty$. Then we claim that $\{M_{t_n} : n \in \mathbf{N}\}$ is a Cauchy sequence. To see this, let $\epsilon > 0$ be given. We must show that there exists an N such that for all $n, m > N$, $\|M_{t_n} - M_{t_m}\|_2 < \epsilon$. But this is easy: Choose N such that $\mathbf{V}M_{t_N} > \lim_{t\to\infty} \mathbf{V}M_t - \delta$ where $\delta > 0$ is yet to be determined, and let $n, m > N$. Then

$$\mathbf{V}(M_{t_N} - M_{t_n}) \leq \delta$$

and therefore $\|M_{t_N} - M_{t_n}\|_2 \leq \sqrt{\delta}$. The same applies if we replace t_n with t_m. The triangle inequality for the $\mathcal{L}_2$ norm then implies that

$$\|M_{t_n} - M_{t_m}\|_2 \leq \|M_{t_N} - M_{t_n}\|_2 + \|M_{t_N} - M_{t_m}\|_2 \leq 2\sqrt{\delta}.$$

So if we choose $\delta < \epsilon^2/4$, we get $\|M_{t_n} - M_{t_m}\|_2 < \epsilon$. Since $\{M_{t_n} : n \in \mathbf{N}\}$ is a Cauchy sequence and the space $\mathcal{L}_2$ is complete, there exists an $M_\infty \in \mathcal{L}_2$ such that $M_{t_n} \to M_\infty$ in mean square. Moreover, it is easy to see that this M_∞ does not depend on the particular sequence $\{t_i : i \in \mathbf{N}\}$.

We omit the proof that the limit is also w.p. 1. This proof uses quite different techniques; see (Williams, 1991). ■

4.6 CONCLUSION

A stochastic process is a family of random variables parameterized by time, e.g. $\{X_t : t \geq 0\}$. For fixed time t, we obtain a random variable $X_t : \Omega \mapsto \mathbf{R}^n$. For fixed realization ω, we obtain a sample path $X.(\omega) : \bar{\mathbf{R}}_+ \mapsto \mathbf{R}^n$, so the theory of stochastic processes can be seen as a construction that allows us to pick a random function of time.

Stochastic processes can be seen as evolving stochastic experiments. They require a large sample space Ω, typically a function space, and a filtration $\{\mathcal{F}_t : t \geq 0\}$ which describes how information is accumulated as the experiment evolves.

In this chapter, we have used Brownian motion as the recurring example of a stochastic process, to make the general concepts more specific. Brownian motion is a fundamental process in the study of diffusions and stochastic differential equations, and it is useful to know its properties in detail. The two most important properties arguably concern the expectation and the variance: The expectation af any increment is 0, so Brownian motion is a martingale. The variance grows linear with time, so that distance scales with the square root of time.

When the Brownian motion is *canonical*, the sample space consists of continuous functions which we identify with the sample path of Brownian motion. Canonical Brownian motion therefore defines a probability measure on the space of continuous functions; phrased differently, a stochastic experiment consisting of picking a random continuous function. A side benefit is that with this construction, all sample paths are continuous by construction.

However, Brownian motion is not differentiable; in contrast, it has infinite total variation. In the following chapters, we shall see that this has profound consequences when we include Brownian motion in differential equations. Instead, it has a very simple quadratic variation: $[B]_t = t$. Since the paths are not differentiable, the quadratic variation is central in the stochastic calculus that we will build, together with the martingale property.

4.7 NOTES AND REFERENCES

Besides Brown, Einstein, and Wiener, early studies of Brownian motion include the work of Louis Bechalier in 1900, whose interest was financial markets, and the study of T.N. Thiele, whose 1880 paper concerned least squares estimation. More in-depth treatments of Brownian motion can be found in (Williams, 1991; Rogers and Williams, 1994a; Øksendal, 2010; Karatzas and Shreve, 1997).

4.8 EXERCISES

Simulation of Brownian Motion

Exercise 4.11 The Brownian Bridge:

1. Write a code which takes Brownian motion $B_{t_1}, B_{t_2}, \ldots, B_{t_n}$ on the time points $0 \leq t_1 < \cdots < t_n$, and which returns a finer partition $s_1, \ldots, s_{2n-1}$ along with simulated values of the Brownian motion $B_{s_1}, \ldots, B_{s_{2n-1}}$. Here, the finer partition includes all mid-points, i.e.

$$s_1 = t_1, \quad s_2 = \frac{t_1 + t_2}{2}, \quad s_3 = t_2, \quad s_4 = \frac{t_2 + t_3}{2}, \quad \ldots, s_{2n-1} = t_n.$$

2. Use this function iteratively to simulate Brownian motion on the interval $[0, 1]$ in the following way: First, simulate $B_0 = 0$ and $B_1 \sim N(0, 1)$. Then, conditional on this, simulate $B_0, B_{1/2}, B_1$ using your function. Then, conditionally on these, simulate $B_0, B_{1/4}, B_{1/2}, B_{3/4}, B_1$. Continue in this fashion until you have simulated Brownian motion with a temporal resolution of $h = 1/512$. Plot the resulting trajectory.

Exercise 4.12 The Brownian Bridge (again): Yet another way of simulating a Brownian bridge uses a basic result regarding conditional simulation in Gaussian distributions. For two jointly Gaussian random variables (X, Y), we can simulate X from the conditional distribution given Y as follows:

1. Compute the conditional mean, $\mathbf{E}\{X|Y\}$.

2. Sample $(\bar{X}, \bar{Y})$ from the joint distribution of (X, Y).

3. Return $\tilde{X} = \mathbf{E}\{X|Y\} - \mathbf{E}\{\bar{X}|\bar{Y}\} + \bar{X}$.

Check that the conditional distribution of this $\tilde{X}$ given Y is identical to the conditional distribution of X given Y.

Then write a function which inputs a partition $0 = t_0 < t_1 < \cdots < t_n = T$ and a value of B_T, and which returns a sample path of the Brownian bridge $B_0, B_{t_1}, \ldots, B_T$ which connects the two end points. Test the function by computing 1000 realizations of $\{B_{nh} : n \in \mathbf{N}, nh \in [0, 2]\}$ with $h = 0.01$ and for which $B_2 = 1$ and plotting the mean and variance as function of time.

Exercise 4.13 The Wiener Expansion: Here, we simulate Brownian motion through frequency domain. First, we simulate harmonics with random amplitude. Then, we add them to approximate white noise. Finally we integrate to approximate Brownian motion. For compactness, we use complex-valued Brownian motion $\{B_t^{(1)} + iB_t^{(2)} : 0 \leq t \leq 2\pi\}$, where $\{B_t^{(1)}\}$ and $\{B_t^{(1)}\}$ are independent standard Brownian motions.

1. Write a code that generates, for a given $N \in \mathbf{N}$, $2N+1$ complex-valued random Gaussian variables $\{V_k : k = -N, \ldots, N\}$ such that the real and imaginary parts are independent and standard Gaussians. Take e.g. $N = 16$. Then generate an approximation to white noise as

$$W_t = \frac{1}{\sqrt{2\pi}} \sum_{k=-N}^{N} V_k e^{ikt}$$

Check by simulation that the real and imaginary parts of W_t are independent and distributed as $N(0, (2N+1)/(2\pi))$. Here $t \in [0, 2\pi]$ is arbitrary; try a few different values. Visual verification suffices.

2. Verify the claims of independence and distribution theoretically.

3. Evaluate W_t on a regular partition of the time interval $[0, 2\pi]$ and plot the empirical autocovariance function. Comment on the degree to which it resembles a Dirac delta.

4. From the simulated noise signal $\{W_t\}$, compute a sample path of approximate Brownian motion by integration:

$$B_t = \int_0^t W_s \, ds.$$

The integration of each harmonic should preferably be done analytically, i.e.,

$$\int_0^t e^{i0s} \, ds = t, \quad \int_0^t e^{iks} \, ds = \frac{1}{ik}(e^{ikt} - 1) \text{ for } k \neq 0.$$

Plot the real part of the sample path.

5. Write a code which inputs N and some time points $t_i \in [0, 2\pi]$, and which returns a sample of (real-valued) Brownian motion evaluated on those time points, using the calculations above. Verify the function by simulating 1,000 realizations of $(B_1, B_{1.5}, B_2, B_5)$ and computing the empirical mean and the empirical covariance matrix.

Convergence

Exercise 4.14: Show that Brownian motion with drift diverges almost surely, i.e. $\liminf_{t \to \infty} (B_t + ut) = \infty$ for $u > 0$. *Hint:* Use the law of the iterated logarithm.

Exercise 4.15 Convergence w.p. 1, but not in $\mathcal{L}_2$: Let $\{B_t : t \geq 0\}$ be Brownian motion and define, for $t \geq 0$

$$X_t = \exp(B_t - \frac{1}{2}t).$$

Factbox: [The log-normal distribution] A random variable Y is said to be log-normal (or log-Gaussian) distributed with location parameter μ and scale parameter $\sigma > 0$, i.e.,

$$Y \sim LN(\mu, \sigma^2)$$

if $X = \log Y$ is Gaussian with mean μ and variance σ^2.

Property	Expression
Mean $\mathbf{E}Y$	$\exp(\mu + \frac{1}{2}\sigma^2)$
Variance $\mathbf{V}Y$	$(\exp(\sigma^2) - 1)\exp(2\mu + \sigma^2)$
C.d.f. $F_Y(y)$	$\Phi(\sigma^{-1}(\log(y) - \mu))$
mea P.d.f. $f_Y(y)$	$\sigma^{-1}y^{-1}\phi(\sigma^{-1}(\log(y) - \mu))$
Median	$\exp(\mu)$
Mode	$\exp(\mu - \sigma^2)$

Show that $X_t \to 0$ almost surely as $t \to \infty$, but that $\mathbf{E}|X_t|^2 \to \infty$. *Note:* This process $\{X_t : t \geq 0\}$ is one example of *geometric Brownian motion*; we will return to this process repeatedly. If the result puzzles you, you may want to simulate a number of realizations of X_t and explain in words how X_t can converge to 0 almost surely while diverging in $\mathcal{L}_2$.

Exercise 4.16 Continuity of Stochastic Processes:

1. Let $\{N_t : t \geq 0\}$ be a Poisson process with unit intensity, i.e., a Markov process $N_0 = 0$ and with transition probabilities given by $N_t|N_s$ being Poisson distributed with mean $t - s$ for $0 \leq s \leq t$. Show that $\{N_t\}$ is continuous in mean square but that almost no sample paths are continuous.

2. Let V be a real-valued random variable such that $\mathbf{E}|V^2| = \infty$ and define the stochastic process $\{X_t : t \geq 0\}$ by $X_t = V \cdot t$. Show that $\{X_t\}$ has continuous sample paths but is not continuous in mean square.

Exercise 4.17 The Precision Matrix of Brownian Motion: If $X = (X_1, X_2, \ldots, X_n)$ is a random variable taking values in $\mathbf{R}^n$ and with variance-covariance matrix $S > 0$, then the *precision matrix* of X is $P = S^{-1}$ (Rue and Held, 2005). Now, let $\{B_t : t \geq 0\}$ be standard Brownian motion.

1. Let $X_i = B_{t_i}$ where $0 < t_1 < \cdots < t_n$. Show that the precision matrix

P is tridiagonal; specifically

$$
P_{ij} = \begin{cases}
\frac{t_2}{t_1(t_2-t_1)} & \text{if } i = j = 1, \\
\frac{1}{t_n-t_{n-1}} & \text{if } i = j = n, \\
\frac{t_{i+1}-t_{i-1}}{(t_{i+1}-t_i)(t_i-t_{i-1})} & \text{if } 1 < i = j < n, \\
-\frac{1}{|t_i-t_j|} & \text{if } i = j \pm 1, \\
0 & \text{else.}
\end{cases}
$$

2. Assume that X is multivariate Gaussian with expectation μ. Show that $1/P_{ii} = \mathbf{V}\{X_i|X_{-i}\}$ and that $\mathbf{E}\{X_i-\mu_i|X_{-i}\} = \sum_{j\neq i} a_j(X_j-\mu_j)$ where $a_j = P_{ij}/P_{ii}$. Here, $X_{-i} = (X_1,\ldots,X_{i-1},X_{i+1},\ldots,X_n)$. *Note:* Precision matrices are most useful in the Gaussian case but can be generalized beyond this in different ways.

3. Combine the previous to show that this precision matrix agrees with the probabilistic graphical model of Brownian motion in Figure 4.1, and the statistics of the Brownian bridge, i.e.:

$$
\mathbf{E}\{B_{t_i}|B_{t_{i-1}}, B_{t_{i+1}}\} = \frac{t_i - t_{i-1}}{t_{i+1} - t_{i-1}} B_{t_{i+1}} + \frac{t_{i+1} - t_i}{t_{i+1} - t_{i-1}} B_{t_{i-1}},
$$

$$
\mathbf{V}\{B_{t_i}|B_{t_{i-1}}, B_{t_{i+1}}\} = \frac{(t_{i+1} - t_i)(t_i - t_{i-1})}{t_{i+1} - t_{i-1}}.
$$

Martingales

Exercise 4.18 Martingales as Random Walks: Let X_i be independent random variables for $i = 1, 2, \ldots$ such that $\mathbf{E}|X_i| < \infty$ and $\mathbf{E}X_i = 0$. Show that the process $\{M_i : i \in \mathbf{N}\}$ given by $M_i = \sum_{j=1}^i X_i$ is a martingale.

Exercise 4.19 Doob's Martingale: Let X be a random variable such that $\mathbf{E}|X| < \infty$, and let $\{\mathcal{F}_t : t \geq 0\}$ be a filtration. Show that the process $\{M_t : t \geq 0\}$ given by $M_t = \mathbf{E}\{X|\mathcal{F}_t\}$ is a martingale w.r.t. $\{\mathcal{F}_t : t \geq 0\}$.

Miscellaneous

Exercise 4.20: Let $\Delta = \{0, 1/n, 2/n, \ldots, 1\}$ be a partition of $[0,1]$, where $n \in \mathbf{N}$, and let V_Δ be the discretized total variation as in (4.2).

1. Show that $\mathbf{E}V_\Delta \sim \sqrt{n}$, (in fact, $\mathbf{E}V_\Delta = \sqrt{2n/\pi}$) so that $\mathbf{E}V_\Delta \to \infty$ as $n \to \infty$.

2. Show that $V_\Delta \to \infty$ w.p. 1 as $n \to \infty$, using

$$\sum_{i=1}^{n} |B_{t_i} - B_{t_{i-1}}|^2 \le \max_i\{|B_{t_i} - B_{t_{i-1}}|\} \cdot \sum_{i=1}^{n} |B_{t_i} - B_{t_{i-1}}|$$

as well as the quadratic variation and continuity of Brownian motion.

Exercise 4.21 The (Second) Arcsine Law: Consider Brownian motion on the time interval [0,1]. Define τ as the last time the process hits 0:

$$\tau = \sup\{t \in [0, 1] : B_t = 0\}$$

1. Show that
$$\mathbf{P}\{\tau \le t\} = \frac{2}{\pi} \arcsin \sqrt{t}$$

for $0 \le t \le 1$. *Hint:* Use reflection as in the proof of Theorem 4.3.2 to relate the probability that the process visits the origin in the interval $(t, 1]$ to the distribution of (B_t, B_1). Then use the result from Exercise 3.14 about two Gaussians having different sign.

2. Estimate the distribution function of τ using Monte Carlo, simulating $N = 1000$ sample paths of $\{B_t : t \in [0, 1]\}$ and reporting the last time the sign is changed. Use a time step of $h = 0.001$. Plot the empirical distribution function and compare with the analytical expression.

Exercise 4.22: Let $\Omega = [0, 1]$, let $\mathcal{F}$ be the Borel algebra on this interval, and let $\mathbf{P}$ be the Lebesgue measure on Ω. Consider the real-valued stochastic processes $\{X_t : t \in [0, 1]\}$ and $\{Y_t : t \in [0, 1]\}$ on this probability space:

$$X_t(\omega) = 0, \quad Y_t(\omega) = \begin{cases} 1 & \text{when } t = \omega, \\ 0 & \text{else.} \end{cases}$$

Show that, $\{X_t\}$ and $\{Y_t\}$ have identical finite-dimensional distributions; we say that $\{X_t\}$ and $\{Y_t\}$ are *versions* of eachother. Note that the sample paths of $\{X_t\}$ are continuous w.p. 1, while those of $\{Y_t\}$ are discontinuous, w.p. 1.

CHAPTER 5

Linear Dynamic Systems

Systems of linear differential equations with exogenous random inputs make an important special case in the theory of stochastic processes. Linear models are important in practical applications, because they quickly give explicit results, based on simple formulas for the mean and for the covariance structure. If a linear model is reasonable, or can be used as a first approximation, then it is typically worthwhile to start the analysis there.

With increasingly strong computers and algorithms, the role of linear systems is no longer as central in science and engineering curricula as it used to be: While earlier generations were forced to simplify dynamic systems in order to analyze them, and therefore often focused on linear models, we can now conduct numerical simulations of larger and more complex systems. As a result, the toolbox of linear systems is no longer rehearsed as eagerly. However, in the case of stochastic systems, simulation is still cumbersome, because we would need to simulate a large number of sample paths before we could say anything about the system at hand with confidence. Then, the analytical techniques of linear systems come to the rescue.

Linear systems can be analyzed in time domain or in frequency domain; the Fourier transform connects to two domains by decomposing the fluctuations of signals into contributions from different frequencies. Since linear systems satisfy the superposition principle, we can consider each frequency in isolation.

Frequency domain methods applied to linear systems also make it possible to give precise meaning to the notion of fast vs. slow signals and dynamics. If a linear system is affected by a noise signal which is fast compared both to the system itself and to the interest of the modeler, then it may be reasonable to approximate the signal with white noise. Doing so, we obtain the first example of a stochastic differential equation. We will see that it is possible to analyse the solution of this equation and give very explicit results for the mean, covariance structure, and spectrum.

DOI: 10.1201/9781003277569-5

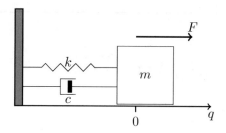

Figure 5.1 A mass-spring-damper system drive by an exogenous force is a simple example of a linear system.

5.1 LINEAR SYSTEMS WITH DETERMINISTIC INPUTS

Let us start by briefly recapitulating the basic theory of linear time-invariant systems driven by additive inputs. A simple example of such a system is a mass connected to a wall with a spring and a damper, and subject to an external force, see Figure 5.1. The governing equations are

$$\frac{dQ_t}{dt} = V_t, \tag{5.1}$$

$$m\frac{dV_t}{dt} = -kQ_t - cV_t + U_t. \tag{5.2}$$

Here, Q_t is the position at time t while V_t is the velocity and U_t is the external force. We use the symbol U_t since external driving inputs are typically denoted U_t in the linear systems literature. The system parameters are the mass m, the spring constant k and the viscous damping coefficient c. The first equation is simply the definition of velocity, while the second equation is Newton's second law, where the total force has contributions from the spring $(-kQ_t$, Hooke's law) and the damper $(-cV_t$, linear viscous damping) in addition to the external force U_t.

State-space formalism is tremendously useful. There, we define the system state $X_t = (Q_t, V_t)^{\top}$ and track the motion of the system in state space; in this case, $\mathbf{R}^2$. The governing equation can be written in vector-matrix notation:

$$\dot{X}_t = AX_t + GU_t. \tag{5.3}$$

Here, the system matrices are

$$A = \begin{bmatrix} 0 & 1 \\ -k/m & -c/m \end{bmatrix} \text{ and } G = \begin{bmatrix} 0 \\ 1/m \end{bmatrix}. \tag{5.4}$$

Writing the equation in state space form is algebraically convenient and a concise shorthand, but also represents a fundamental paradigm to dynamic systems: The state vector summarizes the entire pre-history of the system so that predictions about the future can be based solely on the current state and future inputs, and need not take further elements from the past into account. State space formalism is to dynamic systems what the Markov property (Section 9.1) is to stochastic processes.

For such a linear system with exogenous input, a fundamental property is the impulse response. This is the (fundamental) solution to the system which is at rest before time $t = 0$, i.e., $X_t = 0$ for $t < 0$, and subject to a Dirac delta input, $U_t = \delta(t)$. For the mass-spring-damper system, this means that the mass is standing still before time $t = 0$, $Q_t = V_t = 0$ for $t < 0$. Then, a force of large magnitude is applied over a short time period starting at time 0, effectively changing the momentum instantaneously from 0 to 1. Figure 5.2 shows the impulse response of the position; for correct dimensions, the momentum is changed instantaneously from 0 to 1 Ns. In general, the impulse response of the system (5.3) corresponding to the input $U_t = \delta(t)$ is

$$h(t) = \begin{cases} 0 & \text{for } t < 0, \\ \exp(At)G & \text{for } t \geq 0. \end{cases} \tag{5.5}$$

The matrix exponential $\exp(At)$ is described in the fact box on page 95. For the mass-spring-damper system, it is possible to write up the elements in the matrix exponential in closed form, but this is not possible in general. Rather, there exist powerful numerical algorithms for computing the matrix exponential.

We can use the impulse response to obtain the solution to the linear system (5.3) with the initial condition $X_0 = x_0$, for a general forcing $\{U_t : t \geq 0\}$:

$$X_t = e^{At}x_0 + \int_0^t h(t-s)\, U_s\, ds = e^{At}x_0 + \int_0^t e^{A(t-s)}GU_s\, ds. \tag{5.6}$$

This solution is termed the *superposition* principle: The response at time t to the initial condition only is $\exp(At)x_0$ while the response to a Dirac delta at time s and with strength U_s is $\exp(A(t-s))U_s$. This solution establishes that X_t arises as the linear combination of these responses. The solution can be verified by first noting that the right hand side of (5.6) equals x_0 when $t = 0$, and next differentiating the right hand side, using the Leibniz integral rule for differentiation of an integral:

$$\frac{d}{dt}X_t = Ae^{At}x_0 + e^{A(t-t)}GU_t + \int_0^t Ae^{A(t-s)}GU_s\, ds$$
$$= AX_t + GU_t,$$

as claimed.

5.2 LINEAR SYSTEMS IN THE FREQUENCY DOMAIN

So far we have considered the linear system in the time domain, but frequency domain methods are powerful. There, we decompose all signals into harmonics (i.e., sine and cosine functions or, more conveniently, complex exponentials) using the Fourier transform.

Assume that we apply a force of the form $U_t = U_0 \cos \omega t$ to the mass-spring-damper system, then there is a solution where the system responds with a periodic motion $X_t = U_0 a(\omega) \cos(\omega t + \phi(\omega))$. We will find this solution shortly, in the general case. Note that the frequency of the response is the same as that of the applied force, and that the amplitude $U_0 a(\omega)$ is proportional to the magnitude of the applied force, with the constant of proportionality $a(\omega)$ depending on the frequency ω. Also the phase shift $\phi(\omega)$ depends on frequency ω. It is convenient to write this as

$$X_t = U_0 \mathrm{Re}\left[H(\omega)\exp(i\omega t)\right]$$

where $H(\omega) = a(\omega) \exp(i\phi(\omega))$ is the complex-valued frequency response; i is the imaginary unit.

To find the frequency response for the general linear system (5.3), simply search for solutions of the form $U_t = \exp(i\omega t)$, $X_t = H(\omega) \exp(i\omega t)$:

$$i\omega H(\omega)e^{i\omega t} = AH(\omega)e^{i\omega t} + Ge^{i\omega t}.$$

Then isolate $H(\omega)$:

$$H(\omega) = (i\omega \cdot I - A)^{-1}G,$$

where I is a identity matrix of same dimensions as A. Next, notice that this result coincides with the Fourier transform of the impulse response $h(\cdot)$: Assume that A is stable, i.e., all eigenvalues of A have negative real part, so that $h(t)$ converges to 0 exponentially as $t \to 0$. Then the Fourier transform is

$$H(\omega) = \int_{-\infty}^{+\infty} h(t)e^{-i\omega t}\, dt = \int_0^\infty e^{At}Ge^{-i\omega t}\, dt = (i\omega \cdot I - A)^{-1}G. \quad (5.7)$$

Thus, the frequency response $H(\cdot)$ contains the same information as the impulse response $h(\cdot)$ from (5.5); the two are Fourier pairs.

Finally, if A admits the eigenvalue decomposition $A = T\Lambda T^{-1}$ where Λ is diagonal, then $H(\omega) = T(i\omega \cdot I - \Lambda)^{-1}T^{-1}G$ where the matrix being inverted is diagonal. Therefore, we can view the frequency response as a weighted sum of first-order responses, $1/(i\omega - \lambda_j)$.

The assumption that A is stable deserves a comment: The solution (5.6) includes a response $\exp(At)x_0$ to the initial condition. When A is stable, this element in the solution vanishes as $t \to \infty$ so that the solution eventually becomes independent of the initial condition, and then only the frequency response remains. We can define the frequency response as $(i\omega \cdot I - A)^{-1}G$ even when A is not stable, but in that case the response to the initial condition will grow to dominate the solution, unless the initial condition is chosen with mathematical precision so that its response is 0.

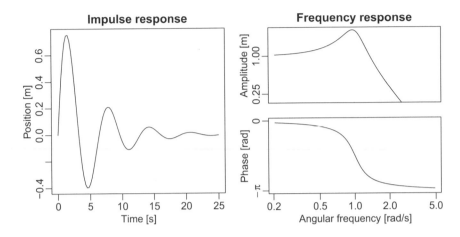

Figure 5.2 Impulse and frequency response for the mass-spring-damper system with parameters $m = 1$ kg, $k = 1$ N/m, and $c = 0.4$ Ns/m. The applied impulse has magnitude 1 Ns. The amplitude response peaks near $\omega = 1$ rad/s, corresponding to the damped oscillations with a period near 2π in the impulse response. At slower frequencies, the response is in phase with the excitation, while at faster frequencies, the response is lagging behind and eventually in counterphase.

For small systems, or when the system matrix A has a particular structure, it is feasible to do the matrix inversion in (5.7) analytically. For the mass-spring-damper system (5.4), we find

$$H(\omega) = \frac{1}{k + ic\omega - m\omega^2} \begin{pmatrix} 1 \\ i\omega \end{pmatrix}.$$

Figure 5.2 shows the frequency response of the position, i.e., the top element of this vector.

In the time domain, we could find the solution for a general forcing $\{U_t : t \geq 0\}$ using the impulse response. The same applies in the frequency domain. Here, it is convenient to consider inputs $\{U_t : t \in \mathbf{R}\}$ defined also for negative times, instead of an initial condition on X_0. If this input $\{U_t\}$ is square integrable, then its Fourier transform $\{U^*(\omega) : \omega \in \mathbf{R}\}$ exists, so $\{U_t\}$ can be decomposed into harmonics. If furthermore A is stable (all eigenvalues have negative real parts), then also the response $\{X_t : t \in \mathbf{R}\}$ will have a Fourier transform $\{X^*(\omega) : \omega \in \mathbf{R}\}$. The two will be related by

$$X^*(\omega) = H(\omega)\, U^*(\omega).$$

This formula expresses that each angular frequency ω can be examined independently; this follows from the superposition principle of linear systems that solutions can be added to form new solutions. The response $X^*(\omega)$ is obtained by multiplying each frequency in the input $U^*(\omega)$ with $H(\omega)$, which specifies the amplification and phase shift of an input with that frequency.

5.3 A LINEAR SYSTEM DRIVEN BY NOISE

Up to this point our models have been deterministic, but we now aim for the situation where the driving input of a dynamic system is a stochastic process. Figure 5.3 shows an example for the mass-spring-damper-system, where the force $\{U_t : t \geq 0\}$ is the realization of a stochastic process; we have simulated this force as well as the response of the mass-spring-damper system (Q_t, V_t). In this example, the driving force $\{U_t\}$ is piecewise constant and changes value at random points in time. These time points of change constitute a Poisson process with mean interarrival time τ. At a point of change, a new force is sampled from a Gaussian distribution with mean 0 and variance σ^2, independently of all other variables.

To simulate this process $\{U_t\}$, we first simulated the interarrival times τ_i for $i = 1, 2, \ldots$, i.e., the time between two subsequent jumps. According to the properties of the Poisson process, these τ_i are independently of eachother, and each follows an exponential distribution with mean τ. We then computed the arrival times T_i recursively by $T_0 = 0$, $T_i = T_{i-1} + \tau_i$. Next, we simulated the levels be $F^{(i)} \sim N(0, \sigma^2)$ for $i = 0, 1, \ldots$, again independently of all other variables. Knowing the time points T_i and the levels $F^{(i)}$, we can plot the piecewise constant sample path $\{U_t(\omega) : t \geq 0\}$ as in Figure 5.3. To reach an expression for U_t at any point in time $t \geq 0$, define the epoch $N_t = \max\{i \geq 0 : T_i \leq t\}$ (i.e., the number of jumps that have occurred before time t) and finally define the force process $U_t = F_{N_t}$.

The sample paths of $\{U_t\}$ are piecewise constant, and therefore we can solve the system dynamics (5.3) as an ordinary differential equation, for each realization. In Figure 5.3, the mean time between jumps in the applied force is 20 seconds, so in most cases the systems falls to rest before the force changes again, and the *step* response of the system is clearly visible.

It is instructive to simulate a driving input and the response it causes, but it is also cumbersome. We would like to have simpler and more general analysis tools for computing the statistics of the response, without knowing the realization of the force, but solely from the statistics of the force. In the following sections, we develop these tools for the first and second order statistics of the response, i.e., the mean value and the covariance structure.

5.4 STATIONARY PROCESSES IN TIME DOMAIN

It is a useful simplification to consider *stationary* stochastic inputs, so we first describe such processes. A stationary stochastic process is one where the statistics do not depend on time; note that this does not at all imply that the sample paths are all constant functions of time. There are several notions of

Factbox: [The matrix exponential] For a square matrix $A \in \mathbf{R}^{n \times n}$, the homogeneous linear system $\dot{X}_t = AX_t$ with initial condition $X_0 = x_0 \in \mathbf{R}^n$ has the unique solution $X_t = \exp(At)x_0$ where $\exp(At)$ is termed the matrix exponential. Here, the matrix exponential $P(t) = \exp(At)$ is itself the unique solution to the matrix differential equation

$$\dot{P}(t) = AP(t), \quad P(0) = I \in \mathbf{R}^{n \times n}.$$

The matrix exponential has the *semigroup* property: $\exp(A(s+t)) = \exp(As)\exp(At)$ for $s, t \in \mathbf{R}$.

In principle, the matrix exponential may be computed through its Taylor series

$$\exp(At) = I + At + \frac{1}{2}(At)^2 + \cdots = \sum_{i=0}^{\infty} \frac{1}{i!}(At)^i$$

but the series converges slowly; it is only useful when t is small. Better algorithms (Moler and Van Loan, 2003) are implemented in good environments for scientific computing. Do not confuse the matrix exponential with element-wise exponential; $(e^A)_{ij}$ does not in general equal $e^{A_{ij}}$. In Matlab or R, compare the two functions exp and expm.

If $A = T\Lambda T^{-1}$, then

$$\exp(At) = T\exp(\Lambda t)T^{-1}$$

and if Λ is diagonal with diagonal elements λ_j, then $\exp(\Lambda t)$ is a diagonal matrix with diagonal elements $\exp(\lambda_j t)$. This may also be written as

$$\exp(At) = \sum_{j=1}^{n} v_j e^{\lambda_j t} u_j$$

where u_j and v_j are left and right eigenvectors of A, normalized so that $u_j v_j = 1$: $Av_j = \lambda_j v_j$, $u_j A = \lambda_j u_j$. v_j is a column in T while u_j is a row in T^{-1}.

Similar results exist when A cannot be diagonalized, using the Jordan canonical form. For example,

$$\exp\left(\begin{bmatrix} \lambda & 1 \\ 0 & \lambda \end{bmatrix} t \right) = \begin{bmatrix} e^{\lambda t} & te^{\lambda t} \\ 0 & e^{\lambda t} \end{bmatrix}.$$

These formulas highlight the central importance of eigenvalues and eigenvectors (including generalized eigenvectors) when solving linear systems of differential equations with constant coefficients.

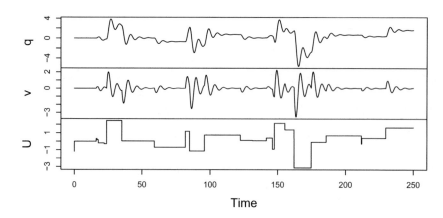

Figure 5.3 The response of the mass-spring-damper system to a force with random steps. Parameters as in Figure 5.2; $\tau = 15$ s. *Top panel:* Position Q [m]. *Middle panel:* Velocity V [m/s]. *Bottom panel:* Force U [N].

stationarity, but for our purpose at this point we only need stationarity of the first and second moment:

Definition 5.4.1 (Second-Order Stationary Process) *A stochastic process* $\{X_t : t \geq 0\}$ *taking values in* $\mathbf{R}^n$ *is said to be second-order stationary, if* $\mathbf{E}|X_t|^2 < \infty$ *for all* $t \geq 0$, *and*

$$\mathbf{E}X_s = \mathbf{E}X_t, \quad \mathbf{E}X_s X_{s+k}^\top = \mathbf{E}X_t X_{t+k}^\top$$

for all $s, t, k \geq 0$.

Second-order stationarity is also referred to as *weak stationarity* or *wide-sense stationarity*. For a second-order stationary process with mean $\bar{x} = \mathbf{E}X_t$, the autocovariance depends only on the time lag k, so we define the autocovariance function $\rho_X : \mathbf{R} \mapsto \mathbf{R}^{n \times n}$ by

$$\rho_X(k) = \mathbf{E}(X_t - \bar{x})(X_{t+k} - \bar{x})^\top.$$

Here t may be chosen arbitrarily. *Careful! The word "autocovariance function" can have slightly different meanings in the literature.* When $X_t = (X_t^{(1)}, \ldots, X_t^{(n)})$ is a vector, i.e. when $n > 1$, the ith diagonal element of $\rho_X(k)$ is the autocovariance of $\{X_t^{(i)}\}$ while an (i, j) off-diagonal element in $\rho_X(k)$ is the covariance $\mathbf{E}(X_t^{(i)} - \bar{x}^{(i)})(X_{t+k}^{(j)} - \bar{x}^{(j)})$.

We now set out to show that the force process $\{U_t : t \geq 0\}$, described in Section 5.3, is second-order stationary, and determine its autocovariance function. To do this, we use a standard trick repeatedly: Include some information, which makes the conditional expectation easy, and then use the tower property (or one of its variants, i.e. the law of total expectation or variance). First, to find the mean and variance of U_t, let $\mathcal{G}$ be the σ-algebra generated by $\{T_i : i \in \mathbf{N}\}$. Then

$$\mathbf{E}\{U_t|\mathcal{G}\} = 0, \quad \mathbf{V}\{U_t|\mathcal{G}\} = \sigma^2,$$

from which we get

$$\mathbf{E}U_t = \mathbf{E}\mathbf{E}\{U_t|\mathcal{G}\} = 0, \quad \mathbf{V}U_t = \mathbf{E}\mathbf{V}\{U_t|\mathcal{G}\} + \mathbf{V}\mathbf{E}\{U_t|\mathcal{G}\} = \mathbf{E}\sigma^2 + \mathbf{V}0 = \sigma^2.$$

To derive the autocovariance function $\rho_U(h) = \mathbf{E}U_t U_{t+h}$ for $h \geq 0$, we condition on $\mathcal{F}_t$, the σ-algebra generated by $\{U_s : 0 \leq s \leq t\}$:

$$\mathbf{E}U_t U_{t+h} = \mathbf{E}(\mathbf{E}\{U_t U_{t+h}|\mathcal{F}_t\}) = \mathbf{E}(U_t \mathbf{E}\{U_{t+h}|\mathcal{F}_t\}).$$

Notice that this connects the problem of determining the autocovariance with that of making predictions. To compute the conditional expectation $\mathbf{E}\{U_{t+h}|\mathcal{F}_t\}$, condition again on the arrival times, i.e. on $\mathcal{G}$. If the force does not change between time t and time $t + h$, then $U_{t+h} = U_t$. On the other hand, if the force changes between time t and time $t+h$, then the conditional expectation of U_{t+h} is 0. That is:

$$\mathbf{E}\{U_{t+h}|\mathcal{F}_t, \mathcal{G}\} = \begin{cases} U_t & \text{if } \{T_i\} \cap [t, t+h] = \emptyset, \\ 0 & \text{else.} \end{cases}$$

Now, the probability that there are no jumps in the time interval is

$$\mathbf{P}[\{T_i\} \cap [t, t+h] = \emptyset|\mathcal{F}_t] = \exp(-h/\tau),$$

since the time between events is exponentially distributed. This hinges on the exponential distribution having no memory, i.e. when we stand at time t and look for the next jump, we do not need to take into account how long time has passed since the previous jump. Combining, we use the Tower property to get

$$\mathbf{E}\{U_{t+h}|\mathcal{F}_t\} = \mathbf{E}[\mathbf{E}\{U_{t+h}|\mathcal{F}_t, \mathcal{G}\}|\mathcal{F}_t] = U_t \exp(-h/\tau),$$

and therefore,

$$\mathbf{E}U_t U_{t+h} = \sigma^2 \exp(-|h|/\tau) = \rho_U(h). \tag{5.8}$$

Here we have used symmetry to obtain also autocovariance at negative lags. Note that $\mathbf{E}U_t U_{t+h}$ does not depend on the time t, but only on the time lag h, so the process $\{U_t\}$ is second order stationary.

The form of the autocovariance is archetypal: It contains a variance, σ^2, and a time constant τ which measures the time scale over which the autocovariance function decays and is therefore termed the *decorrelation time*. See Figure 5.4 (left panel).

5.5 STATIONARY PROCESSES IN FREQUENCY DOMAIN

Just as frequency domain methods are useful for linear systems (Section 5.2), they are also useful for stationary processes. If[1] the autocovariance function

---

[1] According to the Wiener-Khinchin theorem, this requirement can be relaxed.

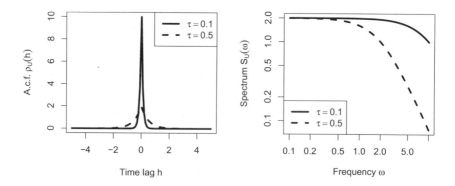

Figure 5.4 Autocovariance function and variance spectrum (in log scales) of the piecewise constant force process $\{U_t : t \geq 0\}$, for two values of the time constant τ. The variance σ^2 is chosen so that $\sigma^2\tau = 1$. Note, in the spectrum, that the *cut-off frequency* $1/\tau$ marks a transition from low-frequency behavior to high-frequency behavior.

ρ_X is $\mathcal{L}_2$, we can define the variance spectrum S_X as its Fourier transform:

$$S_X(\omega) = \int_{-\infty}^{+\infty} \rho_X(t) \, \exp(-i\omega t) \, dt.$$

To justify the name variance spectrum, note that by the inverse Fourier transform, we can compute the autocovariance function from the variance spetrum,

$$\rho_X(t) = \frac{1}{2\pi} \int_{-\infty}^{+\infty} S_X(\omega) \exp(i\omega t) \, d\omega$$

and in particular, with a time lag $t = 0$, we get

$$\mathbf{V}X_t = \rho_X(0) = \frac{1}{2\pi} \int_{-\infty}^{+\infty} S_X(\omega) \, d\omega.$$

We see that the variance spectrum $S_X(\omega)$ decomposes the variance of X_t into contributions from cycles of different frequencies, which justifies the name "variance spectrum".

Remark 5.5.1 *The literature does not agree on where to put the factor 2π. Also, much of the physics and engineering literature measures frequencies in cycles per time ν rather than radians per time ω; i.e., $\omega = 2\pi\nu$. For a scalar process, we can consider positive frequencies only, use $d\omega = 2\pi d\nu$, and obtain*

$$\mathbf{V}X_t = \rho_X(0) = \int_0^{+\infty} 2S_X(2\pi\nu) \, d\nu.$$

In the scalar case, we may replace the complex exponential $\exp(-i\omega h)$ with the cosine $\cos(\omega h)$ since the autocorrelation function is even, and the spectrum is real-valued. This gives the simpler expression

$$S_X(\omega) = \int_{-\infty}^{+\infty} \rho_X(h) \cos(\omega h) \, dh. \tag{5.9}$$

For the particular example of the driving force in Section 5.3, the autocovariance function (5.8) is $\mathcal{L}_2$, so we get the spectrum

$$S_U(\omega) = \int_{-\infty}^{+\infty} \rho_U(h) \exp(-i\omega h) \, dh = \frac{2\sigma^2 \tau}{1 + \omega^2 \tau^2} \tag{5.10}$$

as shown in Figure 5.4 (right panel). This form is archetypal: A low-frequency asymptote $2\sigma^2\tau$ expresses the strength of slow oscillations present in the force, and at high frequencies, there is a roll-off where the contribution of harmonics in the force decays with ω^2. Between the two, there is a cut-off frequency $\omega = 1/\tau$, corresponding to the decorrelation time τ, which indicates the transition from slow to fast modes.

5.6   THE RESPONSE TO NOISE

We now investigate how a linear system responds to a stochastic input $\{U_t\}$.

We consider the general linear system (5.3) where the initial condition $x_0 \in \mathbf{R}^n$ is deterministic, while the input $\{U_t : t \geq 0\}$ is a stochastic process with mean $\bar{u}(t) = \mathbf{E}U_t$ and autocovariance $\rho_U(s,t) = \mathbf{E}(U_s - \bar{u}(s))(U_t - \bar{u}(t))^\top$. We assume that $\{U_t : t \geq 0\}$ is so well-behaved that we can, for each realization of $\{U_t : t \geq 0\}$, compute the corresponding realization of the solution $\{X_t : t \geq 0\}$ by means of the solution formula (5.6).

In this formula (5.6), we can take expectation on both sides. Fubini's theorem allows as to commute expectation and time integration, so we obtain for the mean value $\bar{x}(t) = \mathbf{E}X_t$:

$$\bar{x}(t) = e^{At}x_0 + \int_0^t e^{A(t-s)} G\bar{u}(s) \, ds.$$

Differentiating with respect to time, we obtain an ordinary differential equation governing the mean value:

$$\frac{d}{dt}\bar{x}(t) = A\bar{x}(t) + G\bar{u}(t). \tag{5.11}$$

This equation can be solved uniquely for each initial condition $\bar{x}(0) = x_0$. We see that we can obtain the governing ordinary differential equation for the mean value simply by taking expectation in (5.3).

Next, we aim to obtain the covariance $\rho_X(s,t) = \mathbf{E}(X_s - \bar{x}(s))(X_t - \bar{x}(t))^\top$. Using $\tilde{U}_t = U_t - \bar{u}(t)$ and $\tilde{X}_t = X_t - \bar{x}(t)$ for the deviations of the processes from their mean values, we first write integral formulas for the deviation at time s and t:

$$\tilde{X}_s = \int_0^s e^{Av} G \tilde{U}_{s-v} \, dv \text{ and } \tilde{X}_t = \int_0^t e^{Aw} G \tilde{U}_{t-w} \, dw.$$

Combining the two, and commuting the expectation and integration over time, we obtain

$$\rho_X(s,t) = \mathbf{E}\tilde{X}_s \tilde{X}_t^\top = \int_0^s \int_0^t e^{Av} G \rho_U(s-v, t-w) G^\top e^{A^\top w} \, dw \, dv. \quad (5.12)$$

These integrals may not seem illuminating, but have patience – we will soon see that they lead to a very tractable and explicit result in the frequency domain. Focus on the special case where the input $\{U_t\}$ is stationary and the system is exponentially stable, i.e., all eigenvalues of A have negative real part, so that the effect of the initial condition and old inputs vanishes as time progresses. In this case, there exists a solution $\{X_t : t \geq 0\}$ which is also wide-sense stationary.[2] We focus on this solution.

Writing $\rho_U(t-s)$ for $\rho_U(s,t)$ and $\rho_X(t-s)$ for $\rho_X(s,t)$, we obtain

$$\rho_X(t-s) = \int_0^s \int_0^t e^{Av} G \rho_U(t-s+v-w) G^\top e^{A^\top w} \, dw \, dv$$

and for $s, t \to \infty$ with $l = t - s$ fixed, this converges to

$$\rho_X(l) = \int_0^\infty \int_0^\infty e^{Av} G \rho_U(l+v-w) G^\top e^{A^\top w} \, dw \, dv. \quad (5.13)$$

since $\exp(Av)$ converges exponentially to zero as $v \to \infty$, and since $\rho_U(\cdot)$ is bounded by $\rho_U(0)$.

Now we are ready to jump to frequency domain. Taking Fourier transform of the autocovariance function $\rho_X(\cdot)$, we obtain the variance spectrum

$$S_X(\omega) = \int_{-\infty}^{+\infty} \rho_X(l) \, \exp(-i\omega l) \, dl.$$

It is a standard result for Fourier transforms that convolutions in time domain, such as (5.13), correspond to multiplication in frequency domain. With this, we obtain

$$S_X(\omega) = H(-\omega) \cdot S_U(\omega) \cdot H^\top(\omega). \quad (5.14)$$

Exercise 5.1: Verify the result (5.14).

In case where X_t and U_t are scalar, we get the simpler formula

$$S_X(\omega) = |H(\omega)|^2 \cdot S_U(\omega).$$

In words, the contribution from a given frequency ω to the variance of X_t depends on its presence in $\{U_t\}$ and its magnification through system dynamics.

---

[2]There exist also non-stationary solutions, differing in the initial condition X_0.

5.7   THE WHITE NOISE LIMIT

For the example in Figures 5.2 and 5.3, the impulse response of the mass-spring-damper system displayed damped oscillations with a period near 2π s, since the eigenvalues have magnitude $1\ \text{s}^{-1}$ and are near the imaginary axis. In turn, the driving force was constant of periods of 15 s, on average. In short, the driving force was slow compared to the system dynamics. The stochastic differential equations we are interested in are characterized by the opposite: The driving noise is fast compared to system dynamics. We can, for example, consider the mass-spring-damper system subjected to random forces from collisions with air molecules. The assumption is that there is a separation of time scales where the system evolves on slow time scales, while the force fluctuates on fast time scales. This separation of time scales allows us to simplify the analysis.

Figure 5.5 shows spectra for the situation where the force applied to the mass-spring-damper system is faster than the system dynamics. Specifically, the resonance frequency is still 1 rad/s corresponding to a period of 2π s, but now the mean time between force jumps is $\tau = 0.1$ s.

In terms of the spectra, we see that in the frequency range up to, say, 5 rad/s, we can approximate the spectrum of the driving force with a constant function

$$S_U(\omega) \approx 2\sigma^2\tau \text{ for } \omega \ll 1/\tau.$$

Moreover, we see that the total variance of the response Q_t is not very sensitive to the spectrum of the force F at frequencies larger than 5 rad/s, since the frequency response of the system is small for such high frequencies. Therefore we may as well ignore the details of the spectrum S_U at high frequencies, and approximate $S_U(\omega)$ with the constant $2\sigma^2\tau$ for all frequencies ω. This is called the *white noise approximation*.

Why do we call this white noise? Recall that when characterizing colors, white light has the property that all colors of the rainbow (i.e., all frequencies or wavelengths) are equally present. By analogy, a white noise signal is one where all frequencies contribute equally to the variance or power. Scalar white noise is characterized by one number, its intensity, which we take as the constant value of the variance spectrum. In this example, we approximate the force U_t with white noise with intensity $2\sigma^2\tau$.

White noise signals are an idealization: Such a signal would have infinite variance, since its variance spectrum is not integrable. But it is a useful approximation; when the force changes rapidly compared to the system dynamics, we may approximate the force with a white noise signal. The approximation is valid as long as we operate in the frequency range of the system, i.e., at frequencies $\omega \ll 1/\tau$. We find that the spectrum of the position is well approximated by

$$S_Q(\omega) \approx 2\sigma^2\tau|H(\omega)|^2 \text{ for } \omega \ll 1/\tau.$$

Approximating the force with white noise amounts to letting $\tau \to 0$, but at the same time letting $\sigma^2 \to \infty$ so that the spectrum $S_U(0) = 2\sigma^2\tau$ remains

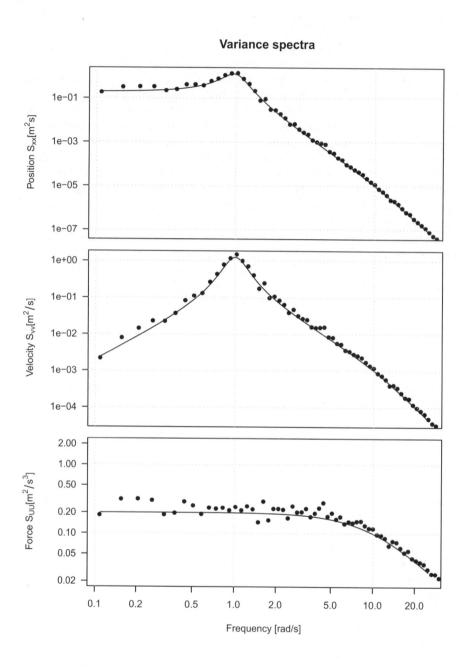

Figure 5.5 Variance spectrum of position, velocity, and force for the mass-spring-damper system. Lines are the analytical expressions. Dots are estimated spectra based on simulation of the process. The average time between jumps in the force is $\tau = 0.1$ s.

constant at frequency 0. At any other frequency ω, we have (pointwise) convergence $S_U(\omega) \to S_U(0)$ as $\tau \to 0$. In terms of the autocovariance function of the driving force, which was

$$\rho_U(h) = \sigma^2 \exp(-|h|/\tau),$$

we see that this corresponds to approximating the autocovariance function with a Dirac delta:

$$\rho_U(h) \to 2\sigma^2\tau \cdot \delta(h).$$

In the limit, $\rho_U(h)$ vanishes for any non-zero h, and therefore the time-domain characterization of white noise is independence, i.e., U_s and U_t are uncorrelated for any $s \neq t$.

5.8 INTEGRATED WHITE NOISE IS BROWNIAN MOTION

In this section, we show the connection between white noise and Brownian motion: Brownian motion can, formally, be regarded as integrated white noise. Stated differently, white noise can – formally – be seen as the derivative of Brownian motion.

First, we investigate the difference quotient of Brownian motion. Let $\{B_t : t \geq 0\}$ be Brownian motion, let a time lag k be given, and define the stochastic process $\{X_t : t \geq 0\}$ by

$$X_t = \frac{1}{k}(B_{t+k} - B_t).$$

This X_t is a difference quotient, and for small k we may think of $\{X_t : t \geq 0\}$ as an approximation to the (non-existing) derivative of Brownian motion.

Exercise 5.2: Show that $\{X_t : t \geq 0\}$ is second order stationary, has mean 0, and the following autocovariance function:

$$\rho_X(h) = \frac{k - |h|}{k^2} \vee 0. \tag{5.15}$$

(Recall our notation that $a \vee b = \max(a, b)$)

The autocorrelation function (5.15) is shown in Figure 5.6. Note that as the time lag k decreases towards 0, the a.c.f. approaches a Dirac delta.[3] This justifies the useful but imprecise statement that the derivative of Brownian motion is delta-correlated.

Figure 5.6 displays also the spectrum of the difference quotient $\{X_t : t \geq 0\}$. The analytical expression for this spectrum is

$$S_X(\omega) = \begin{cases} 2\frac{1-\cos \omega k}{\omega^2 k^2} & \text{for } \omega \neq 0, \\ 1 & \text{for } \omega = 0. \end{cases}$$

---

[3]In the sense of weak convergence of measures, i.e., $\int_{-\infty}^{+\infty} f(h)\rho_X(h)\, dh \downarrow f(0)$ as $h \downarrow 0$ for any continuous function f.

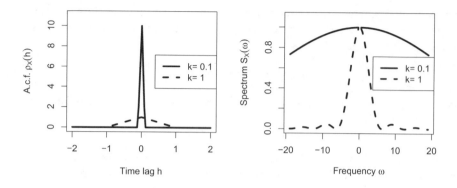

Figure 5.6 Autocorrelation function (left) and variance spectrum (right) of the difference quotient of Brownian motion.

Note that the spectrum at frequency 0 is $S_X(0) = 1$ for any k, since the a.c.f. integrates to 1 for any k. Note also that as $k \to 0$, the spectrum $S_X(\omega)$ converges to the constant 1, for any frequency ω, in agreement with the a.c.f. approaching a Dirac delta. So as the time lag vanishes, $k \to 0$, the spectrum of the difference quotient approaches that of white noise. This motivates the statement "the derivative of Brownian motion is white noise", which is useful but should not be taken too literally since Brownian motion is not differentiable.

Now, conversely, consider a white noise signal $\{U_t\}$ with mean 0 and auto-covariance function $\rho_U(h) = \delta(h)$, which is to say that its variance spectrum is $S_U(\omega) = 1$. The following derivation is purely formal, so try not to be disturbed by the fact that such a signal does not exist! Instead, define the integral process $\{B_t : t \geq 0\}$

$$B_t = \int_0^t U_s \, ds$$

and consider the covariance structure of $\{B_t : t \geq 0\}$. We can apply formula (5.12) with $A = 0$, $G = 1$ to get

$$\rho_B(s,t) = \mathbf{E} B_s B_t = s.$$

In particular, $\mathbf{V} B_t = t$. By stationarity, we get $\mathbf{V}(B_t - B_s) = t - s$ for any $0 < s < t$. *Exercise: Show that the increments of $\{B_t : t \geq 0\}$ are uncorrelated. That is, assume $0 < s < t < v < w$, and show that $\mathbf{E}(B_t - B - s)(B_w - B_v) = 0$.* We see that the mean and covariance structure of $\{B_t : t \geq 0\}$ agrees with our Definition 4.2.1 of Brownian motion. The Gaussianity and continuity of $\{B_t\}$ do not follow from this argument; there we need more properties of the white noise signal $\{U_t\}$. Regardless, this formal calculation justifies the statement "Brownian motion is integrated white noise". Again, this statement is useful but should not be taken too literally since continuous time white noise does not exist as a stochastic process in our sense.

5.9   LINEAR SYSTEMS DRIVEN BY WHITE NOISE

We now return to the linear system driven by noise

$$\dot{X}_t = AX_t + GU_t. \tag{5.16}$$

We are interested in the limiting case where U_t approaches white noise, corresponding to mass-spring-damper example when the mean time τ between jumps tends to 0. We will call this limit a *linear stochastic differential equation in the narrow sense*, because the drift term AX_t is linear in the state, and the noise term GU_t is independent of the state. We say that the noise enters *additively*. Here, we approach this limit indirectly since white noise does not exist as a stochastic process. We first integrate the equation w.r.t. dt to obtain

$$X_t - X_0 = \int_0^t AX_s\, ds + G \int_0^t U_s\, ds.$$

Now, if $\{U_t\}$ approximates white noise, the results of the previous section show that the last integral will approximate Brownian motion

$$X_t - X_0 = \int_0^t AX_s\, ds + GB_t. \tag{5.17}$$

Since this equation involves Brownian motion and not white noise, it does not suffer from the problem that white noise does not exist. We shall see, in the following chapters, that it is the right starting point for a general theory of stochastic differential equations; at that point we will re-write the equation using the notation

$$dX_t = AX_t\, dt + G\, dB_t. \tag{5.18}$$

If the time t is small, we can approximate $\int_0^t AX_s\, ds \approx AX_0 t$, which leads to

$$X_t \approx X_0 + AX_0 t + GB_t.$$

This leads to an Euler-type method, known as the Euler-Maruyama method, for solving the equation approximately recursively:

$$X_{t+h} = X_t + AX_t h + G(B_{t+h} - B_t). \tag{5.19}$$

This algorithm allows us to simulate sample paths of $\{X_t\}$. This is the same algorithm we pursued in Chapter 2 when simulating advective and diffusive transport.

Note that if X_t is Gaussian, then so is X_{t+h}; in fact, the entire process is Gaussian. Let us identify the mean and covariance structure of this X_t. With $\bar{x}_t = \mathbf{E}X_t$, we get from (5.11)

$$\bar{x}_t = \exp(At)\, \bar{x}_0.$$

For the covariance, we obtain from (5.12)

$$\rho_X(s,t) = \int_0^s e^{A(s-v)} GG^\top e^{A^\top(t-v)} \, dv,$$

using that the autocovariance function of the noise is $\rho_U(v,w) = \delta(v-w)$. It is convenient to first look at the variance at time t, $\Sigma(t) = \rho(t,t)$:

$$\Sigma(t) = \int_0^t e^{A(t-v)} GG^\top e^{A^\top(t-v)} \, dv. \tag{5.20}$$

Differentiating with respect to t, we obtain

$$\frac{d}{dt}\Sigma(t) = A\Sigma(t) + \Sigma(t)A^\top + GG^\top. \tag{5.21}$$

This is a linear matrix differential equation, known as the differential Lyapunov equation. Together with the initial condition $\Sigma(0) = 0$ it determines the variance function. See Exercise 5.7 for an alternative derivation, and Exercise 5.8 for methods for finding the solution numerically. With $\Sigma(\cdot)$ in hand, we can find the autocovariance function:

$$\rho(s,t) = \mathbf{E}\tilde{X}_s \tilde{X}_t^\top \tag{5.22}$$
$$= \mathbf{E}(\mathbf{E}\{\tilde{X}_s \tilde{X}_t^\top | X_s\}) \tag{5.23}$$
$$= \mathbf{E}(\tilde{X}_s \mathbf{E}\{X_t^\top | X_s\}) \tag{5.24}$$
$$= \mathbf{E}(\tilde{X}_s \tilde{X}_s^\top e^{A^\top(t-s)}) \tag{5.25}$$
$$= \Sigma(s) \cdot e^{A^\top(t-s)}. \tag{5.26}$$

Here, we have used that the equation (5.11) for the mean also applies to *conditional* expectations, so that $\mathbf{E}\{X_t|X_s\} = \exp(A(t-s))X_s$.

Of special interest are second-order stationary solutions where $\Sigma(t)$ does not depend on t, but satisfies the *algebraic* Lyapunov equation

$$A\Sigma + \Sigma A^\top + GG^\top = 0. \tag{5.27}$$

This Σ is an equilibrium of the differential Lyapunov equation (5.21). It can be shown that this linear matrix equation in Σ has a unique solution if A contains no eigenvalues on the imaginary axis. See Exercise 5.8 for one way to solve it. If A is exponentially stable (all eigenvalues in the open left half plane), then the unique solution Σ is positive semidefinite, and the equation expresses a balance between variance pumped into the system by noise (the term $GG^\top$) and dissipated by the stable system dynamics A. In this case, $\Sigma(t) \to \Sigma$ as $t \to \infty$. This Σ will be positive definite if all linear combinations of states in the system are affected by the noise. This will be the case if G is square and invertible; a weaker and sufficient condition is that the pair (A, G) is *controllable* (see Section 9.11.1).

To elaborate on the stability of A, consider the scalar equation $dX_t = X_t \, dt + dB_t$. The algebraic Lyapunov equation is $2\Sigma + 1 = 0$ so $\Sigma = -1/2!$ The explanation is that the system is unstable, so X_t diverges to infinity, and no steady-state exists.

In summary, in this common situation – a stable system all the dynamics of which are excited by the noise – the process X_t will approach a stationary Gaussian process. The stationary variance is Σ, the unique solution to the algebraic Lyapunov equation (5.27), and the autocovariance function is

$$\rho(h) = \Sigma \exp(A^\top h) \text{ for } h \geq 0. \tag{5.28}$$

For $h < 0$, we use the relationship $\rho_X(-h) = \rho_X^\top(h)$. Note that the matrix exponential $\exp(At)$ determines both the impulse response (5.5) and the autocovariance function of stationary fluctuations. This at the core of so-called *fluctuation-dissipation* theory from statistical physics.

5.10 THE ORNSTEIN-UHLENBECK PROCESS

The simplest example of a linear system driven by white noise $\{U_t\}$ arises when the system is scalar:

$$\dot{X}_t = -\lambda X_t + \sigma U_t$$

where $\lambda, \sigma > 0$. We take $\{U_t\}$ to have unit intensity. This can alternatively be written with the notation of (5.18) as

$$dX_t = -\lambda X_t \, dt + \sigma \, dB_t.$$

This equation is referred to as the Langevin equation. A simulation of this process is seen in Figure 5.7. Its expectation satisfies

$$\bar{x}_t = \mathbf{E}X_t = e^{-\lambda t}\mathbf{E}X_0.$$

The variance $\Sigma(t) = \mathbf{V}X_t$, in turn, satisfies the differential Lyapunov equation $\dot{\Sigma}(t) = -2\lambda\Sigma(t) + \sigma^2$, i.e.,

$$\mathbf{V}X_t = \Sigma(t) = \frac{\sigma^2}{2\lambda} + \left(\Sigma(0) - \frac{\sigma^2}{2\lambda}\right)e^{-2\lambda t}.$$

The stationary solution $\{X_t : t \geq 0\}$ to this equation is called the *Ornstein-Uhlenbeck* process. Its mean is 0 and its variance satisfies the algebraic Lyapunov equation $-2\lambda\mathbf{V}X_t + \sigma^2 = 0$ or $\mathbf{V}X_t = \sigma^2/(2\lambda)$, which gives the autocovariance function

$$\rho_X(h) = \frac{\sigma^2}{2\lambda}e^{-\lambda|h|}.$$

Notice that the form of this a.c.f. coincides with that of the force $\{F_t\}$ from the mass-spring-damper system (equation (5.8) and Figure 5.4); i.e., a two-sided

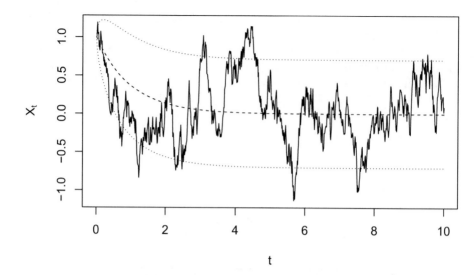

Figure 5.7 A simulation of the Ornstein-Uhlenbeck process (solid) including the expectation (dashed) and plus/minus one standard deviation (dotted). Parameters are $X_0 = 1$, $\lambda = 1$, $\sigma = 1$.

exponential. So the variance spectrum of the Ornstein-Uhlenbeck process has the form (5.10) and is seen in Figure 5.4. However, the Ornstein-Uhlenbeck process is Gaussian while the force $\{F_t\}$ from the mass-spring-damper example was not. The Ornstein-Uhlenbeck process is also referred to as *low-pass filtered white noise*, although this term can also be applied to other processes. It is a fundamental building block in stochastic models in more or less all areas of applications.

5.11 THE NOISY HARMONIC OSCILLATOR

Another basic example of a linear system driven by white noise is the noisy harmonic oscillator. This also serves as a fundamental building block in stochastic models, when you need a stochastic process which is oscillatory and dominated by a specific frequency. The mass-spring-damper system can be seen as a noisy harmonic oscillator, when we subject it to a white noise input. However, an alternative formulation which has a more symmetric form is obtained with the linear stochastic differential equation

$$\dot{X}_t = AX_t + \sigma U_t, \quad \text{or } dX_t = AX_t \, dt + \sigma \, dB_t.$$

Here, $X_t \in \mathbf{R}^2$. The noise process $\{U_t\}$ is two-dimensional; its elements are independent white noise with unit intensity. Correspondingly, $\{B_t : t \geq 0\}$ is two-dimensional standard Brownian motion. The system matrix is

$$A = \begin{bmatrix} -\mu & -k \\ k & -\mu \end{bmatrix}.$$

The parameter k specifies the dominating frequency, while μ specifies the damping and σ scales the process and specifies the variance – see Exercise 5.3 to see exactly how. Simulations of this system are seen in Figure 5.8 for two sets of parameters. In the one case, we take $k = 1$, $\mu = 0.05$, and see quite persistent and regular oscillations. In the second case, $k = 1$, $\mu = 0.5$, the damping is higher, so the oscillations are more irregular. These patterns are also visible in the autocovariance function and in the variance spectrum – note that lower damping implies more sustained oscillations in the a.c.f. and a more defined resonance peak in the variance spectrum.

Exercise 5.3:

1. Show that

$$\exp(At) = e^{-\mu t} \begin{bmatrix} \cos kt & -\sin kt \\ \sin kt & \cos kt \end{bmatrix}.$$

2. Show that the stationary variance of $\{X_t\}$ is

$$\Sigma = \frac{\sigma^2}{2\mu} I.$$

3. Show that the a.c.f. is

$$\rho_X(t) = \frac{\sigma^2}{2\mu} e^{-\mu|t|} \begin{bmatrix} \cos kt & -\sin kt \\ \sin kt & \cos kt \end{bmatrix}.$$

5.12 CONCLUSION

Linear systems of ordinary differential equations driven by random inputs make a tractable class of stochastic dynamic systems. We can determine the mean and autocovariance structure quite explicitly, and even if these two statistics do not fully describe a stochastic process, they may be sufficient for a given purpose. In the stationary case, where systems are stable and we assume that the effect of a distant initial condition has decayed, we obtain explicit formulas, most clearly in frequency domain (5.14): The spectrum of the output is obtained by multiplying the spectrum of the input with the squared frequency response.

When the noise fluctuates fast relative to system dynamics, it may be an advantage to approximate it with white noise. In time domain, this corresponds to approximating the autocovariance function with a Dirac delta, while in frequency domain, it corresponds to approximating the variance spectrum

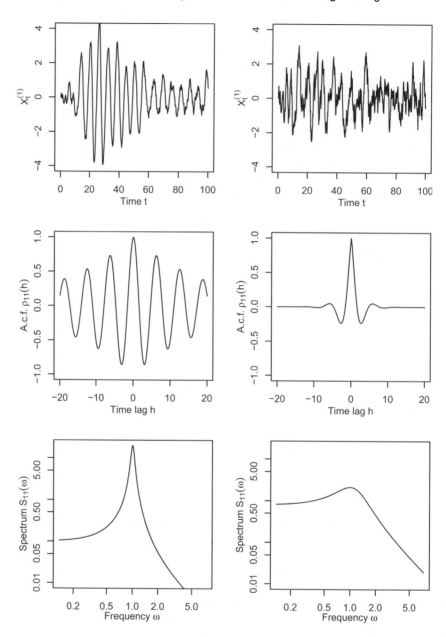

Figure 5.8 The noisy harmonic oscillator with $k = 1$. *Left column:* $\mu = 0.05$, $\sigma^2 = 10$. *Right column:* $\mu = 0.5$, $\sigma^2 = 1$. *Top row:* Simulated sample paths of $\{X_t^{(1)}\}$. *Middle row:* The a.c.f. of $\{X_t^{(1)}\}$. *Bottom row:* The variance spectrum of $\{X_t^{(1)}\}$.

with a constant function. It should be kept in mind that white noise only exists as an idealization. Linear systems drive by white noise is particularly simple to analyze with respect to variance structure; the Lyapunov equation is a key element. Such linear systems are a simple special case of stochastic differential equations. This theory is highly useful in practice.

Frequency domain concepts and techniques are clear and powerful for the case of linear systems, because each frequency can be analyzed in isolation. For the nonlinear systems we study in the following chapters, frequency domain techniques are less directly applicable, so our discussion will focus on time domain. However, when modeling any dynamic system, it is good practice to scrutinize which dynamics are active on which timescales, and which timescales should be resolved by the model. This analysis will give guidance to which dynamics should be considered "slow" and approximated with constants, and which dynamics should be considered "fast" and represented by (possibly filtered) white noise. In this process, frequency domain notions are very useful for framing the discussion, even if the overall model contains nonlinear components.

From the point of view of constructing a theory, it is worth noting that white noise presents a challenge, because it does not exist as a stochastic process *per se* but rather represents a limit. However, we were able to circumvent this problem in two ways: First, we reformulated the differential equation driven by white noise as an *integral* equation where Brownian motion appears. Next, we discretized time with the Euler-Maruyama method. We shall see that these two techniques are also key for non-linear equations, even if they need a bit more effort there.

5.13 EXERCISES

Exercise 5.4 Thermal Noise in an Electrical Circuit: In 1928, John B. Johnson and Harry Nyquist found that random movements of electrons in electrical components have the effect of an extra voltage supply, which can be approximated by a white noise source. For the RC-circuit (resistor-capacitor)

$$R\dot{Q}_t + \frac{Q_t}{C} = V_t$$

where Q_t is the charge in the capacitor, noise in the resistor acts as an external voltage supply $\{V_t\}$ which is white noise. Considering positive frequencies only, its spectrum is $4k_B T R$; with our notation, we have $S_V(\omega) = 2k_B T R$ (compare remark 5.5.1). Here, $k_B = 1.4 \cdot 10^{-23}$ J K^{-1} is the Boltzmann constant, T is the temperature, and R is the resistor.

1. Taking $V_t = \sigma \, dB_t/dt$, where $\{B_t\}$ is standard Brownian motion, find σ. Find the numeric value of σ for $T = 300$ K, $R = 1$ kΩ, $C = 1$ nF.

2. Find a stochastic differential equation which governs Q_t.

3. Find the stationary mean, variance, a.c.f., and variance spectrum, of the charge $\{Q_t\}$ and of the voltage $\{Q_t/C\}$ over the capacitor.

4. Find numerical values for the r.m.s. charge $\{Q_t\}$, the voltage $\{Q_t/C\}$, and their decorrelation time, for the parameters in question 1.

Exercise 5.5: This exercise reproduces Figure 1.1 on page 2 by posing and solving differential equations for the motion in a wind turbine. The equations are

$$
\begin{aligned}
\dot{F}_t &= -\lambda(\bar{f} - F_t) + \sigma\xi_t, \\
\dot{X}_t &= V_t \\
\dot{V}_t &= -kX_t - \mu V_t + F_t
\end{aligned}
$$

where F_t is the force from the wind on the turbine, X_t is the position, V_t is the velocity, and $\{\xi_t\}$ is white noise with unit intensity which drives the force. All quantities are dimensionless; as parameters we take $\lambda = 0.5$, $k = 1$, $\mu = 0.5$, $\bar{f} = 3$, $\sigma = 1$, and the initial conditions are $F_0 = 0.5$, $X_0 = V_0 = 0$.

1. Simulate the noise-free version of the system $(U_t \equiv 0)$ for $t \in [0, 30]$ and plot the force, the position, and the velocity.

2. Include noise and simulate the system with the Euler-Maruyama method. Construct a plot similar to Figure 1.1.

3. Extend the simulation to a longer period, for example $[0, 1000]$. Compute the empirical mean and variance-covariance matrix of force, position, and velocity. Compute the same quantities analytically by solving the algebraic Lyapunov equation (5.27) and compare.

Exercise 5.6: Consider the white noise limit of the wind turbine model from the previous exercise, i.e.,

$$
\dot{X}_t = V_t, \quad \dot{V}_t = -kX_t - \mu V_t + s\,\xi_t.
$$

Here we have set the average force $\bar{f}$ to 0; this corresponds to a shift of origin for the position. Show that in stationarity, the position and velocity are uncorrelated. Next, show that *equipartioning* of energy holds, i.e., the average kinetic energy $\mathbf{E}\frac{1}{2}V_t^2$ equals the average potential energy $\mathbf{E}\frac{1}{2}kX_t^2$.

Exercise 5.7 The Differential Lyapunov Equation Revisited: An alternative derivation of the differential Lyapunov equation (5.21)

$$\dot{\Sigma}(t) = A\Sigma(t) + \Sigma(t)A^\top + GG^\top$$

is as follows: Assume that $\Sigma(t) = \mathbf{V}X_t$ is given. We aim to find $\Sigma(t+h)$. To this end, use the Euler-Maruyama scheme

$$X_{t+h} = X_t + AX_t h + G(B_{t+h} - B_t).$$

Use this to find the variance $\Sigma(t+h)$ of X_{t+h}. Divide with h and let $h \to 0$ to find a differential equation for $\Sigma(t)$.

Exercise 5.8 Numerical Solution of the Differential Lyapunov Equation: Consider again the differential Lyapunov equation (5.21)

$$\dot{\Sigma}(t) = A\Sigma(t) + \Sigma(t)A^\top + GG^\top, \quad \Sigma(0) = Q = Q^\top,$$

governing the variance-covariance matrix for the linear system $dX_t = AX_t\, dt + G\, dB_t$. Here, A, $\Sigma(t)$, $GG^\top$ and Q are n-by-n matrices.

1. Show the following: If there exists a $S = S^\top$ such that $AS + SA^\top + GG^\top = 0$, then the solution can be written

$$\Sigma(t) = S - e^{At}(S - Q)e^{A^\top t}$$

Note: If A is strictly stable, then S is guaranteed to exist, will be non-negative definite, and is the steady-state variance-covariance matrix of the process, i.e., $\Sigma(t) \to S$ as $t \to \infty$.

2. Show that the differential Lyapunov equation can be written as

$$\dot{s}_t = Ms_t + g$$

where s_t is a column vector made from entries of $\Sigma(t)$ by stacking columns on top of each other, g is a column vector made from $GG^\top$ in the same way, and M is the n^2- by- n^2 matrix

$$M = A \otimes I + I \otimes A$$

where $\otimes$ is the Kronecker product, and I is an n-by-n identity matrix.

3. Show that the solution can be written as

$$\begin{pmatrix} g \\ s_t \end{pmatrix} = e^{Pt} \begin{pmatrix} g \\ s_0 \end{pmatrix}$$

where

$$P = \begin{bmatrix} 0 & 0 \\ I & M \end{bmatrix}$$

Hint: Show that (g, s_t) satisfies the linear ODE

$$\begin{pmatrix} \dot{g} \\ \dot{s}_t \end{pmatrix} = \begin{pmatrix} 0 \\ g + Ms_t \end{pmatrix}$$

4. Now, let $\Sigma(t; Q)$ be the solution of the differential Lyapunov condition with non-zero initial condition $\Sigma(0) = Q$. Show that $\Sigma(t; Q) = \exp(At)Q\exp(A^\top t) + S(t; 0)$; i.e., the variance of X_t has two elements: One that stems from the variance of X_0, and one that stems from the noise in the interval $[0, t]$.

These algorithms have been implemented in function dLinSDE in SDEtools.

Exercise 5.9 The Brownian Bridge: Let $\{X_t : 0 \le t \le T\}$ be the solution to the Itô equation

$$dX_t = \frac{b - X_t}{T - t} \, dt + dW_t$$

where $\{W_t\}$ is standard Brownian motion. Show that X_t has the same statistics as the Brownian bridge; i.e., show that $\mathbf{E}X_t = \mathbf{E}\{B_t | B_T = b\}$ and that $\mathbf{V}X_t = \mathbf{V}\{B_t | B_T = b\}$ for $0 \ge t \ge T$.

Exercise 5.10 Vibrating Pearls on a String: Consider n pearls on a string vibrating according to the equation

$$dX_t = V_t \, dt, \quad dV_t = -KX_t \, dt - cV_t \, dt + \sigma dB_t$$

where $X_t, V_t, B_t \in \mathbf{R}^n$. Here, K is the matrix

$$K_{ij} = \begin{cases} 2\kappa & \text{if } i = j, \\ -\kappa & \text{if } i = j \pm 1, \\ 0 & \text{else.} \end{cases}$$

Show that in stationarity, X_t and V_t are uncorrelated, and that the stationary variance of X_t is related to K^{-1} while the stationary variance of V_t is a constant times the identity matrix. Perform a stochastic simulation of the system with $n = 16$, $c = 0.1$, $\kappa = 2$, $\sigma = 1$, $t = 0, 1, 2, \ldots, 900$. Compare the empirical variance with the theoretical. Plot the position of the pearls at a few chosen time points.

II

Stochastic Calculus

CHAPTER 6

Stochastic Integrals and the Euler-Maruyama Method

In Chapter 5, we considered systems of linear differential equations driven by white noise. We now extend this to non-linear systems:

$$\frac{dX_t}{dt} = f(X_t) + g(X_t)\xi_t.$$

Here, $\{\xi_t\}$ is white noise, which we in the last chapter saw could be understood as the time derivative of Brownian motion, $\xi_t = dB_t/dt$, if only formally.

Unfortunately, when the noise intensity $g(x)$ depends on the state x, there is ambiguity in this equation: It can be understood in different ways. So our first challenge is to define exactly how this equation should be understood. The root of the problem is that the white noise process $\{\xi_s\}$ is not a stochastic process in the classical sense, but represents a limit; details in the limiting procedure affect the result.

To resolve this ambiguity, we will make use of the same two techniques which we applied to linear systems: We integrate the equation to obtain an integral equation in which Brownian motion appears. This introduces a stochastic integral, which we need to define; the way we do this is motivated by the Euler-Maruyama method.

The study of these stochastic integrals is technically challenging, because they hinge on small-scale fluctuations. It is useful to build intuition by simulating the solutions of equations. Here, we give a number of such examples.

6.1 ORDINARY DIFFERENTIAL EQUATIONS DRIVEN BY NOISE

We now motivate our framework for stochastic differential equation and stochastic integrals, skipping technicalities. Consider a non-linear ordinary

DOI: 10.1201/9781003277569-6

differential equation $\dot{X}_t = f(X_t)$ perturbed by a driving noise signal $\{\xi_t : t \geq 0\}$:

$$\frac{dX_t}{dt} = f(X_t, t) + G\xi_t, \quad X_0 = x, \tag{6.1}$$

where G is a constant noise intensity. We assume that $\{\xi_t\}$ is smooth enough that this equation has a solution in the classical sense, for each realization. Our interest is now the limit where $\{\xi_t\}$ approaches white noise (see Section 5.8), but we have to approach this limit carefully, since white noise is not a stochastic process in the classical sense. We first integrate the left and right hand sides of the equation over the time interval $[0, t]$, finding:

$$X_t - X_0 = \int_0^t f(X_s, s) \, ds + G \int_0^t \xi_s \, ds.$$

As $\{\xi_t\}$ approaches white noise, the last integral approaches Brownian motion, and the equation becomes

$$X_t - X_0 = \int_0^t f(X_s, s) \, ds + GB_t.$$

Now, the elusive white noise process no longer appears, and in stead we have an *integral* equation in which Brownian motion $\{B_t\}$ enters as a driving term. This is a key observation, even if we at this point cannot say anything about solving this integral equation.

Next, we turn to the more general case, where the noise intensity is allowed to vary with the state and with time:

$$\frac{dX_t}{dt} = f(X_t, t) + g(X_t, t)\xi_t, \quad X_0 = x. \tag{6.2}$$

Repeating the approach in the previous, integration with respect to time leads to the term

$$\int_0^t g(X_s, s)\xi_s \, ds$$

so we need to assign a meaning to this integral in the limit where $\{\xi_t\}$ approaches white noise. One way forward is to pursue an Euler-type algorithm for time discretization, the *Euler-Maruyama scheme*, for the equation (6.2):

$$X_{t+h} = X_t + f(X_t, t) \, h + g(X_t, t)(B_{t+h} - B_t). \tag{6.3}$$

To motivate this scheme, first notice that with no noise ($g \equiv 0$), this is the explicit Euler method for the ordinary differential equation $dX_t/dt = f(X_t, t)$. Next, for the equation $dX_t/dt = \xi_t$ ($f \equiv 0, g \equiv 1$), the scheme leads to $X_t = X_0 + B_t$, which agrees with Brownian motion being integrated white noise (Section 5.8). The scheme (6.3) generalizes the way we simulated the motion of particles subject to advection and (constant) diffusion in Section 2.6. The scheme is explicit and therefore simple to implement and analyze.

Despite these arguments, we emphasize that the Euler-Maruyama scheme is a choice and that there are alternatives; notably, the Stratonovich approach which we will discuss in Section 6.8.

Now, partition the time interval $[0, t]$ into subintervals, $0 = t_0 < t_1 < \cdots < t_n = t$. When we apply the Euler-Maruyama scheme (6.3) to each subinterval, we find the state at the terminal time t:

$$X_t = X_0 + \sum_{i=0}^{n-1} (X_{t_{i+1}} - X_{t_i})$$

$$= X_0 + \sum_{i=0}^{n-1} f(X_{t_i}, t_i) \cdot (t_{i+1} - t_i) + \sum_{i=0}^{n-1} g(X_{t_i}, t_i)(B_{t_{i+1}} - B_{t_i}). \quad (6.4)$$

We can now consider the limit of small time steps. As before, we use the notation $|\Delta| = \max\{t_i - t_{i-1} : i = 1, \ldots, n\}$. With this notation, in the limit $|\Delta| \to 0$, the first sum converges to an integral:

$$\sum_{i=0}^{n-1} f(X_{t_i}, t_i) \cdot (t_{i+1} - t_i) \to \int_0^t f(X_s, s) \, ds.$$

To deal with the second term, $\sum_{i=0}^{n-1} g(X_{t_i}, t_i)(B_{t_{i+1}} - B_{t_i})$, which we have not seen before, we definine the *Itô integral* with respect to Brownian motion:

$$\int_0^t g(X_s, s) \, dB_s = \lim_{|\Delta| \to 0} \sum_{i=0}^{n-1} g(X_{t_i}, t_i)(B_{t_{i+1}} - B_{t_i}). \quad (6.5)$$

Then we obtain the solution X_t, in the limit of small time steps:

$$X_t = X_0 + \int_0^t f(X_s, s) \, ds + \int_0^t g(X_s, s) \, dB_s. \quad (6.6)$$

It is customary to write this in differential form:

$$dX_t = f(X_t, t) \, dt + g(X_t, t) \, dB_t. \quad (6.7)$$

This can either be understood as a shorthand for the integral equation (6.6), as an "infinitesimal version" of the Euler-Maruyama scheme (6.3), or simply as the differential equation (6.2) where we have (formally) multiplied with dt and used $\xi_t = dB_t/dt$.

To summarize, a differential equation driven by white noise can be rewritten as an *integral* equation. Here, a new *Itô integral* appears that integrates with respect to Brownian motion. The solution can be approximated using the Euler-Maruyama discretization.

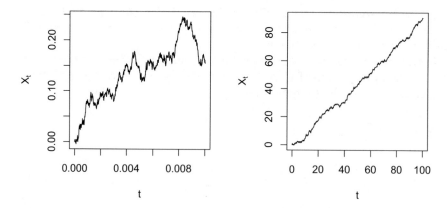

Figure 6.1 Simulation of Brownian motion with drift with $u = \sigma = 1$. *Left panel:* $t \in [0, 0.01]$. *Right panel:* $t \in [0, 100]$.

6.2 SOME EXEMPLARY EQUATIONS

Before we go deep into the mathematical construction of the Itô integral and the resulting stochastic differential equations, it is useful to explore some simple models, primarily by means of simulation using the Euler-Maruyama scheme (6.3). Together with the Ornstein-Uhlenbeck process (Section 5.10) and the noisy harmonic oscillator (Section 5.11), the following examples give an idea of the types of dynamics these models display.

6.2.1 Brownian Motion with Drift

The simplest stochastic differential equation which contains both drift and noise, is:

$$dX_t = u \, dt + \sigma \, dB_t.$$

For an initial condition $X_0 = x_0$, it is easy to see that process $\{X_t : t \geq 0\}$ given by $X_t = x_0 + ut + \sigma B_t$ satisfies this equation; simply verify that it satisfies the Euler-Maruyama scheme *exactly*, for any discretization of time. This process corresponds to the solution (2.10) of the advection-diffusion equation (2.9) with constant flow u and diffusivity $D = \sigma^2/2$. A simulation is seen in Figure 6.1. Notice that depending on the time scale, the bias ut may dominate over the random walk $\sigma \, dB_t$, or vice versa (compare Section 2.4).

6.2.2 The Double Well

Figure 6.2 shows a simulation of the process $\{X_t\}$ given by the stochastic differential equation

$$dX_t = (rX_t - qX_t^3) \, dt + \sigma \, dB_t. \tag{6.8}$$

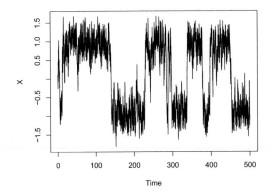

Figure 6.2 Simulation of the double well model with $r = 1$, $q = 1$, $\sigma = 0.5$, $X_0 = 0$. The computational time step is 0.01.

This process combines a nonlinear drift $f(x) = rx - qx^3$ with additive noise, $g(x) = \sigma$. When r, q and σ are positive parameters, as in the simulation, this is called the *double well model*. The reason is that the drift, $f(x) = rx - qx^3$, can be seen as the negative derivative of a potential function $u(x) = \frac{1}{2}rx^3 - \frac{1}{4}qx^4$, which has two "potential wells", i.e., local minima, at $x = \pm\sqrt{r/q}$. The drift takes the process towards the nearest of these potential wells, but the noise perturbs it away from the center of the wells, and occasionally, the process transits from one well to the other. As long as the process stays in one well, the sample paths look quite similar to the Ornstein-Uhlenbeck process. There is no known analytical solution to this stochastic differential equation. The model is an archetype of a bistable system.

6.2.3 Geometric Brownian Motion

We now consider the simplest stochastic differential equation in which the noise intensity $g(x)$ actually varies with the state:

$$dX_t = rX_t \, dt + \sigma X_t \, dB_t, \quad X_0 = x.$$

This equation is called *wide-sense linear* and its solution $\{X_t : t \geq 0\}$ is called *Geometric Brownian motion*. Figure 6.3 shows simulations obtained with two sets of parameters; one that (loosely) leads to stochastic exponential decay, and another which leads to stochastic exponential growth.

In the next chapter, we will use stochastic calculus to verify that the solution to this equation is

$$X_t = x \exp\left(rt - \frac{1}{2}\sigma^2 t + \sigma B_t\right).$$

Notice that the solution is linear in the initial condition x. In particular, if $X_0 = 0$, then $X_t = 0$ for all $t \geq 0$. This happens because both the drift

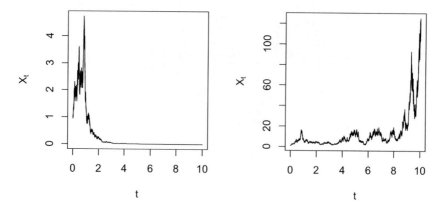

Figure 6.3 Simulation of geometric Brownian motion. *Left panel:* $r = -0.5$, $\sigma = 1$. *Right panel:* $r = 1$, $\sigma = 1$.

$f(x) = rx$ and the noise intensity $g(x) = \sigma x$ vanish at the equilibrium point $x = 0$.

Since geometric Brownian motion is the simplest case with state-varying noise intensity, we will return to this process repeatedly, to illustrate the theory. Geometric Brownian motion may be used in biology to describe population growth, in finance to describe the dynamics of prices, and in general in any domain where noisy exponential growth or decay is needed.

6.2.4 The Stochastic van der Pol Oscillator

As an example of two coupled stochastic differential equations, consider the van der Pol oscillator with additive noise:

$$dX_t = V_t \, dt, \quad dV_t = \mu(1 - X_t^2)V_t \, dt - X_t \, dt + \sigma \, dB_t. \qquad (6.9)$$

Without noise, the van der Pol system is a reference example of a nonlinear oscillator: When the position X_t is near 0, there is negative damping, so if the process starts near the equilibrium at the origin, the state (X_t, V_t) will spiral out. When the position $|X_t|$ is outside the unit interval, the damping becomes positive, so that oscillations remain bounded, and a limit cycle appears. Here, we add noise in the simplest form, i.e., additive noise on the velocity. A simulation is seen in Figure 6.4 where we start at the origin. Noise perturbs the state away from the equilibrium of the drift, and the state quickly spirals out. It then approaches a limit cycle, but is continuously perturbed from it.

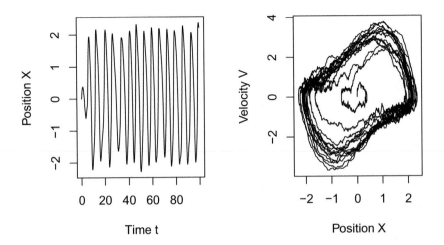

Figure 6.4 The stochastic van der Pol oscillator with $\mu = 1$, $\sigma = 1/2$. *Left panel:* The position X_t vs. time. *Right panel:* Trajectories in the phase plane (X_t, V_t).

6.3 THE ITÔ INTEGRAL AND ITS PROPERTIES

We now turn to the mathematical construction of the Itô integral (6.5). Let $(\Omega, \mathcal{F}, \{\mathcal{F}_t : t \geq 0\}, \mathbf{P})$ be a filtered probability space and let $\{B_t : t \geq 0\}$ be Brownian motion on this space.

Theorem 6.3.1 (Itô integral; $\mathcal{L}_2$ Version) *Let $0 \leq S \leq T$ and let $\{G_t : S \leq t \leq T\}$ be a real-valued stochastic process, which has left-continuous sample paths, which is adapted to $\{\mathcal{F}_t\}$, and for which $\int_S^T \mathbf{E}|G_t|^2 \, dt < \infty$. Then the limit*

$$I = \lim_{|\Delta| \to 0} \sum_{i=1}^{\#\Delta} G_{t_{i-1}}(B_{t_i} - B_{t_{i-1}}) \ (limit \ in \ mean \ square) \qquad (6.10)$$

exists. Here $\Delta = \{S = t_0 < t_1 < t_2 < \cdots < t_n = T\}$ is a partition of $[S, T]$. We say that $\{G_t\}$ is $\mathcal{L}_2$ Itô integrable, and that I is the Itô integral:

$$I = \int_S^T G_t \, dB_t.$$

We outline the proof of this theorem in Section 6.11.1. Notice that we evaluate the integrand $\{G_t\}$ at the left end of each sub-interval $[t_{i-1}, t_i]$, in agreement with the explicit Euler-Maruyama method of Section 6.1, and, in particular, (6.5). The Itô integral can be seen as an operator that combines the two inputs. the *integrand* $\{G_t\}$ and the *integrator* $\{B_t\}$, to produce the output, the *Itô integral* $\{I_t\}$ where $I_t = \int_0^t G_s \, dB_t$ (Figure 6.5).

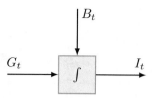

Figure 6.5 The Itô integral seen as an operator that, for each $t \geq 0$, produces $I_t = \int_0^t G_s \, dB_s$, from $\{G_s : 0 \leq s \leq t\}$ and $\{B_s : 0 \leq s \leq t\}$.

Biography: Kiyosi Itô (1915–2008)

 Born in Japan, he held positions Princeton, Aarhus, and Cornell, as well as at several Japanese universities. He made fundamental contributions to the theory of stochastic processes, besides his seminal work on what we now call Itô calculus, where he focused on sample paths and the stochastic differential equations they satisfy. Between 1938 and 1950 he developed the stochastic integral and the associated calculus, inspired by the work of Kolmogorov, Lévy, and Doob.

The definition suggests how to approximate an Itô integral numerically: Fix the partition Δ and compute the sum in the definition. The following code from the package SDEtools assumes that $\{G_t\}$ and $\{B_t\}$ have been tabulated on time points $t_1, t_2, \ldots, t_n$. It then computes and returns the integrals $\int_{t_1}^{t_i} G_t \, dB_t$ for $i = 1, \ldots, n$, based on the approximation that the integrand is piecewise constant.

```
itointegral <- function(G,B)
                c(0,cumsum(head(G,-1)*diff(B)))
```

Now we can state a number of properties of the Itô integral:

Theorem 6.3.2 *Let $0 \leq S \leq T \leq U$; let $\{F_t : 0 \leq t \leq U\}$ and $\{G_t : 0 \leq t \leq U\}$ be $\mathcal{L}_2$ Itô integrable on $[0, U]$ with respect to $\{B_t\}$. Then the following holds:*

1. *Additivity:* $\int_S^U G_t \, dB_t = \int_S^T G_t \, dB_t + \int_T^U G_t \, dB_t$.

2. *Linearity:* $\int_S^T aF_t + bG_t \, dB_t = a \int_S^T F_t \, dB_t + b \int_S^T G_t \, dB_t$ *when* $a, b \in \mathbf{R}$.

3. *Measurability:* $\int_S^T G_t \, dB_t$ *is* $\mathcal{F}_T$-*measurable.*

4. *Continuity:* $\{I_t\}$ *is continuous in mean square and can be taken to have continuous sample paths.*

5. *The martingale property: The process* $\{I_t : 0 \leq t \leq U\}$, *where* $I_t = \int_0^t G_s \, dB_s$, *is a martingale (w.r.t.* $\{\mathcal{F}_t : t \geq 0\}$ *and* $\mathbf{P}$*).*

6. *The Itô isometry:* $\mathbf{E}|\int_S^T G_t \, dB_t|^2 = \int_S^T \mathbf{E}|G_t|^2 \, dt.$

Let us briefly discuss these properties: We would expect any integral to be additive and linear; recall that (probability) measures and expectations are (Chapter 3). We also expect the integral to depend continuously on the upper limit. Measurability is natural from the point of view of information flow: The device in Figure 6.5 needs to know the signals $\{G_s : 0 \leq s \leq t\}$ and $\{B_s : 0 \leq s \leq t\}$ in real time to compute I_t, and then knows the result of this computation. That is, we assume that $\{G_t\}$ and $\{B_t\}$ are $\{\mathcal{F}_t\}$-adapted, and conclude that the same applies to $\{I_t\}$.

What is left are the two most noticeable of these properties: The martingale property and the Itô isometry. The martingale property says that the expected value of a future integral is always 0:

$$\mathbf{E}\left\{\int_S^T G_t \, dB_t \middle| \mathcal{F}_S\right\} = 0.$$

In turn, the Itô isometry establishes the variance of this integral. These two properties are specific to the Itô integral, so they represent a choice. A major argument in favor of Itô's way of defining the integral, is that it leads to these two properties.

Proof: We outline the proof only; see e.g. (Øksendal, 2010). We derive the mean and variance of the Itô integral under the additional assumption there is a $K > 0$ such that $\mathbf{E}|G_t|^2 < K$ for all t; see Section 6.6.1 for how to relax this assumption. Assume that $\{G_t\}$ is Itô integrable on the interval $[S, T]$. Fix the partition $\Delta = \{S = t_0 < t_1 < \cdots < t_n = T\}$ and consider the sum in (6.10), i.e.,

$$I_\Delta = \sum_{i=1}^n G_{t_{i-1}}(B_{t_i} - B_{t_{i-1}}).$$

Each of the terms in this sum has expectation 0; to see this, first condition on $\mathcal{F}_{t_{i-1}}$ and then use the law of total expectation. Hence, $\mathbf{E}I_\Delta = 0$.

Next, any two distinct terms in the sum are uncorrelated: Take $i < j$ and consider the conditional expectation:

$$\mathbf{E}\left\{G_{t_{i-1}}G_{t_{j-1}}(B_{t_i} - B_{t_{i-1}})(B_{t_j} - B_{t_{j-1}})\middle|\mathcal{F}_{t_{j-1}}\right\}.$$

Here, all terms are $\mathcal{F}_{t_{j-1}}$-measurable, except the last Brownian increment, so this conditional expectation equals 0. The law of total expectation then implies that also the unconditional expectation is 0, so that the two terms are indeed uncorrelated. Thus,

$$
\begin{aligned}
\mathbf{E}I_\Delta^2 &= \sum_{i=1}^{n} \mathbf{E}G_{t_{i-1}}^2 (B_{t_i} - B_{t_{i-1}})^2 \\
&= \sum_{i=1}^{n} \mathbf{E}\mathbf{E}\left\{ G_{t_{i-1}}^2 (B_{t_i} - B_{t_{i-1}})^2 \Big| \mathcal{F}_{t_{i-1}} \right\} \\
&= \sum_{i=1}^{n} \mathbf{E}G_{t_{i-1}}^2 (t_i - t_{i-1}).
\end{aligned}
$$

Since $I_\Delta \to I$ in mean square by assumption, this shows that $\mathbf{E}I = 0$ and $\mathbf{E}I^2 = \int_S^T \mathbf{E}G_t^2 \, dt$.

Linearity follows from I_Δ being linear in the integrand. For measurability, note first that I_Δ is obviously $\mathcal{F}_T$-measurable for each partitioning Δ. Now write the limit I as a sum, say, $I = X + Y$, where X is $\mathcal{F}_T$-measurable and Y is in the orthogonal complement. Then $\mathbf{E}(I_\Delta - I)^2 = \mathbf{E}(I_\Delta - X)^2 + \mathbf{E}Y^2$ so we conclude that $I_\Delta \to X$ in mean square and $\mathbf{E}Y^2 = 0$. To see that additivity holds, notice that if $\{G_t\}$ is Itô integrable on an interval $[S, U]$, then the contributions from the subintervals $[S, T]$ and $[T, U]$ are orthogonal, so that $\{G_t\}$ is Itô integrable on subintervals.

This implies that I_t is defined for all $t \in [0, U]$. To see that $\{I_t\}$ is a martingale, note first that I_T is $\mathcal{F}_T$-measurable and that $\mathbf{E}|I_T| < \infty$. Inspecting our previous argument regarding the sum in I_Δ, conditioning on $\mathcal{F}_S$ and using the tower property rather than the law of total expectation, we see that $\mathbf{E}\{\int_S^T G_t \, dB_t | \mathcal{F}_S\} = 0$. Since S and T are arbitrary, this implies that $\{I_t\}$ is a martingale with respect to $\{\mathcal{F}_t\}$.

Continuity of $\{I_t\}$ in mean square follows from the Itô isometry: Note that $I = I_T - I_S$; now let $T \to S$, then $\mathbf{E}|I|^2 \to 0$. See (Øksendal, 2010) for the proof that the sample paths of $\{I_t\}$ can be taken to be continuous. ■

6.4   A CAUTIONARY EXAMPLE: $\int_0^T B_s \, dB_s$

Itô integrals do not behave like Riemann integrals, and this has far-reaching consequences. To demonstrate this with an example, consider integrating Brownian motion with respect to itself, i.e.,

$$
\int_0^t B_s \, dB_s.
$$

Following Theorem 6.3.1, we approximate this integral numerically by discretization, using a uniform partition of the time interval $[0, t]$ with mesh (time step) $h > 0$:

$$
0 = t_0 < t_1 < t_2 < \cdots < t_n = t \text{ with } t_i = ih, \ hn = t.
$$

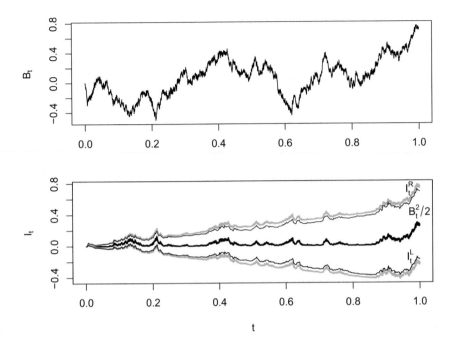

Figure 6.6 *Top panel:* A realization of Brownian motion B_t for $t \in [0, 1]$. *Bottom panel:* Left hand (I_t^L) and right hand (I_t^R) approximations of $\int_0^t B_s \, dB_s$. Thick grey lines use a fine grid ($h = 2^{-13}$) while thin black lines use a coarse grid ($h = 2^{-9}$). Included is also $B_t^2/2$ (thick solid line).

Let I_t^L denote the approximation to the integral based on this partition:

$$I_t^L = \sum_{i=1}^{n} B_{t_{i-1}}(B_{t_i} - B_{t_{i-1}}).$$

In Riemann integration, it does not matter where in the interval $[t_{i-1}, t_i]$ we evaluate the integrand, when we afterwards consider the fine time discretization limit $h \to 0$. To see if this also applies to stochastic integrals, let I_t^R be the corresponding approximation using the right endpoint:

$$I_t^R = \sum_{i=1}^{n} B_{t_i}(B_{t_i} - B_{t_{i-1}}).$$

Figure 6.6 shows the two approximations, as functions of t, and for two values of h. Note that the difference between the two discretizations appears to grow with time, but does not seem very sensitive to the time step h. In fact, it is easy to see that

$$I_t^R - I_t^L = \sum_{i=1}^{n} (B_{t_i} - B_{t_{i-1}})^2. \tag{6.11}$$

Taking expectation, and using that increments in Brownian motion satisfy $\mathbf{E}(B_t - B_s)^2 = |t - s|$, we get that $\mathbf{E}(I_t^R - I_t^L) = t$. Moreover, as the discretization becomes finer, the difference between the two discretizations is exactly the quadratic variation of Brownian motion: $I_t^R - I_t^L \to [B]_t = t$, a.s.

We conclude that the Itô integral is affected by our choice that we evaluate the integrand at the left side. This is contrast to Riemann integrals. The difference is due to the small-scale fluctuations in Brownian motion, i.e., that the quadratic variation $\sum_{i=1}^n (B_{t_i} - B_{t_{i-1}})^2$ does not vanish in the limit of fine partitions, while the variation $\sum_{i=1}^n |B_{t_i} - B_{t_{i-1}}|$ diverges (Section 4.3).

Moreover, if we try to apply the elementary calculus and solve the integral by substitution $u = B_t$, this would lead us to believe that $\int_0^t B_s \, dB_s = \int_0^{B_t} u \, du = \frac{1}{2}u^2 \big|_0^{B_t} = \frac{1}{2}B_t^2$, using that $B_0 = 0$. If you are uncomfortable with this substitution, then go ahead and pretend that B_t is differentiable with derivative ξ_t, rewrite the integral as $\int_0^t B_s \xi_s \, ds$, and then do the substitution $u = B_t$, $du = \xi_t \, dt$. This analytical "result" for the integral is also included in Figure 6.6. Note that it lies perfectly in between the two numerical approximations. In fact, this analytical expression corresponds to the *Stratonovich* integral (Section 6.8) where the integrand is evaluated at the midpoint.

Although we see that standard calculus does not agree with the numerical discretization, it does provide a clue which we can exploit to evaluate the approximation I_t^L. Manipulating the sum, we find

$$I_t^L = \sum_{i=1}^n B_{t_{i-1}} \cdot (B_{t_i} - B_{t_{i-1}})$$

$$= \frac{1}{2}\sum_{i=1}^n (B_{t_i} + B_{t_{i-1}}) \cdot (B_{t_i} - B_{t_{i-1}}) - \frac{1}{2}\sum_{i=1}^n (B_{t_i} - B_{t_{i-1}}) \cdot (B_{t_i} - B_{t_{i-1}})$$

$$= \frac{1}{2}\sum_{i=1}^n (B_{t_i}^2 - B_{t_{i-1}}^2) - \frac{1}{2}\sum_{i=1}^n (B_{t_i} - B_{t_{i-1}})^2$$

$$= \frac{1}{2}B_t^2 - \frac{1}{2}\sum_{i=1}^n (B_{t_i} - B_{t_{i-1}})^2$$

Letting the mesh $|\Delta|$ go to zero, and using the quadratic variation of Brownian motion $[B]_t = t$, we find

$$\int_0^t B_s \, dB_s = \lim_{|\Delta| \to 0} I_t^L = \frac{1}{2}B_t^2 - \frac{1}{2}t. \tag{6.12}$$

It may seem like a painstaking effort needed to compute what should be a simple integral; luckily, in Chapter 7 we will learn how to reach the result (6.12) using calculus, which is much more efficient.

Combining with (6.11) we find that $I_t^R \to \frac{1}{2}B_t^2 + \frac{1}{2}t$, so that the average of the two approximations $\frac{1}{2}(I_t^L + I_t^R)$ approaches $\frac{1}{2}B_t^2$.

In summary: In the Itô integral, we choose to evaluate the integrand at the left side of every sub-interval. This choice matters, in contrast to the Riemann integral: If we evaluate the integrand differently, we get a different integral. Moreover, when we evaluate the integrand in either of the end points, we cannot solve the integral by substitution using the classical chain rule of calculus. Therefore we need to develop a stochastic calculus (Chapter 7).

6.5 ITÔ PROCESSES AND SOLUTIONS TO SDES

The Itô integral $\int_0^t G_s \, dB_s$ allows us to define what it means to add white noise to a differential equation. We can now specify the class of processes in which we search for solutions to stochastic differential equations: These involve an Itô integral with respect to Brownian motion, as well as an integral with respect to time.

Definition 6.5.1 (Itô process; $\mathcal{L}_2$ version) *Given a filtered probability space $(\Omega, \mathcal{F}, \{\mathcal{F}_t\}, \mathbf{P})$, we say that a process $\{X_t : t \geq 0\}$ given by*

$$X_t = X_0 + \int_0^t F_s \, ds + \int_0^t G_s \, dB_s \qquad (6.13)$$

is an $\mathcal{L}_2$ Itô process, provided that the initial condition X_0 is $\mathcal{F}_0$-measurable (for example, deterministic), and that $\{F_t\}$ and $\{G_t\}$ are adapted and have left-continuous sample paths and locally integrable variance.

We use the shorthand

$$dX_t = F_t \, dt + G_t \, dB_t$$

for such a process; this shorthand does not refer to the initial position X_0, which is required to fully specify the process. We call $\{F_t\}$ the *drift* of the process, and $\{G_t\}$ the *intensity*.

Proposition 6.5.1 *Let $\{X_t\}$ be an $\mathcal{L}_2$ Itô process, then*

$$\mathbf{E}X_t = \mathbf{E}X_0 + \int_0^t \mathbf{E}F_s \, ds$$

and the quadratic variation is

$$[X]_t = \int_0^t |G_s|^2 \, ds.$$

Proof: Assume first that $F_t = F_0$ and $G_t = G_0$ where F_0 and G_0 are bounded $\mathcal{F}_0$-measurable random variables. Then $X_t = X_0 + F_0 t + G_0 B_t$ so the expectation $\mathbf{E}X_t$ is immediate. For the quadratic variation, consider a partition Δ

of $[0, t]$; then

$$\sum_{i=1}^{\#\Delta} |X_{t_i} - X_{t_{i-1}}|^2 = \sum_{i=1}^{\#\Delta} |F_0(t_i - t_{i-1}) + G_0(B_{t_i} - B_{t_{i-1}})|^2$$

$$= F_0^2 \sum_{i=1}^{\#\Delta} (t_i - t_{i-1})^2 + G_0^2 \sum_{i=1}^{\#\Delta} (B_{t_i} - B_{t_{i-1}})^2$$

$$+ 2F_0 G_0 \sum_{i=1}^{\#\Delta} (t_{i-1} - t_i)(B_{t_i} - B_{t_{i-1}}).$$

In the limit $|\Delta| \to 0$, the first term involving F_0^2 vanishes. The second term converges to $G_0^2[B]_t = G_0^2 t$. The third term, conditional on $\mathcal{F}_0$ and for a fixed partition Δ, is Gaussian with mean 0 and variance $4F_0^2 G_0^2 \sum (\Delta t_i)^3$, so converges to 0 in mean square as $|\Delta| \to 0$.

Next, the results hold also if $\{F_t\}$ and $\{G_t\}$ are elementary processes; applying the previous to the each sub-interval where the integrands are constant. Finally, we approximate $\{F_t\}$ and $\{G_t\}$ with elementary processes and see that the results carry over to the limit. ■

If the time step h is small and the integrands are continuous, then the conditional distribution of X_{t+h} given $\mathcal{F}_t$ is approximately normal with mean and variance given by

$$\mathbf{E}\{X_{t+h}|\mathcal{F}_t\} = X_t + F_t \cdot h + o(h), \quad \mathbf{V}\{X_{t+h}|\mathcal{F}_t\} \approx |G_t|^2 \cdot h + o(h).$$

Thus, F_t determines the incremental mean while G_t determines the incremental variance.

Although Itô processes are not tied up to the application of particles moving in fluids, it is useful to think of an Itô process as the position of such a particle. In that case, $\{F_t\}$ is the random "drift" term responsible for the mean change in position of $\{X_t\}$, while $\{G_t\}$ is the random intensity of the unbiased random walk resulting from collisions with fluid molecules.[1]

Our main motivation for defining Itô processes is that they serve as solutions for stochastic differential equations:

Definition 6.5.2 (Solution of a stochastic differential equation)   *We say that the stochastic process $\{X_t\}$ satisfies the (Itô) stochastic differential equation*

$$dX_t = f(X_t, t) \, dt + g(X_t, t) \, dB_t$$

where $\{B_t\}$ is Brownian motion on a filtered probability space $(\Omega, \mathcal{F}, \{\mathcal{F}_t : t \geq 0\}, \mathbf{P})$ if $\{X_t\}$ is an Itô process

$$dX_t = F_t \, dt + G_t \, dB_t$$

---

[1]It is tempting to equate F_t with the bulk fluid flow at the position of the particle, and relate G_t to the diffusivity at the particle, but this is only true when the diffusivity in constant in space. We shall return to this issue later, in Section 9.5.

where $F_t = f(X_t, t)$ and $G_t = g(X_t, t)$. In that case we call $\{X_t\}$ an Itô diffusion.

The theory of stochastic differential equations operate also with other definitions of solutions, and then, the one given here is denoted a *strong solution*. We consider only such strong solutions.

6.6 RELAXING THE $\mathcal{L}_2$ CONSTRAINT

Theorem 6.3.1, which established the Itô integral, assumed that the process $\{G_t : t \geq 0\}$ had locally integrable variance, $\int_0^t \mathbf{E}|G_s|^2 \, ds < \infty$. This assumption allowed us to use the machinery of $\mathcal{L}_2$ spaces when proving convergence. Although this assumption holds in the majority of applications we are interested in, it is somewhat restrictive and becomes tedious, because we must constantly check the variance of the integrands. It turns out that we can relax the assumption and only require that almost all sample paths of $\{G_t\}$ are square integrable:

$$\mathbf{P}\left(\forall t \geq 0 : \int_0^t |G_s|^2 \, ds < \infty\right) = 1. \tag{6.14}$$

Notice that this holds if the sample paths of $\{G_t\}$ are continuous. The Itô integral can be extended to cover also these integrands; see (Øksendal, 2010; Karatzas and Shreve, 1997; Rogers and Williams, 1994b). The technique is to approximate $\{G_t\}$ with bounded processes $\{G_t^{(n)}\}$ given by $G_t^{(n)} = G_t \mathbf{1}\{|G_t| \leq n\}$. Each of these $\{G_t^{(n)}\}$ are Itô integrable in the sense of Theorem 6.3.1 and their Itô integrals $I^{(n)} = \int_S^T G_s^{(n)} \, dB_s$ converge in probability. We define the Itô integral $I = \int_S^T G_s \, dB_s$ as the limit.

The resulting Itô integral may not have a well-defined expectation, because tail contributions do not vanish. Consider for example the constant integrand $G_t \equiv G = \exp(X^2)$ where X is a standard Gaussian variable which is $\mathcal{F}_0$-measurable (and therefore independent of the Brownian motion $\{B_t\}$). Note that $\mathbf{E}G = \infty$. We then get

$$I_t = \int_0^t G_s \, dB_s = e^{X^2} B_t.$$

and therefore $\mathbf{E}|I_t| = \mathbf{E}G \cdot \mathbf{E}|B_t| = \infty$. So $\mathbf{E}I_t$ is not defined and the process $\{I_t = \int_0^t G_s \, dB_s\}$ is not a martingale. Moreover, the Itô isometry only applies in the sense that $\mathbf{E}|I_t|^2 = \infty = \int_0^t \mathbf{E}|G_s|^2 \, ds$. To get an operational theory, we *localize* the integral: We introduce an outer bound $R > 0$ and the stopping time $\tau_R = \inf\{t \geq 0 : |I_t| > R\}$. Then, the stopped process $\{I_{t \wedge \tau_R}\}$ is a martingale and we say that $\{I_t\}$ is a *local* martingale. This technique of localization is then used throughout. This allows us to generalize also the notion of Itô processes:

Definition 6.6.1 (Itô process; general version) *With the setup in Definition 6.5.1, we say that $\{X_t\}$ is an Itô process, if X_0 is $\mathcal{F}_0$-measurable, $\{F_t\}$ and $\{G_t\}$ are adapted and have left-continuous sample paths.*

Karatzas and Shreve (1997) relax the assumption of left-continuity even further.

6.7 INTEGRATION WITH RESPECT TO ITÔ PROCESSES

We introduced the notation

$$dX_t = F_t \, dt + G_t \, dB_t$$

for an Itô process as a shorthand for

$$X_t - X_0 = \int_0^t F_s \, ds + \int_0^t G_s \, dB_s.$$

However, the notation suggests that dX_t, dt and dB_t are objects belonging to the same class. For example, provided $G_t > 0$, we can re-write formally

$$\frac{1}{G_t} \, dX_t - \frac{F_t}{G_t} \, dt = dB_t$$

which would then be a shorthand for

$$\int_0^t \frac{1}{G_s} \, dX_s - \int_0^t \frac{F_s}{G_s} \, ds = B_t - B_0.$$

This, however, requires that we can integrate not only with respect to time t and Brownian motion $\{B_t : t \geq 0\}$, but also with respect to an Itô process $\{X_t : t \geq 0\}$. It turns out that we can indeed extend the definition of the Itô integral, so that we can also integrate with respect to Itô processes:

$$\int_0^t H_s \, dX_s = \lim_{|\Delta| \to 0} \sum_{i=0}^{\#\Delta} H_{t_i}(X_{t_{i+1}} - X_{t_i}).$$

The limit needs only be in probability, but will often be in mean square. The conclusion from this is that the stochastic differential equation $dX_t = F_t \, dt + G_t \, dB_t$ can be seen as an equation among integrators. One particular Itô process, which we will repeatedly use as an integrator, is the quadratic variation process $[X]_t$.

It is possible to define Itô integrals with respect to a larger class of processes known as *semimartingales*; see (Karatzas and Shreve, 1997) or (Rogers and Williams, 1994b).

Exercise 6.1 Numerical integration w.r.t. an Itô process: Choose your favorite drift term $\{F_t\}$ and noise intensity $\{G_t\}$. Then, construct numerically a sample path of the Itô process $\{X_t\}$ with $dX_t = F_t \, dt + G_t \, dB_t$. Then, reconstruct the driving Brownian motion numerically, i.e., solve $dW_t = (G_t)^{-1}(dX_t - F_t \, dt)$. Does $\{W_t\}$ equal $\{B_t\}$?

Biography: Ruslan Leontievich Stratonovich (1930–1997)
Stratonovich was born and lived in Moscow. His work focused on the mathematics of noise in physics and engineering. Besides the stochastic calculus which is centered around the Stratonovich integral, his most important contribution was a general technique for filtering in non-linear dynamic systems, which includes the Kalman-Bucy filter as a special case.

6.8   THE STRATONOVICH INTEGRAL

The Itô integral involves Riemann sums where the integrand is evaluated at the left end-point of each sub-interval, and we saw in Section 6.4 that this choice has an impact on the resulting integral. Therefore, we could define an entire family of stochastic integrals, parametrized by where we choose to evaluate the integrand. In this family, the most prominent member beside the Itô integral is the Stratonovich integral, where we evaluate the integrand at the mid-point, or equivalently use the trapezoidal rule

$$\int_0^t G_s \circ dB_s = \lim_{|\Delta| \to 0} \sum_{i=1}^{\#\Delta} \frac{1}{2} \left(G_{t_{i-1}} + G_{t_i} \right) \left(B_{t_i} - B_{t_{i-1}} \right)$$

where $\Delta = \{0 = t_0 < t_1 < \cdots < t_n = t\}$ is a partition of the interval $[0, t]$ and the limit is in mean square.

Example 6.8.1 *With $G_t = B_t$, we get*

$$\int_0^t B_s \circ dB_s = \lim_{|\Delta| \to 0} \sum_{i=1}^{\#\Delta} \frac{1}{2} \left(B_{t_i} + B_{t_{i-1}} \right) \left(B_{t_i} - B_{t_{i-1}} \right)$$

$$= \lim_{|\Delta| \to 0} \sum_{i=1}^{\#\Delta} \frac{1}{2} \left(B_{t_i}^2 - B_{t_{i-1}}^2 \right)$$

$$= \frac{1}{2} B_t^2.$$

This example is interesting, because it suggests that Stratonovich integrals obey the same rules of calculus as normal Riemann integrals. This is indeed the case, as we will see in the next chapter.

Exercise 6.2 Numerical Stratonovich integration: Write a function, which takes as input a time partition $t_0 < t_1 < \cdots < t_n$, as well as the integrand $\{G_{t_i} : i = 0, \ldots, n\}$ and the Brownian motion $\{B_{t_i} : i = 0, \ldots, n\}$ sampled at these time points, and which returns (an approximation to) the Stratonovich integral $\int_{t_0}^{t_n} G_t \circ dB_t$ sampled at the same time points. Verify the function by computing the integral $\int_0^1 B_t \circ dB_t$ and comparing it to the theoretical result $\frac{1}{2} B_1^2$, as in Figure 6.6.

How does the Stratonovich integral relate to the Itô integral? The difference between the two integrals originate from the different treatment of fine-scale fluctuations in the two processes, the integrator $\{B_t\}$ and the integrand $\{G_t\}$. To quantify this, we introduce:

Definition 6.8.1 (Cross-variation) *Let $\{X_t\}$ and $\{Y_t\}$ be two real-valued stochastic processes. We define their cross-variation as the limit in probability*

$$\langle X, Y \rangle_t = \lim_{|\Delta| \to 0} \sum_{i=1}^{\#\Delta} (X_{t_i} - X_{t_{i-1}})(Y_{t_i} - Y_{t_{i-1}})$$

whenever this limit exists.

Comparing the discrete-time approximations of the Itô and the Stratonovich integral, we see that

$$\int_0^t G_s \circ dB_s - \int_0^t G_s \, dB_s = \frac{1}{2} \lim_{|\Delta| \to 0} \sum_{i=1}^{n} (G_{t_i} - G_{t_{i-1}})(B_{t_i} - B_{t_{i-1}})$$

$$= \frac{1}{2} \langle G, B \rangle_t. \tag{6.15}$$

For example, with $G_t = B_t$, we can compare Example 6.8.1 and Section 6.4 and find

$$\int_0^t B_s \circ dB_s - \int_0^t B_s \, dB_s = \frac{1}{2}B_t^2 - \left(\frac{1}{2}B_t^2 - \frac{1}{2}t\right) = \frac{1}{2}t$$

which agrees with $\langle B, B \rangle_t = [B]_t = t$.

If the integrand is of bounded total variation, then the cross-variation vanishes, $\langle G, B \rangle_t = 0$, and the Itô and Stratonovich integral are identical.

Our focus will be mostly on the Itô integral, primarily because of its martingale property, which is connected to the explicit Euler-Maruyama method (6.3). This property is key in the theoretical development. However, there are two reasons why the Stratonovich integral is a popular alternative to the Itô integral in applications, besides the perhaps natural and symmetric choice of the mid-point:

1. As we shall see in Chapter 7, the stochastic calculus that results from the Stratonovich integral appears simpler and closer to ordinary (deterministic) calculus. An example of this is the result in Example 6.8.1 that $\int_0^t B_s \circ dB_s = B_t^2/2$, as we would expect from a naive application of standard rules of calculus. This simplicity is a particular advantage when the application involves many coordinate transformations. We will also make use of the Stratonovich integral in some examples to compare with deterministic results.

2. When exogenous noise $\{\xi_t\}$ drives an ordinary differential equation, and this noise is band-limited but approximates white noise, then the Stratonovich integral emerges as the limit. The following exercise gives an example.

Exercise 6.3 The Stratonovich integral and band-limited noise: For each $N \in \mathbf{N}$, let $\{B_t^{(N)} : t \geq 0\}$ be a stochastic process with smooth sample paths, such that $B_t^{(N)} \to B_t$ in mean square, for each t. For example, $\{B_t^{(N)}\}$ may be the approximation of $\{B_t\}$ based on frequencies $-N, -N+1, \ldots, N$ as in Exercise 4.13. Verify that

$$\int_0^t B_s^{(N)} \, dB_s^{(N)} \to \int_0^t B_s \circ dB_s$$

in mean square, as $N \to \infty$.

6.9 CALCULUS OF CROSS-VARIATIONS

We have seen that the properties of stochastic integrals hinge critically on the small-scale fluctuations of the processes (the integrand and the integrator) and how we treat them. Specifically, the quadratic variation of Brownian motion is the reason the Riemann integral is insufficient (Section 6.4). Recall the key result (Theorem 4.3.1) that the quadratic variation of Brownian motion $\{B_t\}$ is $[B]_t = t$. Moreover, the difference between the Itô and the Stratonovich integrals equals the cross-variation between the integrand and the integrator (Section 6.8).

The two descriptors, the quadratic variation $[X]_t$ of a process $\{X_t\}$, and the cross-variation $\langle X, Y \rangle_t$ between two processes $\{X_t\}$ and $\{Y_t\}$, both quantify small-scale fluctuations, and they are tightly coupled: We have $\langle X, X \rangle_t = [X]_t$ and, in turn, $\langle X, Y \rangle_t = ([X+Y]_t - [X-Y]_t)/4$, which can be verified by simply writing out the terms for a given partition of $[0, t]$.

Two basic results for cross-variations are given in the following exercises.

Exercise 6.4: Let $\{T_t\}$ be the deterministic process with $T_t = t$. Show that $[T]_t = \langle T, B \rangle_t = 0$.

Exercise 6.5: Let $\{B_t : t \geq 0\}$ and $\{W_t : t \geq 0\}$ be independent Brownian motions on the same filtered probability space. Show that $\langle B, W \rangle_t = 0$.

In stochastic calculus, we often end up integrating with respect to the cross-variation of two Itô processes $\{L_t\}$ and $\{M_t\}$, i.e., consider integrals of the form $\int_0^t H_t \, d\langle L, M \rangle_t$. It is convenient to use the notational convention

$$dL_t \, dM_t = d\langle L, M \rangle_t.$$

This notation also reflects how we would approximate the Itô integral when discretizing time:

$$\int_0^t H_t \, d\langle L, M\rangle_t \approx \sum_{i=1}^n H_{t_{i-1}}(L_{t_i} - L_{t_{i-1}})(M_{t_i} - M_{t_{i-1}}).$$

With this notation, Exercises 6.4 and 6.5 showed that

$$(dt)^2 = 0, \quad dt \, dB_t = 0, \quad (dB_t)^2 = dt, \quad dB_t \, dW_t = 0. \tag{6.16}$$

A reasonable question is if we ever need terms that are cubic in the increments. The answer is "no":

Exercise 6.6: Show that $\sum_{i=1}^n (\Delta B_i)^3 \to 0$ in mean square as $|\Delta| \to 0$. Here, as usual, $0 = t_0 < t_1 \cdots < t_n = t$ is a partition of $[0, t]$, and $\Delta B_i = B_{t_i} - B_{t_{i-1}}$.

One way to synthesize these results is through the scaling relationship $B_t \sim \sqrt{t}$: The terms $(dt)^2$, $dt \, dB_t$, and $(dB_t)^3$ are all $o(dt)$ and therefore their sum vanish as the time step goes to zero.

We can now use these rules to simplify sums and products of increments:

Exercise 6.7: Let $\{X_t\}$ and $\{Y_t\}$ be scalar Itô processes given by

$$dX_t = F_t \, dt + G_t \, dB_t, \quad dY_t = K_t \, dt + L_t \, dB_t.$$

Show that $dX_t \, dY_t = G_t L_t \, dt$, i.e. $\langle X, Y\rangle_t = \int_0^t G_s L_s \, ds$. You may assume, for simplicity, that the integrands are bounded.

6.10 CONCLUSION

Stochastic differential equations are most easily understood in term of the Euler-Maruyama scheme; the Itô integral appears in the limit of vanishing time steps. The key feature of this integral, that we evaluate the integrand at the left hand side of sub-intervals, is consistent with that the Euler-Maruyama method is explicit.

The Euler-Maruyama method is a highly useful way of simulating solutions to stochastic differential equations, even if it does not perform impressively from the point of view of numerical analysis (as we shall see in Chapter 8). It is also very useful in the process of constructing the theory.

In the Itô integral, we assume that the integrating device knows the integrand $\{G_t\}$ in real time and does not anticipate the Brownian motion. Stated in the language of measure theory, $\{G_t\}$ is adapted w.r.t. $\{\mathcal{F}_t\}$ while $\{B_t\}$ is Brownian motion w.r.t. $\{\mathcal{F}_t\}$. As a result of these two assumptions, the expected contribution to the integral over a small time step is zero: $\mathbf{E}G_{t_{i-1}}(B_{t_i} - B_{t_{i-1}}) = 0$. This is the reason the Itô integral is a martingale, which is the key property of the integral. The fact that the Itô integral has expectation 0, and a tractable variance which is given by the Itô isometry, makes the integral attractive both from the point of view of modeling and theory.

In turn, the Stratonovich interpretation has a natural symmetry which may seem appealing, and as we shall see in the next chapter, this result of this symmetry is that Stratonovich integrals gives rise to a simpler stochastic calculus than the Itô calculus. However, the Stratonovich integral does not have expectation 0.

We convert between Itô and Stratonovich integrals using the cross-variation of the integrand and the integrator, which measure to which degree small-scale fluctuations in the two processes are correlated. Cross-variations also turn out to be central to the stochastic calculus that we develop in the next chapter.

6.11 NOTES AND REFERENCES

The Itô integral was introduced to the English-reading audience in (Itô, 1944). The Euler-Maruyama method was introduced by Gisiro Maruyama in a brief report in 1953 and later in (Maruyama, 1955). Stratonovich published his integral in English in (Stratonovich, 1966); the same integral appeared in the thesis of D.L. Fisk in 1963. The exposition in this chapter is based on (Karatzas and Shreve, 1997; Øksendal, 2010; Mao, 2008).

6.11.1 The Proof of Itô Integrability

Here, we outline the proof of Theorem 6.3.1; see (Karatzas and Shreve, 1997), (Øksendal, 2010) or (Mao, 2008). We first consider a particularly simple class of integrands, namely the *elementary processes*, which have bounded and piecewise constant sample paths:

Definition 6.11.1 (Elementary process) *A process $\{G_t : S \leq t \leq T\}$ is said to be* elementary, *if:*

1. *There exists a bound $K > 0$, such that $|G_t| \leq K$ for all $t \in [S, T]$.*

2. *There exists a deterministic sequence $S = t_0 < t_1 < \cdots < t_n = T$, such that the sample paths of G_t are constant on each interval $[t_{i-1}, t_i)$. Thus, there exists a sequence of random variables $\{\gamma_i : i = 1, \ldots, n\}$ such that*

$$G_t = \gamma_i \text{ when } t_{i-1} \leq t < t_i.$$

Elementary processes play the same role in the theory of stochastic integrals as simple random variables do in the theory of expectation (Section 3.3). First, elementary processes are clearly Itô integrable: If $\{G_t\}$ is elementary as in the definition, then

$$\int_S^T G_t \, dB_t = \sum_{i=1}^n G_{t_{i-1}}(B_{t_i} - B_{t_{i-1}}).$$

Next, we use elementary processes to approximate integrands in a much larger class:

Definition 6.11.2 *The real-valued stochastic process $\{G_t : t \geq 0\}$ is said to be progressively measurable w.r.t. $\{\mathcal{F}_t\}$ if $G : [0, t] \times \Omega \mapsto \mathbf{R}$ is measurable w.r.t. the product-σ-algebra of $\mathcal{B}([0, t])$ and $\mathcal{F}_t$, for each $t \geq 0$. Here $\mathcal{B}([0, t])$ is the Borel algebra on $[0, t]$.*

This requirement guarantees, for example, that the integral $\int_0^t |G_s|^2 \, ds$ is a well-defined random variables which is $\mathcal{F}_t$-measurable, i.e., its realized value is known at time t. Progressively measurable processes are necessarily adapted. Conversely, an adapted process with left-continuous sample paths is progressiveluy measurable (Karatzas and Shreve, 1997).

Lemma 6.11.1 *Let $\{G_t : S \leq t \leq T\}$ be a progressively measurable stochastic process such that $\int_S^T \mathbf{E}G_t^2 \, dt < \infty$. Then there exists a sequence $\{G_t^{(i)}\}$ of elementary processes such that $\int_S^T \mathbf{E}|G_t^{(i)} - G_t|^2 \, dt \to 0$ as $i \to \infty$.*

See (Karatzas and Shreve, 1997), (Øksendal, 2010) or (Mao, 2008) for the proof of this lemma - first we approximate $\{G_t\}$ with a bounded process, then we approximate this bounded process with a Lipschitz continuous process, and finally we approximate this continuous process with a piecewise constant one, i.e. an elementary process.

Next, the Itô isometry says that the Itô integral depends continuously on the integrand: When $\{G_t^{(i)}\}$ is a sequence of elementary processes that converge to $\{G_t\}$ in the sense of the lemma, then the Itô integrals $\int_S^T G_t^{(i)} \, dB_t$ converge in mean square. The limit is the integral $\int_S^T G_t \, dB_t$.

6.11.2   Weak Solutions to SDE's

Our notion of solutions, as in Definition 6.5.2, is termed a *strong* solution: We start with a probability space $(\Omega, \mathcal{F}, \mathbf{P})$, a filtration $\{\mathcal{F}_t\}$, Brownian motion $\{B_t\}$, and seek an Itô process $\{X_t\}$ so that the stochastic differential equation is satisfied. The literature also operates with a *weak* notion of solutions, where we are just given the functions f and g and asked to find a filtered probability space as well as two adapted processes $\{B_t, X_t\}$ which satisfy the stochastic differential equation and such that $\{B_t\}$ is Brownian motion. See e.g. (Karatzas and Shreve, 1997).

6.12   EXERCISES

Exercise 6.8: Give a probabilistic interpretation to the integral

$$\int_0^\infty t^2 \, dF_t \text{ with } F_t = 1 - \exp(-\lambda t)$$

and state its value.

Exercise 6.9: Consider the Itô integral (compare Section 6.4)

$$I_t = \int_0^t B_s \, dB_s = \frac{1}{2}B_t^2 - \frac{1}{2}t.$$

Use the properties of Gaussian variables (Exercise 3.13) to determine the mean and variance of I_t, and verify that this agrees with the properties of Itô integrals. *Hint:* Exercise 3.13 contains a result for the moments of Gaussian variables.

Exercise 6.10 Stochastic resonance: Consider the double well model (6.8) with periodic forcing

$$dX_t = (X_t - X_t^3 + A\cos\omega t) \, dt + \sigma \, dB_t.$$

Take $A = 0.12$, $\omega = 10^{-3}$, and $\sigma = 0.25$. Simulate the system using the Euler-Maruyama method on the time interval $[0, 10^4]$ using a time step of 0.01. Repeat with $\sigma = 0.1$ and $\sigma = 0.5$. *Note:* This phenomenon was observed and analyzed by (Benzi et al., 1981).

Exercise 6.11: Let $\{X_t\}$ be the Ornstein-Uhlenbeck process given by

$$dX_t = -\lambda X_t \, dt + \lambda \, dB_t, \quad X_0 = 0,$$

and define $\{Y_t\}$ by

$$Y_t = \int_0^t X_s \, ds.$$

1. Using $\lambda = 100$, simulate a sample path of $\{B_t\}$, $\{X_t\}$ and $\{Y_t\}$ on the time interval $t \in [0, 10]$, using a time step of 0.001. Plot $\{Y_t\}$ and $\{B_t\}$.

2. Why is $\{Y_t\}$ so close to $\{B_t\}$? Show, for example, that $Y_t - B_t = -X_t/\lambda$, so that in stationarity, $\mathbf{E}(Y_t - B_t)^2 = 1/(2\lambda)$.

3. In a new window, plot the sample paths of the following Itô processes:

$$\int_0^t B_s \, dB_s, \quad \int_0^t Y_s \, dY_s, \quad \int_0^t B_s \circ dB_s, \quad \int_0^t Y_s \circ dY_s.$$

Explain the findings.

4. In a new window, plot

$$\int_0^t Y_s \, dB_s, \quad \int_0^t B_s \, dY_s, \quad \int_0^t Y_s \circ dB_s, \quad \int_0^t B_s \circ dY_s.$$

Explain the findings.

Exercise 6.12 Mean and variance in the Cox-Ingersoll-Ross process:
Consider the Itô equation

$$dX_t = \lambda(\xi - X_t)\, dt + \gamma\, \sqrt{X_t}\, dB_t, \quad X_0 = x > 0,$$

where $\{B_t\}$ is standard Brownian and λ, ξ, γ are positive parameters. We take for granted that a unique solution to this equation exists and can be approximated with the Euler-Maruyama method; this $\{X_t\}$ is the so-called *Cox-Ingersoll-Ross process*. Write down difference equations for the mean and variance of X_t and derive differential equations by passing to the limit of small time steps. Conclude on the stationary mean, variance, and autocovariance function of $\{X_t\}$.

CHAPTER 7

The Stochastic Chain Rule

The Itô integral is the key component in the theory of diffusion processes and stochastic differential equations, but we rarely compute Itô integrals explicitly. In stead, we almost invariably use rules of calculus, just as we do when analyzing deterministic models. However, since Brownian motion and Itô processes are non-differentiable, the ordinary rules of calculus do not apply, so in this chapter, we develop the corresponding stochastic calculus. The most important element in this calculus is Itô's lemma, a stochastic version of the chain rule, which in its simplest form states that if $\{X_t : t \geq 0\}$ is a scalar Itô process, and $\{Y_t : t \geq 0\}$ is a new scalar stochastic process defined by $Y_t = h(X_t)$, then $\{Y_t\}$ is again an Itô process. Although Itô's lemma generalizes the chain rule, it involves second order derivatives of h as well as the quadratic variation of $\{X_t\}$; terms, which are unfamiliar to ordinary calculus.

In this chapter, once we have stated Itô's lemma, we give examples of applications of the result. The main applications are:

- To solve stochastic differential equations analytically, in the few situations where this is possible.

- To change coordinates. For example, the equation may describe the motion of a particle in Cartesian or polar coordinates.

- To find the dynamics of quantities that are derived from the state, such as the energy of a particle.

Itô's lemma tells us how to transform the dependent variable. We also give formulas for how to transform the independent variable, i.e., change the scale of time.

DOI: 10.1201/9781003277569-7

7.1   THE CHAIN RULE OF DETERMINISTIC CALCULUS

Let us first recapitulate the well-known chain rule from ordinary (deterministic) calculus. Let $\{X_t : t \geq 0\}$ be a real-valued function with derivative $\frac{dX_t}{dt} = \dot{X}_t$, and let $h : \mathbf{R} \times \mathbf{R} \mapsto \mathbf{R}$ be a differentiable function with partial derivatives $\dot{h}$ and h'. Define

$$Y_t = h(t, X_t)$$

then, according to the chain rule, $\{Y_t : t \geq 0\}$ is differentiable with derivative

$$\dot{Y}_t = \dot{h}(t, X_t) + h'(t, X_t)\,\dot{X}_t.$$

Formally, we may multiply this with dt and obtain (omitting arguments)

$$dY_t = \dot{h}\,dt + h'\,dX_t.$$

We can rewrite these two equations in integral form:

$$Y_t = Y_0 + \int_0^t \left[\dot{h}(s, X_s) + h'(s, X_s)\dot{X}_s \right] ds$$

$$= Y_0 + \int_0^t \dot{h}(s, X_s)\,ds + \int_0^t h'(s, X_s)\,dX_s.$$

This is a slightly unusual way to write the chain rule, but for our purpose it has the advantage that it is similar to the way we write stochastic differential equations. This makes comparison easier.

7.2   TRANSFORMATIONS OF RANDOM VARIABLES

Since stochastic calculus concerns transformations of Itô processes, it is useful to recap how scalar random variables and their statistics behave under transformations. Let X be a scalar random variable and define $Y = h(X)$ where $h : \mathbf{R} \mapsto \mathbf{R}$ is smooth. Figure 7.1 displays an example.

Recall that the medians $\bar{x}$ and $\bar{y}$ of X and Y satisfy $\bar{y} = h(\bar{x})$ whenever $h(\cdot)$ is monotonic. For the expectations, the relationship is not as simple, but textbooks on statistics use the following approximation, where μ, σ^2, and f are the mean, variance, and p.d.f. of X:

$$\mathbf{E}Y = \int_{-\infty}^{+\infty} f(x)h(x)\,dx$$

$$\approx \int_{-\infty}^{+\infty} f(x)[h(\mu) + h'(\mu) \cdot (x - \mu) + \frac{1}{2}h''(\mu) \cdot (x - \mu)^2]\,dx$$

$$= h(\mu) + \frac{1}{2}h''(\mu) \cdot \sigma^2.$$

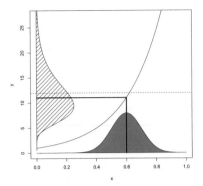

Figure 7.1 Two random variables $X \sim N(0.6, 0.1)$ and $Y = \exp 4X$. The gray and shaded region displays the p.d.f. of X and Y, rescaled. Solid lines indicate medians; the dashed line indicates $\mathbf{E}Y$.

Here, we have Taylor expanded h around μ, discarding cubic and higher order terms, which is valid when f resembles a Gaussian on a region where h resembles a parabola. Notice that the variance of X, in concert with the curvature of h, together yield a contribution to the expectation of Y.

As for the deviations from the mean, we can use the linear approximation

$$Y - \mathbf{E}Y \approx h'(\mu) \cdot (X - \mu)$$

which, combined with the expression for $\mathbf{E}Y$, gives

$$Y \approx h(\mu) + \frac{1}{2}h''(\mu)\sigma^2 + h'(\mu) \cdot (X - \mu).$$

For stochastic processes, we shall now see that this approximation is pivotal, when applied to the *increments* of the original and transformed processes.

7.3 ITÔ'S LEMMA: THE STOCHASTIC CHAIN RULE

The deterministic chain rule of Section 7.1 required that $\{X_t\}$ is smooth, so it needs not apply when $\{X_t\}$ is Brownian motion or an Itô process. Indeed, Section 6.4 indicated that the chain rule failed to helps us compute $\int_0^t B_s \, dB_s$. In stead we have the following chain rule of stochastic calculus:

Theorem 7.3.1 (Itô's lemma) *Let $\{X_t : t \geq 0\}$ be an Itô process as in Definition 6.6.1, taking values in $\mathbf{R}^n$ and given by*

$$dX_t = F_t \, dt + G_t \, dB_t$$

where $\{B_t : t \geq 0\}$ is d-dimensional Brownian motion. Let $h : \mathbf{R}^n \times \mathbf{R} \mapsto \mathbf{R}$ be differentiable w.r.t. time t and twice differentiable w.r.t. x, with continuous derivatives and define $Y_t = h(X_t, t)$. Then $\{Y_t\}$ is an Itô process given by

$$dY_t = \dot{h}\, dt + \nabla h\, dX_t + \frac{1}{2}dX_t^\top \mathbf{H}h\, dX_t \tag{7.1}$$

$$= \dot{h}\, dt + \left(\nabla h\, F_t + \frac{1}{2}\mathrm{tr}G_t^\top \mathbf{H}h\, G_t \right) dt + \nabla h\, G_t\, dB_t. \tag{7.2}$$

Here, $\dot{h} = \partial h / \partial t(X_t, t)$ is the partial derivative with respect to time, while ∇h is the spatial gradient and $\mathbf{H}h = \partial^2 h / \partial x^2$ is the Hessian matrix containing double derivatives w.r.t. spatial coordinats. We omit the arguments (X_t, t) for clarity. We view ∇h as a row vector, so that $\nabla h\, dX_t$ is the inner product.

Itô's lemma gives two expressions for dY_t: The first, (7.1) looks like (and essentially is) a Taylor expansion of

$$Y_t + dY_t = h(X_t + dX_t, t + dt)$$

around the point (X_t, t), where we keep first order terms in dt and up to second order terms in dX_t. We need second order terms in dX_t because dX_t generally scales with $\sqrt{dt}$, so that $(dX_t)^2$ scales as dt (proposition 6.5.1).

The second form follows from the first, using arithmetic for the increments and the crossvariation (Section 6.9): We insert $dX_t = F_t\, dt + G_t\, dB_t$ and $dX_t dX_t^\top = G_t G_t^\top\, dt$. We then use linear algebra: Recall that the trace of an n-by-n matrix is $\mathrm{tr}P = P_{11} + \cdots + P_{nn}$ and that $\mathrm{tr}PQ = \mathrm{tr}QP$ when $P \in \mathbf{R}^{n \times m}$, $Q \in \mathbf{R}^{m \times n}$. So

$$dX_t^\top \mathbf{H}h\, dX_t = \mathrm{tr}[\mathbf{H}h\, G_t G_t^\top\, dt] = \mathrm{tr}[G_t^\top \mathbf{H}h\, G_t]\, dt.$$

This form, (7.2), shows that the drift of $\{Y_t\}$ involves three terms: The terms $\dot{h}\, dt + \nabla h\, F_t\, dt$ are what we would expect from the usual chain rule (Section 7.1). The "extra" drift term $\mathrm{tr}(G_t^\top \mathbf{H}h\, G_t)/2\, dt$ has the same form as the approximation we considered in Section 7.2. In turn, the noise intensity of $\{Y_t\}$, i.e. $\nabla h\, G_t\, dB_t$, also has the form that is to be expected from Section 7.2.

Proof: We outline the proof in the simple case where h does not depend on time t and $\{X_t\}$ is scalar. First, assume also that the triple derivative h''' is bounded, and that $\{F_t\}$ and $\{G_t\}$ are deterministic and constant, $F_t = F$, $G_t = G$. Let $\Delta = (t_0, t_1, \ldots, t_n)$ be a partition of $[0, t]$ and write

$$Y_t = Y_0 + \sum_{i=0}^{n-1} \Delta Y_i \text{ with } \Delta Y_i = Y_{t_{i+1}} - Y_{t_i}$$

To evaluate the increment ΔY_i, Taylor expand the function h around X_{t_i}:

$$\Delta Y_i = h'(X_{t_i})\, \Delta X_i + \frac{1}{2}h''(X_{t_i})\, (\Delta X_i)^2 + R_i$$

where $\Delta X_i = X_{t_{i+1}} - X_{t_i}$ and the residual R_i can be written in the mean value (or Lagrange) form

$$R_i = \frac{1}{6}h'''(\xi_i)(\Delta X_i)^3.$$

Here, ξ_i is, for each i, a random variable between $X_{t_{i-1}}$ and X_{t_i}]. Since $h'''(\xi_i)$ is bounded, it follows from a minor extension of Exercise 6.6 that the sum of these residuals vanish in mean square as $|\Delta| \to 0$.

The sum $h'(X_{t_i})\,\Delta X_i$ converges to the Itô integral $\int_0^t h'(X_s)\,dX_s$. As for the sum of the terms $\frac{1}{2}h''(X_{t_i})\,(\Delta X_i)^2$, it converges to the integral

$$\int_0^t \frac{1}{2}h''(X_s)\,d[X]_s$$

where $[X]_t$ is the quadratic variation process of $\{X_t\}$. With the notation $(dX_s)^2 = d[X]_s$ from Section 6.9, we get the first form (7.1):

$$Y_t = Y_0 + \int_0^t h'(X_s)\,dX_s + \int_0^t \frac{1}{2}h''(X_s)\,(dX_s)^2.$$

With the rules for manipulating variations (Section 6.9, Exercise 6.7), we can rewrite this as

$$Y_t = Y_0 + \int_0^t h'(X_s)F_s\,ds + \int_0^t h'(X_s)G_s\,dB_s + \int_0^t \frac{1}{2}h''(X_s)\,G_s^2\,dt,$$

i.e., the second form (7.2). Now relax the simplifying assumptions: First, the argument holds also if F and G are random but $\mathcal{F}_0$-measurable and bounded. Next, if $\{F_t\}$ and $\{G_t\}$ are elementary processes, then apply the previous argument to each sub-interval where they are constant; then, the conclusion still stands. Now, if $\{F_t\}$ and $\{G_t\}$ are as in the theorem, approximate them with elementary processes to see that the conclusion remains. Finally, if $h'''(x)$ is not bounded, then approximate h on a bounded domain with a function with bounded triple derivative, and stop the process $\{X_t\}$ upon exit from this domain. Then, the conclusion remains for the stopped processes. Let the bounded domain grow to cover the entire state space to see that the conclusion stands. This only sketches the proof; see e.g. (Øksendal, 2010; Mao, 2008). ■

Example 7.3.1 *We confirm the result (6.12) that $\int_0^t B_s\,dB_s = (B_t^2 - t)/2$: Take $X_t = B_t$ and $Y_t = h(X_t, t)$ with $h(x, t) = (x^2 - t)/2$. Then $\dot{h} = -1/2$, $h' = x$, $h'' = 1$, and*

$$dY_t = \dot{h}\,dt + h'\,dX_t + \frac{1}{2}h''\,(dX_t)^2 = -dt/2 + B_t\,dB_t + dt/2 = B_t\,dB_t.$$

Since $Y_0 = 0$, we conclude that $Y_t = (B_t^2 - t)/2 = \int_0^t B_s\,dB_s$.

We have stated Itô's lemma for a *scalar-valued* function h. When the function h maps the state X_t to a vector $Y_t \in \mathbf{R}^m$, i.e., $h : \mathbf{R}^n \times \mathbf{R} \mapsto \mathbf{R}^m$, we may apply Itô's lemma to each coordinate of Y_t at a time. I.e., we apply the previous formula to $Y_t^{(i)} = h_i(X_t, t)$, for $i = 1, \ldots, m$.

Example 7.3.2 *Let $\{X_t : t \geq 0\}$ be a vector-valued Itô process given by $dX_t = F_t \, dt + G_t \, dB_t$ and let $Y_t = TX_t$ where T is a matrix. Then*

$$dY_t = TF_t \, dt + TG_t \, dB_t.$$

Being even more specific, let $\{X_t\}$ satisfy the linear SDE $dX_t = AX_t \, dt + G \, dB_t$ and assume that T is square and invertible. Then $\{Y_t\}$ satisfies the linear SDE

$$dY_t = TAT^{-1}Y_t \, dt + TG \, dB_t.$$

Example 7.3.3 *[A stochastic product rule] Let $\{X_t\}$ and $\{Y_t\}$ be two scalar Itô processes and define $\{Z_t\}$ by $Z_t = X_t Y_t$. We aim to write $\{Z_t\}$ as an Itô process. To this end, introduce $h(x, y) = xy$, so that*

$$\nabla h = [y \quad x], \quad \mathbf{H}h = \begin{bmatrix} 0 & 1 \\ 1 & 0 \end{bmatrix}.$$

We get

$$dZ_t = \nabla h \cdot \begin{pmatrix} dX_t \\ dY_t \end{pmatrix} + \begin{pmatrix} dX_t \\ dY_t \end{pmatrix}^\top \frac{1}{2} \mathbf{H}h \begin{pmatrix} dX_t \\ dY_t \end{pmatrix} = Y_t \, dX_t + X_t \, dY_t + dX_t \, dY_t$$

where, as usual, $dX_t \, dY_t = d\langle X, Y \rangle_t$. I.e., we have the product rule

$$d(X_t Y_t) = X_t \, dY_t + Y_t \, dX_t + dX_t \, dY_t.$$

More specifically, assume that

$$dX_t = F_t \, dt + G_t \, dB_t, \quad dY_t = K_t \, dt + L_t \, dB_t$$

where $\{B_t\}$ is multivariate Brownian motion; then $dX_t \, dY_t = G_t L_t^\top \, dt$ and

$$d(X_t Y_t) = (X_t K_t + Y_t F_t + G_t L_t^\top) \, dt + (X_t L_t + Y_t G_t) \, dB_t.$$

7.4   SOME SDE'S WITH ANALYTICAL SOLUTIONS

Itô's lemma applies to transformations of and Itô process $\{X_t\}$, i.e., the integrands $\{F_t\}$ and $\{G_t\}$ can be arbitrary stochastic processes, as long as they satisfy the technical requirements. However, in most of the applications we are interested in, the Itô process $\{X_t\}$ is a solution to a stochastic differential equation, i.e., an Itô diffusion:

$$dX_t = f(X_t, t) \, dt + g(X_t, t) \, dB_t.$$

In this case, Itô's lemma has the form

$$dY_t = \dot{h}\ dt + \nabla h\ f\ dt + \frac{1}{2}\mathrm{tr}[gg^\mathsf{T}\mathbf{H}h]\ dt + \nabla h\ g\ dB_t$$

when $Y_t = h(X_t, t)$; we assume that h is $C^{2,1}$ and have again omitted the arguments (X_t, t). The drift term is so important, and appears so frequently, that we introduce a special symbol for it, namely the second order linear differential operator L given by

$$Lh = \nabla h\ f + \frac{1}{2}\mathrm{tr}[gg^\mathsf{T}\mathbf{H}h]$$

whenever $h(\cdot, t)$ is C^2 for each t. This operator works along the spatial coordinates x and can be applied both when $h(x, t)$ depends on time, and to a function $h(x)$ which does not depend on time. We will later introduce the name "backward Kolmogorov operator" for L.

An important application of Itô's lemma is to find and verify analytical solutions to stochastic differential equations. This is possible only for a small class of equations, but these special cases play a prominent role due to their tractability.

Exercise 7.1 Geometric Brownian Motion: In Section 6.2.3, we claimed that the scalar stochastic differential equation

$$dY_t = rY_t\ dt + \sigma Y_t\ dB_t, \quad Y_0 = y.$$

has the solution

$$Y_t = y\ \exp((r - \frac{1}{2}\sigma^2)t + \sigma B_t).$$

Show that this is true. Next, find the mean, mean square, and variance of Y_t, using the properties of the log-normal distribution. *Hint:* For example, take $X_t = (r - \sigma^2/2)t + \sigma B_t$ and $Y_t = h(t, X_t)$ with $h(t, x) = y\exp(x)$. *Note:* See Exercise 7.23 for a partial extension to the multivariate case.

Exercise 7.2: Show that $Y_t = \tanh B_t$ satisfies

$$dY_t = -Y_t(1 - Y_t^2)\ dt + (1 - Y_t^2)\ dB_t, \quad Y_0 = 0.$$

Exercise 7.3: Show that $Y_t = (y^{1/3} + \frac{1}{3}B_t)^3$ satisfies

$$dY_t = \frac{1}{3}Y_t^{1/3}\ dt + Y_t^{2/3}\ dB_t, \quad Y_0 = y.$$

7.4.1 The Ornstein-Uhlenbeck Process

The narrow-sense linear stochastic differential equation

$$dX_t = -\lambda X_t \, dt + \sigma \, dB_t, \quad X_0 = x, \tag{7.3}$$

has the solution which would expected from ordinary calculus:

$$X_t = xe^{-\lambda t} + \int_0^t \sigma e^{-\lambda(t-s)} dB_s. \tag{7.4}$$

To verify this, we cannot use Itô's lemma directly, because the integrand in the Itô integral depends on the upper limit t. To circumvent this difficulty, we introduce a transformed version of X_t, namely the process Y_t given by

$$Y_t = h(t, X_t) = e^{\lambda t} X_t \text{ where } h(t, x) = e^{\lambda t} x$$

Itô's lemma assures that if X_t satisfies the original SDE (7.3), then $\{Y_t\}$ satisfies

$$
\begin{aligned}
dY_t &= \dot{h} \, dt + h' \, dX_t + \frac{1}{2} h'' \, dX_t^2 \\
&= \lambda e^{\lambda t} X_t \, dt + e^{\lambda t}(-\lambda X_t \, dt + \sigma \, dB_t) \\
&= e^{\lambda t} \sigma \, dB_t.
\end{aligned}
$$

This stochastic differential equation can easily be solved for Y_t, using $Y_0 = h(0, X_0) = x$:

$$Y_t = x + \int_0^t e^{\lambda s} \sigma dB_s.$$

We can now back-transform to find the solution to the original equation:

$$X_t = e^{-\lambda t} Y_t = xe^{-\lambda t} + \int_0^t e^{-\lambda(t-s)} \sigma dB_s$$

The solution (7.4) is called the Ornstein-Uhlenbeck process and was introduced by Uhlenbeck and Ornstein in 1930 as a model for the velocity of a molecule under diffusion. Compared to Brownian motion it has the advantage that it predicts finite velocities! It is used frequently in applications, whether in physics, engineering, biology or finance. Analyzing the solution, using the martingale property of the Itô integral, we see that (compare section 5.10)

$$\mathbf{E} X_t = e^{-\lambda t} x.$$

To find the variance of X_t, we use the Itô isometry:

$$
\begin{aligned}
\mathbf{V} X_t &= \mathbf{V} \left(\int_0^t e^{-\lambda(t-s)} \sigma \, dB_s \right) \\
&= \int_0^t e^{-2\lambda(t-s)} \sigma^2 \, ds \\
&= \begin{cases} \frac{\sigma^2}{2\lambda}(1 - \exp(-2\lambda t)) & \text{if } \lambda \neq 0, \\ \sigma^2 t & \text{if } \lambda = 0. \end{cases}
\end{aligned}
$$

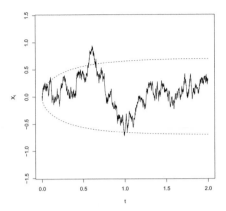

Figure 7.2 A realization of the Ornstein-Uhlenbeck process $\{X_t\}$ given by the SDE $dX_t = -\lambda X_t\, dt + \sigma\, dB_t$ (7.3) (solid) with $X_0 = 0$, $\lambda = 1$, $\sigma = 1$. Included is also plus/minus the standard deviation of X_t (dashed).

This agrees with what we found studying the Lyapunov equation (Section 5.9 and Exercise 5.7). A realization of the integral is seen in Figure 7.2. Since X_t arises as a linear combination of Gaussian random variables, X_t is also Gaussian, so its distribution is determined by the mean and variance.

The Ornstein-Uhlenbeck process can be generalized to vector processes, and to the case of time-varying parameters:

Exercise 7.4: Consider a vector process $\{X_t \in \mathbf{R}^n : t \geq 0\}$ which satisfies the narrow-sense linear SDE

$$dX_t = (AX_t + w_t)\, dt + G\, dB_t$$

with initial condition $X_0 = x$, where the external input w_t is deterministic and of bounded variation. Show that X_t can be written as

$$X_t = e^{At}x + \int_0^t e^{A(t-s)}(w_s\, ds + GdB_s).$$

State the mean, variance and distribution of X_t. *Hint:* Follow the reasoning for the Ornstein-Uhlenbeck process in the previous, i.e., start by defining $Y_t = h(t, X_t)$ with $h(t, x) = \exp(-At)x$.

Exercise 7.5: Consider the scalar time-varying equation

$$dX_t = \lambda_t X_t\, dt + \sigma_t\, dB_t$$

with initial condition $X_0 = x$, where $\{\lambda_t : t \geq 0\}$ and $\{\sigma_t : t \geq 0\}$ are deterministic functions of bounded variation. Show that the solution is

$$X_t = e^{F_t} x + \int_0^t e^{F_t - F_s} \sigma_s \, dB_s$$

where

$$F_t = \int_0^t \lambda_s \, ds.$$

In summary, for linear stochastic differential equations, there is a unique solution which has a closed-form expression.

7.5   DYNAMICS OF DERIVED QUANTITIES

A second common application of Itô's lemma is the situation where a dynamic system evolves according to a stochastic differential equation, and we are interested in some prescribed function of interest defined on state space. We give two examples.

7.5.1   The Energy in a Position-Velocity System

Consider a physical system with a position $\{X_t\}$ and a velocity $\{V_t\}$, which satisfy two coupled stochastic differential equations

$$dX_t = V_t \, dt, \quad dV_t = -u'(X_t) \, dt - \mu V_t \, dt + \sigma \, dB_t.$$

Here, $u(\cdot)$ is a potential, so that $-u'(x)$ is the (mass specific) force acting on the system. There is linear dampling given by μ, and stochastic forces $\sigma \, dB_t$. This system generalizes the mass-spring-damper system of Chapter 5, where the potential was the energy in the spring, $u(x) = kx^2/(2m)$.

Define the potential, kinetic, and total energies in the system:

$$U_t = u(X_t), \quad T_t = \frac{1}{2} V_t^2 \quad E_t = U_t + T_t.$$

Then with Itô's lemma, these can be written as Itô processes

$$dU_t = u'(X)V_t \, dt$$

$$dT_t = V_t \, dV_t + \frac{1}{2}(dV_t)^2 = \left[-V_t u'(X_t) - \mu V_t^2 + \frac{1}{2}\sigma^2 \right] \, dt + \sigma V_t \, dB_t$$

$$dE_t = dU_t + dT_t = \left[\frac{1}{2}\sigma^2 - \mu V_t^2 \right] \, dt + \sigma V_t \, dB_t.$$

The noise gives rise to fluctuations in the energy through the term $\sigma V_t \, dB_t$: The random impulse $\sigma \, dB_t$ increases the kinetic and total energy, if it has the same sign as V_t; otherwise it decreases the energy. The presence of noise also gives rise to an expected increase in total energy through the term $\frac{1}{2}\sigma^2 \, dt$. Thus, the stochastic forces pump energy into the system.

We now examine stationary solutions to these equations, i.e., the statistics are invariant to time translations (Section 5.4). With an appeal to physics we would expect stationary solutions to exist for some potentials $u(\cdot)$. We will later call such potentials *confining*; for now, we simply assume stationarity. Then, the drift must have expectation 0. For the energy E_t, this implies

$$\mathbf{E}\left[\frac{1}{2}\sigma^2 - \mu V_t^2\right] = 0 \Leftrightarrow \mathbf{E}\frac{1}{2}V_t^2 = \frac{\sigma^2}{4\mu}$$

In stationarity, the expected kinetic energy is therefore $\sigma^2/(4\mu)$ which expresses a balance between what is supplied by the noise and what is dissipated through friction. For quadratic potentials, we can show as in Exercise 5.6 that *equipartitioning* holds, i.e., the expected potential energy equals the expected kinetic energy. We consider stationary processes further in Section 9.8.

7.5.2 The Cox-Ingersoll-Ross Process and the Bessel Processes

Consider $n \geq 2$ independent Ornstein-Uhlenbeck processes $\{X_t^{(i)} : t \geq 0\}$, for $i = 1, \ldots, n$, which each satisfy

$$dX_t^{(i)} = -\mu X_t^{(i)} \, dt + \sigma \, dB_t^{(i)}$$

where $B_t = (B_t^{(1)}, \ldots, B_t^{(n)})$ is n-dimensional standard Brownian motion. Now form the sum of squares:

$$Y_t = \sum_{i=1}^{n} |X_t^{(i)}|^2.$$

Then Itô's lemma gives, since the processes $\{X_t^{(i)}\}$ are independent

$$dY_t = 2\sum_{i=1}^{n} X_t^{(i)} \, dX_t^{(i)} + \sum_{i=1}^{n} |dX_t^{(i)}|^2 = (-2\mu Y_t + n\sigma^2) \, dt + 2\sigma \sum_{i=1}^{n} X_t^{(i)} \, dB_t^{(i)}.$$

This does not seem very tractable, due to the noise term, which couples the n Brownian motions with the n state variables. However, we can make progress by defining $\{W_t : t \geq 0\}$ by $W_0 = 0$ and

$$\sqrt{Y_t} \, dW_t = \sum_{i=1}^{n} X_t^{(i)} \, dB_t^{(i)},$$

then $\{W_t\}$ is a continuous martingale with quadratic variation given by

$$Y_t \, d[W]_t = \sum_{i=1}^{n} |X_t^{(i)}|^2 \, dt$$

or simply $d[W]_t = dt$. Thus, $\{W_t\}$ is Brownian motion. Here, we have ignored the singularity where $Y_t = 0$. To summarize,

$$dY_t = (n\sigma^2 - 2\mu Y_t)\,dt + 2\sigma\,\sqrt{Y_t}\,dW_t \ .$$

We see that we have arrived at an equation which governs the sum of squares, which does not make reference to the individual states $X_t^{(i)}$ or the individual Brownian motions $B_t^{(i)}$. The process $\{Y_t\}$ is referred to as the Cox-Ingersoll-Ross process (Cox et al., 1985), even if it was considered earlier by Feller (1951). A more general formulation of this process, when $\mu \neq 0$, is the Itô equation

$$dY_t = \lambda(\xi - Y_t)\,dt + \gamma\,\sqrt{Y_t}\,dW_t \tag{7.5}$$

and we can relate the two formulations by $\lambda = 2\mu$, $\xi = n\sigma^2/\lambda$, $\gamma = 2\sigma$. Note that the more general formulation does not require n to be integer or greater than or equal to 2. The Cox-Ingersoll-Ross process finds applications in statistical physics, where it can describe fluctuations in energy, in economy, where it can describe fluctuations in interest rates, as well as in mathematical biology, where it can model fluctuations in abundances of populations. Exercise 6.12 found the stationary mean, variance and autocovariance of this process.

With $\mu = 0$ we obtain the so-called squared Bessel process $\{Y_t = \sigma^2 |B_t|^2 : t \geq 0\}$ given by

$$dY_t = n\sigma^2\,dt + 2\sigma\sqrt{Y_t}\,dW_t,$$

and taking the square root, we obtain the Bessel process $\{Z_t = \sqrt{Y_t} = \sigma|B_t| : t \geq 0\}$, which satisfies the SDE

$$dZ_t = \frac{n-1}{2}\frac{1}{Z_t}\sigma^2\,dt + \sigma\,dW_t.$$

Exercise 7.6: Derive the equation governing $\{Z_t\}$ using Itô's lemma.

These processes earn their names because a Bessel function appears in their transition probabilities. Notice that the squared Bessel process has a singularity in its noise intensity at $y = 0$, while the Bessel process has a singularity in its drift. Also these processes are meaningful when n is non-integer, and when $n < 2$, but the singularity needs to be addressed.

7.6 COORDINATE TRANSFORMATIONS

A final frequent application of Itô's lemma is to change coordinates in the underlying state space. Arguably, we have used this already in Section 7.4, when we established that geometric Brownian motion solves a wide-sense linear stochastic differential equation: There, we changed coordinates from natural to logarithmic (Exercise 7.1). We now give a few more examples.

7.6.1 Brownian Motion on the Circle

Brownian motion on the circle is a process taking values on the unit circle in the plane. In polar coordinates (r, θ), the process is easy to describe: The radius r is constant and equal to 1, while the angle θ is Brownian motion, $\Theta_t = B_t$. Transforming to Cartesian coordinates, we find:

$$X_t = \cos \Theta_t = \cos B_t, \quad Y_t = \sin \Theta_t = \sin B_t.$$

Itô's lemma yields that $(X_t, Y_t) = (\cos B_t, \sin B_t)$ satisfies

$$\begin{pmatrix} dX_t \\ dY_t \end{pmatrix} = \begin{pmatrix} -\sin B_t \ dB_t - \frac{1}{2} \cos B_t \ dt \\ \cos B_t \ dB_t - \frac{1}{2} \sin B_t \ dt \end{pmatrix}$$

and substituting the $\cos B_t = X_t$, $\sin B_t = Y_t$, we can rewrite this

$$\begin{pmatrix} dX_t \\ dY_t \end{pmatrix} = \begin{pmatrix} -Y_t \\ X_t \end{pmatrix} dB_t - \frac{1}{2} \begin{pmatrix} X_t \\ Y_t \end{pmatrix} dt. \tag{7.6}$$

This is a wide-sense linear Itô stochastic differential equation governing the processes $\{(X_t, Y_t) : t \geq 0\}$, i.e., the motion in Cartesian coordinates. It is useful to think about this geometrically: The dB_t-term is orthogonal to the position (X_t, Y_t), as we would expect of motion along a tangent. However, since the dB_t-term is of order $\sqrt{dt}$, it acts to project the particle along the tangent which would increase the radius. To balance this, dt-term contracts the particle towards the origin.

Exercise 7.7: Use the stochastic differential equation to show that $\mathbf{E} \cos B_t = \exp(-t/2)$.

7.6.2 The Lamperti Transform

Coordinate transformations allow us to express system dynamics in different coordinate systems, but we should obtain the same results regardless of the choice of coordinate systems. This raises the question if some coordinate system is more convenient for a given purpose. In the previous example, an argument in favor of polar coordinates is that the diffusivity is constant in this coordinate system. We say that the noise is *additive*. As we shall see later, statistics and numerics are simpler in this additive case. We can therefore ask if it is possible to change coordinates so that in the transformed system, the noise is additive? If so, we say that the transformation is a *Lamperti* transform. Consider the scalar stochastic differential equation

$$dX_t = f(X_t) \ dt + g(X_t) \ dB_t$$

and assume that $g(x) > 0$ for all x. Then define the Lamperti transformed process $\{Y_t\}$ by $Y_t = h(X_t)$ where

$$h(x) = \int^x \frac{1}{g(v)} \ dv. \tag{7.7}$$

The lower integration limit is arbitrary; h is any antiderivative. We get

$$h'(x) = \frac{1}{g(x)}, \quad h''(x) = -\frac{g'(x)}{g^2(x)},$$

and so Itô's lemma yields

$$dY_t = h'(X_t)\, dX_t + \frac{1}{2} h''(X_t)\, g^2(X_t)\, dt$$

$$= \frac{f(X_t)}{g(X_t)}\, dt + dB_t - \frac{1}{2} g'(X_t)\, dt$$

$$= \left[\frac{f(h^{-1}(Y_t))}{g(h^{-1}(Y_t))} - \frac{1}{2} g'(h^{-1}(Y_t)) \right] dt + dB_t.$$

Note that in the Lamperti transformed coordinate Y_t, the noise is additive, as required. Here, the noise intensity is 1. We will use the term "Lamperti transform" as long as the noise is additive in the transformed system, even if the intensity is not 1, because it is sometimes convenient to allow a variance parameter.

Exercise 7.8 Geometric Brownian Motion: Consider the SDE

$$dX_t = rX_t\, dt + \sigma X_t\, dB_t.$$

Find a Lamperti transform and write the SDE in the transformed coordinate.

7.6.3 The Scale Function

For a scalar process, it is also possible to transform the coordinate so that the transformed process is driftless, i.e., a martingale. Specifically, consider again the general scalar Itô equation

$$dX_t = f(X_t)\, dt + g(X_t)\, dB_t$$

where we assume, in addition to Lipschitz continuity, that g is bounded away from 0, i.e., there exits an $\epsilon > 0$ such that $g(x) > \epsilon$ for all x. Now, introduce the transformed coordinate $Y_t = s(X_t)$ where we have yet to determine the transform s. Then $\{Y_t\}$ is an Itô process with

$$dY_t = \left[f(X_t)s'(X_t) + \frac{1}{2} g^2(X_t)s''(X_t) \right] dt + s'(X_t)g(X_t)\, dB_t.$$

We now aim to choose the transform $s : \mathbf{R} \mapsto \mathbf{R}$ so that the drift vanishes:

$$fs' + \frac{1}{2} g^2 s'' = 0.$$

Introduce $\phi = s'$, then the equation in ϕ is

$$f\phi + \frac{1}{2} g^2 \phi' = 0$$

which has the solution

$$\phi(x) = \exp\left(-\int_{x_0}^{x} \frac{2f(y)}{g^2(y)} \, dy\right)$$

where the reference point x_0 is arbitrary; i.e., $\log\phi$ is any antiderivative of $-2f/g^2$. From ϕ we may then find the transformation s through

$$s(x) = \int_{x_1}^{x} \phi(y) \, dy$$

where also x_1 is arbitrary. The two arbitrary constants x_0 and x_1 correspond to every affine transformation $Z_t = aY_t + b$ also being driftless.

The resulting s is known as the *scale function*; it appears also, for example, in the analysis of boundary points. Transforming the state so that the drift vanishes, i.e., so that the transformed process is a martingale, is in some sense complementary to the Lamperti transform: The former simplifies the drift term, while the latter simplifies the noise intensity term.

Exercise 7.9: Consider Brownian motion with drift, i.e., $dX_t = \mu \, dt + \sigma \, dB_t$. Determine the scale function s and the governing equation for the transformed coordinate Y_t, and explain in words how $\{Y_t\}$ can be driftless when $\{X_t\}$ has constant drift.

Exercise 7.10: Consider geometric Brownian motion $\{X_t\}$ given by the stochastic differential equation $dX_t = rX_t \, dt + \sigma X_t \, dB_t$. Show that the scale function is

$$s(x) = \begin{cases} \frac{1}{\nu} x^\nu & \text{when } \nu := 1 - 2r/\sigma^2 \neq 0, \\ \log x & \text{when } \nu = 0. \end{cases} \tag{7.8}$$

As a financial application of this scale function for geometric Brownian motion, assume that X_t is the price of a stock at time t. For simplicity, consider a situation where the discount rate is 0, so that money is borrowed or lent without interest. Then a stock typically offers an expected positive return $(r > 0)$ to compensate for the risk of a loss. To explain this quantitatively, assume that the *utility* of owning the stock to an investor is

$$Y_t = s(X_t)$$

where s is the scale function of the process. Then, the utility $\{Y_t\}$ is a martingale. In particular, the expected future utility equals the current utility; buying the stock is no better and no worse than hiding the money in the mattress. Thus, the price is *fair*. With $r > 0$, the utility is an increasing concave (decelerating) function of the price: The richer the investor is, the happier she is, but the first euro brings more joy than the second. For a given utility function, the model predicts that more volatile stocks (σ^2 larger) have higher average yields (r larger). In utility theory, the particular form (7.8) is known as *isoelastic* or *constant relative risk aversion*.

7.7   TIME CHANGE

Itô's lemma allows us to change the coordinate system used to describe the state space, i.e., the dependent variable. In some situations it is equally desirable to change coordinates on the time axis, i.e., the independent variable. In a deterministic setting, the chain rule applies to both situations, but this is not so in the stochastic setting.

A first example is the simple task of rescaling time. Consider a stochastic differential equation

$$dX_t = f(X_t) \, dt + g(X_t) \, dB_t$$

and assume that we want to express this in rescaled time $u = \alpha t$. To this end, define first $Y_u = X_t = X_{u/\alpha}$. Next, we aim to define a new Brownian motion $\{W_u : u \leq 0\}$ which is standard in rescaled time. Recall the scaling of Brownian motion in Section 4.3 to find

$$W_u = \alpha^{1/2} B_{u/\alpha}.$$

Now, $\mathbf{V}W_u = \alpha \mathbf{V}B_{u/\alpha} = u$, so $\{W_u\}$ is standard. We then get

$$dY_u = \alpha^{-1} f(Y_u) \, du + \alpha^{-1/2} g(Y_u) \, dW_u.$$

This can be used e.g. when changing units from hours to seconds; then we would take $\alpha = 3600$ s/hour. Another frequent use of this is to transform between dimensional and non-dimensional models:

Example 7.7.1 (Dimensionless Stochastic Logistic Growth) *Consider the stochastic logistic growth equation from the introduction (page 6)*

$$dX_t = rX_t(1 - X_t/K) \, dt + \sigma X_t \, dB_t \tag{7.9}$$

where r, K and σ are positive parameters, as is the initial condition $X_0 = x$. All quantities have dimensions; e.g., we measure time in seconds and the abundance X_t in grams. We rescale X_t to express it in units of K, and rescale time to express it in units of $1/r$, obtaining the process Y_u given by

$$Y_u = \frac{X_{u/r}}{K}.$$

The stochastic differential equation for $\{Y_u\}$ is obtained with Itô's lemma - which is simple here, because the map $y = h(x) = x/K$ is linear - and with the time change formula. We find

$$dY_u = Y_u(1 - Y_u) \, du + \frac{\sigma}{\sqrt{r}} Y_u \, dW_u$$

where $\{W_u\}$ is dimensionless standard Brownian motion with u being dimensionless time. We see that the original model contains two scale parameters, the carrying capacity K and the time scale $1/r$ of growth, and one dimensionless parameter, the noise level $\sigma/\sqrt{r}$.

Random Time Change

We now generalize to dynamic and random time change. For example, a chemical or biological process may run with a speed which depends on temperature, with the temperature evolving randomly, so we may wish to introduce a clock that itself runs with a speed that depends on temperature.

To this end, define a new transformed time $\{U_t : t \geq 0\}$ by

$$dU_t = H_t \, dt, \quad U_0 = 0$$

where $\{H_t : t \geq 0\}$ gives the speed ratio between new and old time. We require that $\{H_t\}$ is a continuous $\{\mathcal{F}_t\}$-adapted process which is positive and bounded away from zero. Then $t \mapsto U_t$ defines an invertible map from $\mathbf{R}_+$ onto itself, for each realization. Let $T_u = \inf\{t \geq 0 : U_t \geq u\}$ be the inverse map, so that we can transform back and forth between original time t and transformed time u, for each realization. Next, define the process $\{W_u : u \geq 0\}$ by

$$W_u = \int_0^{T_u} \sqrt{H_t} \, dB_t = \int_0^{\infty} \mathbf{1}(t \leq T_u) \sqrt{H_t} \, dB_t.$$

We now argue heuristically that $\{W_u : u \geq 0\}$ is standard Brownian motion with respect to the filtration $\{\mathcal{G}_u : u \geq 0\}$ where $\mathcal{G}_u = \mathcal{F}_{T_u}$: Assume $0 < u < v$ and that $v - u$ is small. Then

$$W_v - W_u \approx \sqrt{H_{T_u}}(B_{T_v} - B_{T_u})$$

and $T_v - T_u$ is approximately $(v - u)/H_{T_u}$. Conditional on $\mathcal{G}_u$, this increment is therefore Gaussian with mean 0 and variance $v - u$, which implies that $\{W_u : u \geq 0\}$ is standard Brownian motion.

Now, let an Itô process $\{X_t : t \geq 0\}$ be given by $dX_t = F_t \, dt + G_t \, dB_t$, then we consider the process $\{Y_u : u \geq 0\}$ given by $Y_u = X_{T_u}$, i.e., in transformed time. This process satisfies the Itô stochastic differential equation

$$dY_u = \frac{F_{T_u}}{H_{T_u}} \, du + \frac{G_{T_u}}{\sqrt{H_{T_u}}} \, dW_u.$$

Note that, as always, different rescalings in the time integral and the Itô integral, due to the scaling properties of Brownian motion.

As an example, recall that we discussed rescaled Brownian motion in logarithmic time in the context of the Law of the Iterated Logarithm (Theorem 4.3.4; page 74). The following exercise establishes the properties of this process using the method we just developed.

Exercise 7.11: Let $\{B_t : t \geq 0\}$ be standard Brownian motion and define, for $t > 0$

$$Y_u = \frac{1}{\sqrt{t}} B_t$$

where $t = \exp(u)$. Show that $\{Y_u : u \geq 0\}$ is governed by

$$dY_u = -\frac{1}{2} Y_u \, du + dW_u$$

i.e., $\{Y_u : u \geq 0\}$ is an Ornstein-Uhlenbeck process. Show also that Y_0 is distributed according to the stationary distribution, i.e., $\{Y_u\}$ is stationary.

When the Itô process $\{X_t\}$ is a solution of a stochastic differential equation, it may happen that the speed ratio H_t depends on the state X_t itself. So consider a process $\{X_t : t \geq 0\}$ given by the stochastic differential equation

$$dX_t = f(X_t) \, dt + g(X_t) \, dB_t$$

and consider the time change $dU_t = h(X_t) \, dt$. Then the process $\{Y_u : u \geq 0\}$ satisfies the Itô stochastic differential equation

$$dY_u = \frac{f(Y_u)}{h(Y_u)} \, du + \frac{g(Y_u)}{\sqrt{h(Y_u)}} \, dW_u \tag{7.10}$$

Example 7.7.2 (Kinesis) *Given a spatially varying diffusivity $D(x) > 0$ (with $x \in \mathbf{R}^d$) we can define a* kinesis *process $\{X_t : t \geq 0\}$ as the solution to the equation*

$$dX_t = \sqrt{2D(X_t)} \, dB_t$$

where $\{B_t : t \geq 0\}$ is d-dimensional standard Brownian motion. This process X_t is unbiased (i.e., each coordinate is a martingale) but it will spend more time in regions where the diffusivity is low. Define the time change $dU_t = 2D(X_t) \, dt$, then we have

$$dY_u = \frac{\sqrt{2D(X_{T_u})}}{\sqrt{2D(X_{T_u})}} \, dW_u = dW_u$$

i.e., $\{Y_u\}$ is Brownian motion. In other words, the process $\{X_t\}$ is a random time change of Brownian motion: We can think of a Brownian particle that carries a clock, which runs with a speed that depends on the position.

Exercise 7.12: Let $\{X_t\}$ be given by the stochastic logistic growth model

$$dX_t = X_t(1 - X_t) \, dt + \sigma X_t \, dB_t$$

and define transformed time by $dU_t = X_t \, dt$. Show that the time-transformed process $\{Y_u\}$ satisfies

$$dY_u = (1 - Y_u) \, du + \sigma \sqrt{Y_u} \, dW_u,$$

i.e., $\{Y_u\}$ is a Cox-Ingersoll-Ross process, at least as long as the process is bounded away from the singularity at $x = 0$. (It may occur that U_t remains bounded when $t \to \infty$)

Exercise 7.13: Consider a scalar diffusion process $\{X_t : t \geq 0\}$ given by $dX_t = f(X_t)\,dt + g(X_t)\,dB_t$ and let $s(x)$ be the associated scale function. Let $\{Y_u : u \geq 0\}$ be a time-changed process given by $dU_t = h(X_t)\,dt$. Show that $s(\cdot)$ is also the scale function of $\{Y_u\}$.

7.8 STRATONOVICH CALCULUS

Our emphasis is on the Itô integral and Itô's interpretation of stochastic differential equations, but we now briefly summarize the corresponding results for the Stratonovich integral and the Stratonovich interpretation.

Theorem 7.8.1 (The Chain Rule of Stratonovich Calculus) *Let $\{X_t : t \geq 0\}$ be an Itô process given in terms of a Stratonovich integral:*

$$dX_t = F_t\,dt + G_t \circ dB_t$$

and consider the image of this process under a smooth map h:

$$Y_t = h(t, X_t).$$

Then $\{Y_t\}$ is an Itô process given in terms of the Stratonovich integral:

$$dY_t = \frac{\partial h}{\partial t}\,dt + \frac{\partial h}{\partial x} \circ dX_t$$

$$= \frac{\partial h}{\partial t}\,dt + \frac{\partial h}{\partial x} F_t\,dt + \frac{\partial h}{\partial x} G_t \circ dB_t.$$

Notice that this is what we would expect if we naively applied the chain rule from deterministic calculus.

Proof: Aiming to write $\{Y_t\}$ in terms of a Stratonovich integral, we make a detour over the Itô calculus. First, recall that a Stratonovich integral can be written in terms of an equivalent Itô integral

$$\int_0^t G_s \circ dB_s = \int_0^T G_s\,dB_s + \frac{1}{2}\langle G, B \rangle_t$$

where $\langle G, B \rangle_t$ is the cross-variation between the processes $\{G_t\}$ and $\{B_t\}$. Therefore, we can rewrite $\{X_t\}$ in the standard form of an Itô process:

$$dX_t = F_t\,dt + G_t\,dB_t + \frac{1}{2}d\langle G, B \rangle_t.$$

Now Itô's lemma yields:

$$dY_t = \frac{\partial h}{\partial t}\,dt + \frac{\partial h}{\partial x}\,dX_t + \frac{1}{2}\frac{\partial^2 h}{\partial x^2}\,(dX_t)^2$$

We rewrite the middle integral as a Stratonovich integral:

$$dY_t = \frac{\partial h}{\partial t}\,dt + \frac{\partial h}{\partial x} \circ dX_t - \frac{1}{2}d\left\langle \frac{\partial h}{\partial x}, X \right\rangle_t + \frac{1}{2}\frac{\partial^2 h}{\partial x^2}\,(dX_t)^2. \tag{7.11}$$

Exercise 7.18 shows that the two last terms cancel, so we reach the final result:

$$dY_t = \frac{\partial h}{\partial t} \, dt + \frac{\partial h}{\partial x} \circ dX_t.$$

■

The conversion between Itô integrals and Stratonovich integrals is not very explicit, due to the term $\langle G, B \rangle_t$. But in the case of stochastic differential equations, we can get more explicit results:

Proposition 7.8.2 *Consider a process $\{X_t \in \mathbf{R}^n : t \geq 0\}$ which satisfies the Stratonovich stochastic differential equation*

$$dX_t = f_S(X_t) \, dt + g(X_t) \circ dB_t \ . \tag{7.12}$$

Then this process also satisfies an Itô equation

$$dX_t = f_I(X_t) \, dt + g(X_t) \, dB_t \tag{7.13}$$

with the same noise intensity g but different drift term f_I. In the case of scalar Brownian motion, the relationship between the Itô drift f_I and the Stratonovich drift f_S is:

$$f_I(x) = f_S(x) + \frac{1}{2} \frac{\partial g}{\partial x}(x) \, g(x) \tag{7.14}$$

Here, $\partial g / \partial x$ is the Jacobian of g. In the case of m-dimensional Brownian motion

$$dX_t = f_S(X_t) \, dt + \sum_{k=1}^{m} g_k(X_t) \circ dB_t^{(k)} = f_I(X_t) \, dt + \sum_{k=1}^{m} g_k(X_t) \, dB_t^{(k)}$$

we have the relationship between two drift terms:

$$f_I(x) = f_S(x) + \frac{1}{2} \sum_{k=1}^{m} \frac{\partial g_k}{\partial x}(x) \, g_k(x)$$

or, written out explicitly element-wise:

$$f_{I,i}(x) = f_{S,i}(x) + \frac{1}{2} \sum_{k=1}^{m} \sum_{j=1}^{n} \frac{\partial g_{ik}}{\partial x_j}(x) \, g_{jk}(x).$$

Proof: We first rewrite X_t in terms of an Itô integral:

$$X_t = X_0 + \int_0^t f_S(X_t) \, dt + \sum_{k=1}^{m} g_k(X_t) \circ dB_t^{(k)}$$

$$= X_0 + \int_0^t f_S(X_t) \, dt + \sum_{k=1}^{m} g_k(X_t) \, dB_t^{(k)} + \sum_{k=1}^{m} \frac{1}{2} \langle g_k(X), B^{(k)} \rangle_t$$

For the last term, we get

$$d\langle g_k(X), B^{(k)}\rangle_t = \frac{\partial g_k}{\partial x} d\langle X, B^{(k)}\rangle_t = \frac{\partial g_k}{\partial x} g_k \, dt.$$

Inserting this in the integral for X_t, we get

$$X_t = X_0 + \int_0^t f_S(X_t) \, dt + \sum_{k=1}^m g(X_t) \, dB_t + \frac{1}{2} \frac{\partial g^{(k)}}{\partial x}(X_s)g^{(k)}(X_s) \, ds$$

or

$$dX_t = \left[f_S(X_t) + \frac{1}{2} \sum_{k=1}^m \frac{\partial g^{(k)}}{\partial x}(X_s)g^{(k)}(X_s) \right] dt + g(X_t) \, dB_t$$

$$= f_I(X_t) \, dt + g(X_t) \, dB_t$$

as claimed.

■

Example 7.8.1 *Consider the process $\{X_t = \exp B_t : t \geq 0\}$. By the chain rule of Stratonovich calculus, $\{X_t\}$ satisifies the Stratonovich equation $dX_t = X_t \circ dB_t$. With proposition 7.8.2, we find $f_S(x) = 0$, $g(x) = x$ and thus $f_I(x) = \frac{1}{2}x$, so $\{X_t\}$ also satisfies the Itô equation $dX_t = \frac{1}{2}X_t \, dt + X_t \, dB_t$. This is, of course, in agreement with what we find by appplying Itô's lemma.*

With this formula, we can convert back and forth between Itô and Stratonovich equations, i.e., for a given Itô equation, we can find a Stratonovich equation which has the same solution, and conversely.

Since Itô equations and Stratonovich equations describe the same class of processes, we could base the mathematical framework on either the Itô integral or the Stratonovich integral and then transform the results to also apply to the other. In this exposition, we generally prefer the Itô intepretation, because the Itô integral is a martingale, and because the simplest numerical method for simulating sample paths is the explicit Euler-Maruyama method for Itô equations. Despite this choice, there are applications where you may prefer the Stratonovich calculus. The simpler chain rule is a good reason to choose Stratonovich calculus when many transformations will be done, e.g., when changing coordinates.

A different question is if a modeler should prefer Itô or Stratonovich equations to describe a given system. In many situations, we know the drift term from physical laws, so that the starting point for the model is an ordinary differential equation, $dX_t = f(X_t) \, dt$. If we then want to add state-dependent noise, should this be Itô noise, $g(X_t) \, dB_t$, or Stratonovich noise, $g(X_t) \circ dB_t$? A main argument for choosing Itô noise in this situation, could be that the instantaneous rate of change of the mean $\mathbf{E}X_t$ is then given by the original drift term f. On the other hand, the Stratonovich interpretation has the nice property that it appears as the limit, when the noise is band-limited with increasing band-width. This stems from the property that the Stratonovich integral arises as the limit when we approximate the integrator, Brownian motion, with its harmonic expansion (Exercise 6.3). Finally, in some situations we would like the noise to mimic Fickian diffusion. This neither leads directly to Itô or Stratonovich models; see Chapter 9. Of course, if time series data is available for the system, then one can infer if the Itô interpretation or the Stratonovich interpretation – or something third – matches the statistics of the observations best.

Exercise 7.14: Consider again Brownian motion on the circle in Cartesian coordinates, i.e., $X_t = \cos B_t$, $Y_t = \sin B_t$ (compare Section 7.6). Derive Stratonovich equations that govern (X_t, Y_t).

Exercise 7.15 Lamperti Transforming a Stratonovich Equation: Consider the scalar Stratonovich equation

$$dX_t = f(X_t) \, dt + g(X_t) \circ dB_t$$

where $g(x) > 0$ for all x. Verify that to transform to coordinates where the noise is additive, we must use the same Lamperti transform as in the Itô case, i.e., (7.7). Next, write the Itô equation which governs $\{X_t\}$, and Lamperti transform this equation. Do we arrive at the same resulting equation?

7.9 CONCLUSION

In this chapter, we have established the rules of calculus that apply to Itô processes. Most importantly, Itô's lemma states that when we map an Itô process $\{X_t\}$ to obtain $Y_t = h(t, X_t)$, the resulting image $\{Y_t\}$ is again an Itô process and the lemma allows us to compute the drift and intensity of $\{Y_t\}$. Relative to standard calculus, Itô's lemma contains a new term in the drift for $\{Y_t\}$ which combines the curvature of h and the quadratic variation of $\{X_t\}$. It is useful to memorize Itô's lemma as a truncated Taylor series in increments where we keep second order terms $(dX_t)^2$, since dX_t in general scales with $\sqrt{dt}$. We can then add and multiply increments according to the rules of Section 6.9, in particular using the formulas $(dt)^2 = dt \, dB_t = 0$, $(dB_t)^2 = dt$.

   Itô's lemma is the main workhorse when we analyze stochastic differential equations. We have used it to verify analytical solutions of some stochastic differential equations. Although analytical solutions are the expectation rather than the rule, those models for which analytical solutions are known play a predominant role in applications. We have also used Itô's lemma to perform coordinate transformations. From a modeler's point of view, transforming to logarithmic or polar coordinates is routine. In addition, we have shown two coordinate transformations for scalar processes which have special interest: The Lamperti transform, which makes the noise additive in the transformed system, and the scale function, which makes the transformed process a martingale, i.e., driftless. Some analysis questions are more easily answered in these transformed coordinate systems.

We have also shown how to transform time. It is a routine task to change the units of time, if one is explicit about the units and remembers that in standard Brownian motion, variance equals time. The same scaling of the two integrands appear also when time changes randomly, even if this occurs less frequently in applications.

Finally, we have derived the chain rule of Stratonovich calculus, i.e. when Stratonovich integrals are involved. This is a simpler chain rule than Itô's lemma, and closer to the chain rule for standard calculus. This can lead one to prefer the Stratonovich framework in some situations, but the martingale property of the Itô integral is a strong argument in favour of the Itô framework.

7.10 NOTES AND REFERENCES

Itô's lemma appeared in (Itô, 1951a), although the special case of Brownian motion was presented in (Itô, 1944), and in Japanese in 1942. When Stratonovich (1966) introduced his interpretation of the integral, he also developed the calculus, and his discussion of the pros and cons of the two interpretations is still valid. Gard (1988) present more techniques for solving stochastic differential equations analytically than what we have shown here.

7.11 EXERCISES

Exercise 7.16: Let $\{X_t\}$ be geometric Brownian motion, $dX_t = rX_t\,dt + \sigma X_t\,dB_t$, with $X_0 = x > 0$. Let $Y_t = h(X_t)$ where $h(x) = x^p$, $p \in \mathbf{R}$. Show that $\{Y_t\}$ is also geometric Brownian motion and determine its parameters.

Exercise 7.17: Verify that $Y_t = \sinh B_t$ satisfies the Itô SDE

$$dY_t = \frac{1}{2}Y_t\,dt + \sqrt{1 + Y_t^2}\,dB_t.$$

Next, verify through simulation: Simulate a path of Brownian motion and solve for $\{Y_t\}$ using the Euler-Maruyama scheme. Plot the analytical solution, the Euler-Maruyama solution, and the two numerical integrals

$$\int_0^t \frac{1}{2}Y_t\,dt + \int_0^t \sqrt{1 + Y_t^2}\,dB_t, \qquad \int_0^t \cosh B_t\,dB_t + \int_0^t \frac{1}{2}\sinh B_t\,d[B]_t,$$

in the same graph and compare.

Exercise 7.18: Let $\{X_t\}$ and $\{Y_t\}$ be scalar Itô processes and set $Z_t = h(X_t, t)$ where $h : \mathbf{R} \times \mathbf{R} \mapsto \mathbf{R}$ is smooth. Show that

$$d\langle Z, Y \rangle_t = h'(X_t, t)\,dX_t\,dY_t$$

where $h' = \partial h/\partial x$.

Exercise 7.19: Using the product rule in Example 7.3.3, show that

$$tB_t = \int_0^t s \, dB_s + \int_0^t B_s \, ds.$$

Next, find the mean and variance of each of these two integrals, and their covariance. Verify the result using Monte Carlo simulation.

Exercise 7.20: Show that

$$\mathbf{E}\left\{\int_0^t B_s \, ds \middle| B_t\right\} = \mathbf{E}\left\{\int_0^t s \, dB_s \middle| B_t\right\} = \frac{1}{2}tB_t$$

Hint: You may use the properties of the Brownian bridge, and the product rule $d(tB_t) = B_t \, dt + t \, dB_t$.

Exercise 7.21: Let $\{X_t\}$ be geometric Brownian motion, i.e.,

$$dX_t = rX_t \, dt + \sigma X_t \, dB_t,$$

and introduce scaled time U_t with $dU_t = X_t \, dt$. We ignore that the speed ratio is not bounded away from 0. Let the process $\{Y_u\}$ be X_t in scaled time, i.e., $Y_u = X_{\tau_u}$ with $\tau_u = \inf\{t \geq 0 : U_t \geq u\}$. Show that $\{Y_u\}$ satisfies

$$dY_u = r \, du + \sigma\sqrt{Y_u} \, dW_u.$$

i.e., $\{Y_u\}$ is a squared Bessel process. *Note:* If X_t approaches the origin, the rescaled time $\{U_t\}$ slows down. In fact, when the drift is weak, $r < \frac{1}{2}\sigma^2$, X_t converges to 0 as $t \to \infty$ w.p. 1 (compare Exercise 7.1). Then U_t remains bounded, and the process $\{Y_u\}$ is only defined up to it hits the origin. Compare Exercise 7.12.

Exercise 7.22: To see what happens when mapping diffusions through functions that diverge quickly, let $\{B_t\}$ be standard Brownian motion.

1. Define $X_t = \exp B_t^2$. Show that $\mathbf{E}X_t$ is well defined if and only if $t < 1/2$.

2. Define $X_t = \exp\exp B_t$. Show that $\mathbf{E}|X_t| = \infty$ for all $t > 0$.

The point of these examples is that $\{X_t\}$ are Itô processes in both situations, as guaranteed by Itô's lemma, but not $\mathcal{L}_2$ Itô processes.

Exercise 7.23 A Multivariate Wide-Sense Linear Equation: Show that $X_t = \exp(At + GB_t)x$ satisfies the wide-sense linear equation

$$dX_t = (A + \frac{1}{2}G^2)X_t \, dt + GX_t \, dB_t, \quad X_0 = x,$$

if matrices A and G commute. Here $\{B_t\}$ is scalar. *Note:* An example of such a system is Brownian motion on the circle; Section 7.6.1. In general, multivariate wide-sense linear equations do not have known closed-form solutions.

Exercise 7.24 Brownian Motion on the Sphere: Consider the Stratonovich equation in n dimensions:

$$dX_t = (I - \frac{1}{|X_t|^2} X_t X_t^\top) \circ dB_t, \quad X_0 = x \neq 0,$$

where $X_t \in \mathbf{R}^n$ and $\{B_t\}$ is n-dimensional Brownian motion.

1. Show that $|X_t|^2 = |x|^2$, i.e. any sphere $\{x : |x| = r\}$ is invariant for $r > 0$.

2. Show that if $\{X_t\}$ satisfies this equation and U is a rotation matrix $(U^\top U = U U^\top = I)$, then $\{U X_t\}$ satisfies the same equation but with a different Brownian motion.

3. It can be shown that $\{X_t\}$ satisfies the Itô equation

$$dX_t = \frac{1-n}{2|X_t|^2} X_t \, dt + (I - \frac{1}{|X_t|^2} X_t X_t^\top) \, dB_t.$$

Verify that the drift term is correct in the sense that $|X_t|^2$ is constant along trajectories, if $\{X_t\}$ satisfies this Itô equation. Then find a differential equation which $\mathbf{E} X_t$ satisfies and state the solution, i.e., write $\mathbf{E} X_t$ as a function of $\mathbf{E} X_0$. Verify the result with a stochastic simulation.

Exercise 7.25 An Exponential Martingale: Let $\{B_t\}$ be standard Brownian motion as usual and let $\{G_t\}$ be $\mathcal{L}_2$ Itô integrable. Define $\{X_t\}$ by $X_0 = 0$ and $dX_t = G_t \, dB_t - \frac{1}{2} G_t^2 \, dt$, and set $Y_t = \exp X_t$. Show that

$$dY_t = Y_t G_t \, dB_t.$$

Verify the result with numerical simulation of a sample path, using $G_t = \sin B_t + \cos t$. *Note:* Since $\{Y_t\}$ is an Itô integral, it is a local martingale. If $\{G_t\}$ satisfies *Novikov's condition* that $\mathbf{E} \exp(\frac{1}{2} \int_0^t G_s^2 \, ds) < \infty$ for all $t \geq 0$, then $\{Y_t\}$ is a martingale.

CHAPTER 8

Existence, Uniqueness, and Numerics

At this point we can, at least in principle, verify if a given process $\{X_t : t \geq 0\}$ satisfies a given stochastic differential equation $dX_t = f(t, X_t)\, dt + g(t, X_t)\, dB_t$. In most applications, the problem is the inverse: We are given the functions f, g in the model and the initial condition $X_0 = x$, and we aim to *solve* the stochastic differential equation, i.e., find $\{X_t\}$. Before we attempt to solve the equation, whether analytically or numerically, we would like to be assured that there does indeed exist a solution and that it is unique. In this chapter we state existence and uniqueness theorems to this end.

Uniqueness can fail at singularities of f and g. Global existence of solutions can fail through *explosions*, where the state X_t diverges to ∞ in finite time. The existence and uniqueness theorems presented in this chapter work by ruling out these phenomena.

Once we know that a unique solution exists, we often wish to find this solution or at least characterize it. We typically cannot write up the solution explicitly, so numerical simulation of sample paths is important. Even when we *can* write up the solution explicitly, we may learn much about the system by inspecting simulated sample paths. The basic algorithm for numerical simulation of sample paths is the Euler-Maruyama method.

In this chapter, we analyze the performance of the Euler-Maruyama scheme (which turns out to be not very impressive): This concerns how fast individual simulated sample paths converge to the true sample paths, as well as how fast the statistics of sample paths converge to the true ones. These modes of convergence lead to the *strong* and *weak* order. We next discuss some improvements: The Mil'shtein scheme and Heun's method. We finally discuss numerical stability, implicit methods, and design of simulation experiments.

DOI: 10.1201/9781003277569-8

8.1 THE INITIAL VALUE PROBLEM

In this chapter, we consider the initial value problem consisting of the Itô stochastic differential equation

$$dX_t = f(X_t, t)\, dt + g(X_t, t)\, dB_t \tag{8.1}$$

and the initial condition

$$X_0 = x. \tag{8.2}$$

Here, not just the model (f, g) and the initial condition x is given, but also the underlying filtered probability space $(\Omega, \mathcal{F}, \{\mathcal{F}_t\}, \mathbf{P})$, along with $\{B_t : t \geq 0\}$ which is Brownian motion w.r.t. $\{\mathcal{F}_t\}$ and $\mathbf{P}$.

We then ask: Does there exist an Itô process $\{X_t\}$ which satisfies this stochastic differential equation and the initial condition? If so, is it unique? Such an $\{X_t\}$ is called a *strong* solution, and the notion of uniqueness is called *strong* or *pathwise*. There is also a weak notion of solutions and uniqueness, but we consider the strong notion only.

8.2 UNIQUENESS OF SOLUTIONS

We prefer initial value problems with unique solutions, but are not always so lucky. A counterexample from ordinary differential equations is:

8.2.1 Non-Uniqueness: The Falling Ball

A ball is held at rest in a gravity field in space at time $t = 0$. We use X_t to denote its vertical position at time t; the axis is directed downwards. The potential energy of the ball at time t is $-mgX_t$ where m is the mass of the ball and g is the gravitational acceleration. The kinetic energy is $\frac{1}{2}mV_t^2$ where $V_t = dX_t/dt$ is the vertical velocity downwards. Energy conservation dictates

$$\frac{1}{2}mV_t^2 - mgX_t = 0.$$

We can isolate V_t to obtain

$$V_t = \frac{dX_t}{dt} = \sqrt{2gX_t} \tag{8.3}$$

where we have used physical reasoning to discard the negative solution. This is a first order ordinary differential equation in $\{X_t\}$. We may write it in the standard form:

$$\dot{X}_t = f(X_t, t) \text{ where } f(x, t) = \sqrt{2gx}.$$

With the initial condition $X_0 = 0$, one solution is

$$X_t = \frac{1}{2}gt^2.$$

But this solution is not unique: Another solution is $X_t = 0$. In fact, for any non-negative t_0, we can construct a solution

$$X_t = \begin{cases} 0 & \text{for } t \leq t_0, \\ \frac{1}{2}g(t - t_0)^2 & \text{for } t > t_0. \end{cases}$$

The physical interpretation of these solutions is that we hold the ball until time t_0 and then let go. Note that for each parameter t_0, this expression defines a continuously differentiable function of time t, so it is a valid solution in every mathematical sense. Uniqueness fails because when the particle is at standstill, we may apply a force without doing work by holding the particle. So energy conservation does not determine the force and hence the motion.

Mathematically, the problem is that the right hand side of the ODE, $f(x,t) = \sqrt{2gx}$, has a singularity at $x = 0$: the derivative is

$$\frac{\partial f}{\partial x}(x,t) = \sqrt{g/2x}$$

which approaches ∞ as $x \downarrow 0$.

As the following exercise shows, the same phenomenon can occur in a stochastic setting.

Exercise 8.1: Show that, given any deterministic $T \geq 0$, the process

$$X_t = \begin{cases} 0 & \text{for } 0 \leq t \leq T \\ (B_t - B_T)^3 & \text{for } t > T \end{cases}$$

satisfies the Stratonovich stochastic differential equation

$$dX_t = 3|X_t|^{2/3} \circ dB_t, \quad X_0 = 0.$$

We use Stratonovich calculus to get a closer analogy to the deterministic case. There is no straightforward physical interpretation of this example, but mathematically, it contains a similar singularity at the origin: The function g given by $g(x) = 3|x|^{2/3}$ is non-differentiable at $x = 0$. The exercise considers the non-uniqueness that appears when starting at the singularity, but notice that we could also have started away from the singularity (say, $X_0 = 1$), in which non-uniqueness would arise when we hit the singularity.

8.2.2 Local Lipschitz Continuity Implies Uniqueness

Non-uniqueness of solutions should, in general, be avoided in the modeling process. In order to rule out non-uniqueness that appears at singularities, it is standard to require that the model is Lipschitz continuous (see Figure 8.1).

Definition 8.2.1 (Lipschitz continuity) *Let f be a function from one normed space $\mathbf{X}$ to another $\mathbf{Y}$. We say that f is globally Lipschitz continuous if there exists a constant $K > 0$, such that*

$$|f(x_1) - f(x_2)| \leq K|x_1 - x_2|$$

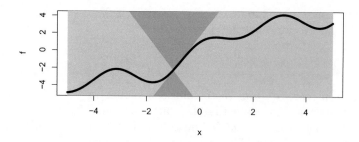

Figure 8.1 Lipschitz continuity. The function $x \mapsto x + \sin x + \cos 2x$ is Lipschitz continuous with Lipschitz constant 4. For each point x, this restricts the graph to a cone, indicated by light gray for the point $x = -1$; the dark gray set is a "forbidden zone".

holds for any $x_1, x_2 \in \mathbf{X}$. We say that f is locally Lipschitz *continuous if for each $x \in \mathbf{X}$ there exists a neighborhood A of x, such that the restriction of f to A is Lipschitz continuous.*

For brevity, we may skip the terms "globally" and "continuous" and just say that a function is Lipschitz.

Our interest is in the finite-dimensional spaces $\mathbf{R}^{n \times m}$ for natural n and m. In this case, it does not matter which norm we choose, since they are all equivalent (i.e., they can be bounded in terms of eachother). Also note that, a C^1 function is necessarily locally Lipschitz, but globally Lipschitz if and only if the derivative is bounded. However, a Lipschitz function does not have to be differentiable; for example the function $x \mapsto |x|$ is Lipschitz but not differentiable at 0.

For time-invariant systems, local Lipschitz continuity is enough to guarantee uniqueness of a solution. For time-varying systems, the requirement is a little bit more elaborate:

Theorem 8.2.1 *Let $T > 0$ and assume that for any $R > 0$ there exist $K_f > 0$, $K_g > 0$, such that*

$$|f(x,t) - f(y,t)| \le K_f \cdot |x - y|, \quad |g(x,t) - g(y,t)| \le K_g \cdot |x - y|$$

whenever $0 \le t \le T$ and $|x|, |y| < R$. Then there can exist at most one Itô process $\{X_t : 0 \le t \le T\}$ which satisfies the initial value problems (8.1), (8.2). That is, if $\{X_t : t \ge 0\}$ and $\{Y_t : t \ge 0\}$ are two Itô processes which both satisfy the initial value problem, then $X_t = Y_t$ for all t, almost surely.

Factbox: [The Grönwall-Bellman inequality] This lemma states that "linear bounds on dynamics imply exponential bounds on solutions". T.H. Grönwall considered a differentiable function $v : [0, \infty) \mapsto \mathbf{R}$ such that

$$v'(t) \leq a(t)v(t)$$

for all $t \geq 0$, where $a : [0, \infty) \mapsto \mathbf{R}$ is continuous, and concluded that

$$v(t) \leq v(0) \exp \left(\int_0^t a(s) \ ds \right)$$

holds for all $t \geq 0$. To see this, write $v'(t) = a(t)v(t) + f(t)$ where $f(t) \leq 0$. Then $v(t) = v(0) \exp \left(\int_0^t a(s) \ ds \right) + \int_0^t \exp(\int_s^t a(u) \ du) f(s) \ ds$. Now note that the second term is non-positive.

R. Bellman considered an integral form, so that v does not need to be differentiable. One version assumes that $a, b, v : [0, \infty) \mapsto \mathbf{R}$ are continuous functions, with a non-negative and b non-decreasing. If v satisfies

$$v(t) \leq b(t) + \int_0^t a(s)v(s) \ ds \text{ for all } t \geq 0,$$

then v also satisfies

$$v(t) \leq b(t) \exp \left(\int_0^t a(s) \ ds \right) \quad \text{for all } t \geq 0.$$

Proof: We first assume that f and g are bounded and globally Lipschitz with constants K_f and K_g, using the Euclidean norm for the vectors x and f and the Frobenius norm $|g|^2 = \sum_{ij} |g_{ij}|^2$ for the matrix g. Let $\{X_t\}$ and $\{Y_t\}$ be Itô processes which both satisfy the stochastic differential equation, but not necessarily the initial condition. Define the processes

$$
\begin{aligned}
D_t &= X_t - Y_t \\
F_t &= f(X_t, t) - f(Y_t, t) \\
G_t &= g(X_t, t) - g(Y_t, t) \\
S_t &= |X_t - Y_t|^2 = |D_t|^2.
\end{aligned}
$$

Then, by the Itô formula, $\{D_t\}$ and $\{S_t\}$ are Itô processes. We get:

$$dD_t = F_t \ dt + G_t \ dB_t$$

and

$$dS_t = 2D_t^\top \, dD_t + |dD_t|^2$$
$$= 2D_t^\top F_t \, dt + 2D_t^\top G_t \, dB_t + \text{tr}[G_t G_t^\top] \, dt.$$

If $\mathbf{E}S_s$ is bounded for $s \in [0, t]$, then the Itô integral is a martingale, so

$$\mathbf{E}S_t = S_0 + \mathbf{E} \int_0^t \left(2 \, D_s^\top F_s + \text{tr}[G_s G_s^\top] \right) \, ds.$$

Now, the Cauchy-Schwarz inequality $|\langle a, b \rangle| \le \|a\| \, \|b\|$ implies that $|D_s^\top F_s| \le |D_s||F_s|$, and with the Lipschitz continuity of f we get $|D_s^\top F_s| \le K_f |D_s|^2$. Similarly, $\text{tr}[G_s G_s^\top] = |G_s|^2 \le K_g^2 |D_s|^2$. We therefore get

$$\mathbf{E}S_t \le S_0 + \int_0^t (2 \, K_f + K_g^2) \mathbf{E}S_s \, ds$$

and in turn the Grönwall-Bellman inequality implies that

$$\mathbf{E}S_t \le S_0 e^{[2K_f + K_g^2]t}.$$

I.e., trajectories diverge at most exponentially from each other with a rate that depends on the Lipschitz constants K_f, K_g. Now, if the two initial conditions coincide, $X_0 = Y_0$, then $S_0 = 0$, and so $\mathbf{E}S_t = 0$ for all t. That is, the solutions X_t and Y_t agree w.p. 1 for each t. By continuity it follows that the two solutions agree for all t, w.p. 1.

Now, if f and g are only locally Lipschitz continuous: For a given R, modify f and g to get $\tilde{f}$ and $\tilde{g}$ given by

$$\tilde{f}(x) = \begin{cases} f(x) & \text{for } |x| \le R \\ f(xR/|x|) & \text{for } |x| > R \end{cases}, \quad \tilde{g}(x) = \begin{cases} g(x) & \text{for } |x| \le R \\ g(xR/|x|) & \text{for } |x| > R \end{cases}.$$

Then $\tilde{f}$ and $\tilde{g}$ are bounded and globally Lipschitz continuous. Now X_t and Y_t satisfy the SDE $dX_t = \tilde{f}(X_t) \, dt + \tilde{g}(X_t) \, dB_t$ and S_t remains bounded up to the stopping time $\tau = \inf\{t \ge 0 : |X_t| > R \vee |Y_t| > R\}$, so it must hold that $X_t = Y_t$ for $t \le \tau$. Since R is arbitrary, we conclude that $X_t = Y_t$ as long as $|X_t| < \infty$, $|Y_t| < \infty$. But since $\{X_t\}$ and $\{Y_t\}$ were assumed to be Itô processes, this holds for all t. ■

8.3 EXISTENCE OF SOLUTIONS

We now ask if the initial value problem (8.1), (8.2) is guaranteed to have a solution. The most interesting reason this may fail, is that a solution exists only up to some (random) point of time, where it *explodes*. It is instructive to begin with a deterministic example:

Example 8.3.1 (Explosion in an Ordinary Differential Equation) *For*

$$\dot{X}_t = 1 + X_t^2, \quad X_0 = 0,$$

there exists a (unique maximal) solution $X_t = \tan t$ which is defined for $0 \leq t < \pi/2$. An explosion occurs at $t = \pi/2$ where $X_t \to \infty$.

Explosions can also occur in continuous-time Markov chains taking discrete values: Consider the "birth" process $\{X_t \in \mathbf{N} : t \geq 0\}$ where the only state transitions are from one state x to $x+1$, which happens with rate x^2. This can model population growth or a nuclear chain reaction. Starting with $X_0 = 1$, let τ_n be the time of arrival to state $n \in \mathbf{N}$, i.e., $\tau_n = \inf\{t \geq 0 : X_t = n\}$. The expected "sojourn time" (i.e., time spent) in state x is $1/x^2$, so τ_n has expectation $\mathbf{E}\tau_n = \sum_{x=1}^{n-1} x^{-2}$ which remains bounded as n increases: $\mathbf{E}\tau_n \to 1.077$ as $n \to \infty$, approximately. Now, define $\tau = \lim_{n\to\infty} \tau_n$. We say that an explosion occurs in finite time if $\tau < \infty$, and see that $\mathbf{E}\tau \approx 1.077$. So an explosion occurs at a finite time, almost surely.

Combining these two examples of explosions, it is not surprising that solutions to stochastic differential equations may explode:

Example 8.3.2 (Explosion in a Stratonovich SDE) *The process $\{X_t = \tan B_t\}$ satisfies the Stratonovich SDE*

$$dX_t = (1 + X_t^2) \circ dB_t \tag{8.4}$$

until the stopping time τ given by

$$\tau = \inf\{t : |B_t| \geq \frac{\pi}{2}\}.$$

The time τ is the (random) time of explosion. See Figure 8.2. Note: *To verify that $\{X_t\}$ satisfies this SDE, we use the chain rule of Stratonovich calculus. However, the chain rule assumes the process to be well defined for all $t \geq 0$. The way around this obstacle is again to localize: Introduce the stopping time $T = \inf\{t \geq 0 : |X_t| \geq R\}$; then the stopped process $\{X_{t \wedge T} : t \geq 0\}$ is well-behaved, and $X_T = \tan B_T = R$. Then let $R \to \infty$.*

8.3.1 Linear Bounds Rule Out Explosions

In some applications, explosions are an important part of the dynamics, and we may wish to know when and how the system explodes. However, at this point we prefer our stochastic differential equations to have solutions defined on the entire time axis, so we aim to rule out explosions. The following theorem is broad enough to apply in many applications:

Theorem 8.3.1 *Let the Itô process $\{X_t : 0 \leq t \leq T\}$ satisfy the initial value problem (8.1), (8.2) for $t \in [0, T]$ where $T > 0$. If (f, g) satisfy the bound*

$$x^\top f(x, t) \leq C \cdot (1 + |x|^2), \quad |g(x, t)|^2 \leq C \cdot (1 + |x|^2)$$

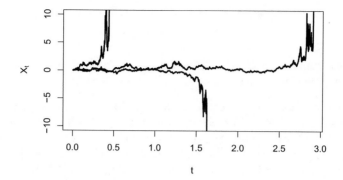

Figure 8.2 Explosion at a random time. Three sample paths of the process (8.4), up to the random time τ of explosion.

for $C > 0$, all $x \in \mathbf{R}^n$, and all $t \in [0, T]$, then

$$\mathbf{E}|X_t|^2 \le (x_0^2 + 3Ct)e^{3Ct}.$$

In particular, $\mathbf{E}|X_t|^2$ is finite and bounded on $[0, T]$.

Proof: Define $S_t = |X_t|^2$. By Itô's lemma, we have

$$
\begin{aligned}
dS_t &= 2X_t^\top \, dX_t + |dX_t|^2 \\
&= 2X_t^\top f(X_t) \, dt + \mathrm{tr}[g^\top(X_t, t)g(X_t, t)] \, dt + 2X_t^\top g(X_t) \, dB_t.
\end{aligned}
$$

We localize to ensure that the Itô integral is a martingale (skipping details) and get

$$\mathbf{E}S_t = S_0 + \mathbf{E}\int_0^t 2X_s^\top f(X_s) + \mathrm{tr}[g^\top(X_s, s)g(X_s, s)] \, ds$$

and, with the bounds on the functions f and g,

$$\mathbf{E}S_t \le S_0 + \int_0^t 3C(1 + \mathbf{E}S_s) \, ds.$$

Now the claim follows from the Grönwall-Bellman inequality with $v(t) = \mathbf{E}S_t$, $b(t) = S_0 + 3Ct$, and $a(s) = 3C$. ■

The condition $x^\top f(x) \le C(1 + |x|^2)$ deserves an explanation. It holds if f is globally Lipschitz continuous, but that would be too restrictive to require, since many models of interest are not. For example, the ordinary differential equation $\dot{X}_t = -X_t^3$ is not globally Lipschitz, but the equation still admits a solution which is defined for all $t \ge 0$, for every initial condition $X_0 = x$. Indeed, this model satisfies the conditions of Theorem 8.3.1, since $g(x) = 0$ and $xf(x) \le 0$ hold for all x. The condition $x^\top f(x) \le C(1 + |x|^2)$ allows the function f to contain superlinear growth terms, as long as these are directed towards the origin.

Theorem 8.3.1 is not an existence theorem, because it assumes a solution and then states a bound on that solution. However, it turns out that it is exactly this bound that allows us to conclude that a solution exists:

Theorem 8.3.2 *Consider the initial value problems (8.1), (8.2) and let $T > 0$. Assume that the functions (f, g) satisfies the local Lipschitz condition for uniqueness in Theorem 8.2.1, and the linear growth bounds in Theorem 8.3.1 which rule out explosions. Then there exists a unique Itô process $\{X_t : 0 \leq t \leq T\}$ which satisfies the initial value problem.*

We omit the proof of this theorem; see, e.g. (Mao, 2008). Briefly, the outline of the proof is as follows: First, we assume that the functions f, g are globally Lipschitz. Then we use the method of successive approximations, also known as Picard iteration, to construct the solution. This iteration starts with $X_t^{(1)} = x$ and employs the recursion

$$X_t^{(n+1)} = x + \int_0^t f(X_s^{(n)}, s) \ ds + \int_0^t g(X_s^{(n)}, s) \ dB_s.$$

First, we use the bounds on f and g to show that each iterate remains bounded in $\mathcal{L}_2$. Then, we show that the sequence is a Cauchy sequence, hence convergent in $\mathcal{L}_2$. Finally, we show that the limit satisfies the stochastic differential equations. If the model is not globally Lipschitz, then we approximate the functions f, g with globally Lipschitz functions as in the proof of Theorem 8.2.1. This implies that a solution exists until the stopping time where it escapes any bounded sphere; i.e., up to explosion. But Theorem 8.3.1 rules out such an explosion. We conclude that the solution is defined at all times.

Exercise 8.2 Picard Iteration for the Wide-Sense Linear Stratonovich Equation: Consider the Stratonovich SDE

$$dX_t = rX_t \ dt + \sigma X_t \circ dB_t \tag{8.5}$$

and the initial condition $X_0 = 1$. Conduct the Picard iteration and show that at each step in the iteration, the solution $X_t^{(n)}$ is the truncated Taylor series of $\exp(rt + \sigma B_t)$.

8.4 NUMERICAL SIMULATION OF SAMPLE PATHS

We have already made use of the simplest algorithm for numerical simulation of sample paths, viz. the Euler-Maruyama scheme. In this section, we analyze the performance of this scheme and show two improvements, the Mil'shtein scheme and the Heun scheme.

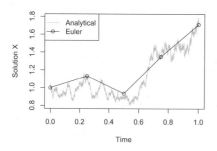 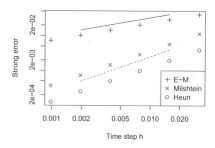

Figure 8.3 Geometric Brownian motion and its discretization. The parameters are $r = 0.5$, $\sigma = 0.5$, $x = 1$, $T = 1$. *Left panel:* A single sample path of the analytical solution evaluated on a very fine time grid (grey), and the Euler-Maruyama approximation with a coarse time step, $h = 0.25$. *Right panel:* The mean absolute error of the Euler-Maruyama scheme, the Mil'shtein scheme, and the Heun scheme, plotted against the time step h. The errors are based on 100,000 realizations. Included are also straight lines corresponding to strong orders of 0.5 and 1.0.

8.4.1   The Strong Order of the Euler-Maruyama Method for Geometric Brownian Motion

We assess the performance of the Euler-Maruyama method through simulation. Consider geometric Brownian motion given by the Itô equation:

$$dX_t = rX_t\, dt + \sigma X_t\, dB_t, \quad X_0 = x$$

for which the unique solution is $X_t = x \exp((r - \frac{1}{2}\sigma^2)t + \sigma B_t)$. We approximate it with the Euler-Maruyama method

$$X_{t+h}^{(h)} = X_t^{(h)} + rX_t^{(h)}h + \sigma X_t^{(h)}\left(B_{t+h} - B_t\right), \quad X_0^{(h)} = x.$$

We use the superscript $X_t^{(h)}$ to emphasize that the approximation of X_t is based on the time step h. We fix the terminal time T and measure the error $X_T - X_T^{(h)}$, which is a random variable. Figure 8.3 (left panel) shows one realization of geometric Brownian motion and its Euler-Maruyama discretization.

To assess the error, we first simulate a large number ($N = 10^5$) of realizations of Brownian motion on a fine temporal grid ($h = 2^{-10}$). For each realization, we compute the analytical solution, the Euler-Maruyama approximation, and the absolute error $|X_T - X_T^{(h)}|$. Next, we sub-sample the Brownian motion on a coarser grid, omitting every other time step, so increasing the time step with a factor 2. We repeat the computations of the Euler-Maruyama discretized solution and the error, and subsampling to ever coarser grids. If the analytical solution had not been available, we would have compared the coarser simulations with the finest one.

Figure 8.3 (right panel) shows the resulting mean absolute error plotted against the time step. It is a double logarithmic plot, so that power relationships display as straight lines. The line with a slope of 0.5 corresponds to a square root scaling, which shows good agreement with experimental results:

$$\mathbf{E}|X_T - X_T^{(h)}| \sim \sqrt{h}.$$

In fact, as we will see in the following, this is the theoretical prediction. We say that the Euler-Maruyama scheme, in general, has *strong order 0.5*.

8.4.2 Errors in the Euler-Maruyama Scheme

Let us investigate how the errors in the Euler-Maruyama scheme arise and accumulate. We consider a scalar SDE

$$X_t = X_0 + \int_0^t f(X_s)\ ds + \int_0^t g(X_s)\ dB_s, \quad X_0 = x,$$

governing the Itô diffusion $\{X_t\}$, and its Euler-Maruyama discretization $\{X_t^{(h)}\}$ with time step h, where the first time step is given by

$$X_h^{(h)} = x + f(x)\ h + g(x)\ B_h.$$

The local error, introduced during the first time step $[0, h]$, is:

$$X_h - X_h^{(h)} = \int_0^h f(X_s) - f(x)\ ds + \int_0^h g(X_s) - g(x)\ dB_s.$$

Define $\{F_s = f(X_s) - f(x)\}$ and $\{G_s = g(X_s) - g(x)\}$ to be the two integrands in this expression, then $F_0 = G_0 = 0$, and Itô's lemma gives us:

$$dF_s = f'(X_s)\ dX_s + \frac{1}{2}f''(X_s)\ (dX_s)^2$$
$$= f'(X_s)f(X_s)\ ds + f'(X_s)g(X_s)\ dB_s + \frac{1}{2}f''(X_s)g^2(X_s)\ ds$$

and

$$dG_s = g'(X_s)\ dX_s + \frac{1}{2}g''(X_s)\ (dX_s)^2$$
$$= g'(X_s)f(X_s)\ ds + g'(X_s)g(X_s)\ dB_s + \frac{1}{2}g''(X_s)g^2(X_s)\ ds.$$

We can now make the following approximations for these processes $\{F_s\}$ and $\{G_s\}$:

$$F_s \approx f'(x)f(x)s + f'(x)g(x)B_s + \frac{1}{2}f''(x)g^2(x)s,$$
$$G_s \approx g'(x)f(x)s + g'(x)g(x)B_s + \frac{1}{2}g''(x)g^2(x)s.$$

Here, we have omitted the error terms and used the symbol $\approx$ to indicate this. Inserting in the integrals for the error, we find

$$X_h - X_h^{(h)} \approx [f'f + \frac{1}{2}f''g^2] \int_0^h s \, ds$$

$$+ f'g \int_0^h B_s \, ds$$

$$+ [g'f + \frac{1}{2}g''g^2] \int_0^h s \, dB_s$$

$$+ g'g \int_0^h B_s \, dB_s$$

where we have omitted the argument x of $f(x)$, $f'(x)$, etc. We can now assess the size of each term. The first integral is deterministic, $\int_0^h s \, ds = \frac{1}{2}h^2$. The second, $\int_0^h B_s \, ds$ is a Gaussian random variable with mean 0 and variance $h^3/3$, and the same applies to the third, $\int_0^h s \, dB_s$ (Exercise 7.19). Finally, for the last integral, we have $\int_0^h B_s \, dB_s = (B_h^2 - h)/2$ (compare equation (6.12)) which has expectation 0 and variance $h^2/2$ (Exercise 3.13). Of these three stochastic terms, when the time step h is small, the last term will dominate since it has lower order, assuming that the coefficient $g'g$ does not vanish.

Our next task is to assess how local errors, introduced during the first time step and subsequent ones, propagate to the end of the simulation, $t = T$. A thorough treatment can be found in (Kloeden and Platen, 1999; Milstein and Tretyakov, 2004); here, we provide an outline. We focus on the dominating term in the local error, $g'g \, (B_h^2 - h)/2$, which has mean 0 and variance $O(h^2)$. This term is amplified or attenuated over the time interval $[h, T]$ by system dynamics; we established an upper bound for the amplification when proving uniqueness (Theorem 8.2.1). Later local errors, introduced in time intervals $[h, 2h]$, $[2h, 3h]$, etc, are uncorrelated. When the entire simulation interval $[0, T]$ is divided into $n = T/h$ subintervals, the total error will therefore have a variance which scales as $O(nh^2) = O(h)$. The error itself is therefore $O(h^{1/2})$, in agreement with the simulation in Section 8.4.1: The Euler-Maruyama scheme has strong order $1/2$.

Note that the error term we have considered includes the factor $g'(x)$. This implies that when the noise is additive ($g(x)$ is a constant function of x), this error term vanishes. The next term turns out to have order 1, as in the Euler scheme for deterministic differential equations. This suggests to apply the Euler-Maruyama scheme to a Lamperti-transformed version of the equation. This is indeed a good idea, in general.

8.4.3 The Mil'shtein Scheme

This error analysis of the Euler-Maruyama scheme also immediately suggests an improvement: Since the dominating term in the local error is $g'g(B_h^2 - h)/2$, we can simply include this in the discrete approximation. This leads to the *Mil'shtein scheme for a scalar equation* (Milstein and Tretyakov, 2004):

$$X_{t+h} = X_t + f(X_t)\,h + g(X_t, t)\Delta B + \frac{1}{2}g'(X_t)g(X_t)\left[(\Delta B)^2 - h\right]$$

with $\Delta B = B_{t+h} - B_t$. The last term can be seen as a correction term, which has expectation zero and variance proportional to h^2, conditional on X_t. The strong order of the scheme is 1.

For a multivariate SDE with multiple noise terms,

$$dX_t = f(X_t)\,dt + \sum_{k=1}^{m} g_k(X_t)\,dB_t^{(k)},$$

the Mil'shtein scheme can be generalized to

$$X_{t+h} = X_t + f\,h + \sum_{k=1}^{m} g_k \Delta B^{(k)} + \sum_{k,l=1}^{m} \nabla g_k\, g_l \int_t^{t+h} B_s^{(l)}\,dB_s^{(k)}. \qquad (8.6)$$

Here, ∇g_k is the Jacobian of g_k, i.e., a matrix with (i,j)-entry $\partial g_i / \partial x_j$. All functions f, g and ∇g are evaluated at X_t. The integrals of the different Brownian motions w.r.t. each other complicate the implementation. However, when there is just a single noise term, or when the cross terms vanish (e.g., *diagonal* or *commutative* noise; see Section 8.6.1), then it is easy to implement the scheme, since we know $\int_t^{t+h} B_s^{(l)}\,dB_s^{(l)}$ from (6.12).

Example 8.4.1 *Consider Brownian motion on the unit circle* $\{(X_t, Y_t)\}$, *given by the Itô stochastic differential equation*

$$dX_t = -\frac{1}{2}X_t\,dt - Y_t\,dB_t, \qquad dY_t = -\frac{1}{2}Y_t\,dt + X_t\,dB_t$$

for which the solution corresponding to $X_0 = 1$, $Y_0 = 0$ *is*

$$X_t = \cos B_t, \qquad Y_t = \sin B_t.$$

To construct the Mil'shtein scheme for this process, we first identify ∇g:

$$\nabla g = \begin{bmatrix} 0 & -1 \\ 1 & 0 \end{bmatrix}$$

and thus $(\nabla g)g = (-x, -y)$. *The Mil'shtein scheme is then*

$$\begin{pmatrix} X_{t+h} \\ Y_{t+h} \end{pmatrix} = \begin{pmatrix} X_t \\ Y_t \end{pmatrix} - \frac{1}{2}\begin{pmatrix} X_t \\ Y_t \end{pmatrix}h + \begin{pmatrix} -Y_t \\ X_t \end{pmatrix}\Delta B - \frac{1}{2}\begin{pmatrix} X_t \\ Y_t \end{pmatrix}((\Delta B)^2 - h)$$

$$= \begin{pmatrix} X_t \\ Y_t \end{pmatrix} + \begin{pmatrix} -Y_t \\ X_t \end{pmatrix}\Delta B - \frac{1}{2}\begin{pmatrix} X_t \\ Y_t \end{pmatrix}(\Delta B)^2$$

See Figure 8.4. Replacing the last term involving $(\Delta B)^2$ *with its expectation, we are back at the Euler-Maruyama scheme, which would return the point* $(1 - h/2, \Delta B)$.

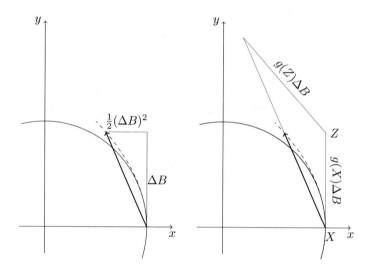

Figure 8.4 The Mil'shtein scheme (left) and the Heun scheme (right) for Brownian motion on the unit circle. One step is shown, starting at position $(X_t, Y_t) = (1, 0)$. Note that the schemes return a point on the supporting parabola (dashed).

8.4.4 The Stochastic Heun Method

The Heun method is a predictor-corrector scheme for ordinary differential equations. It may be generalized to stochastic differential equations, most conveniently in the Stratonovich case. So consider the Stratonovich equation

$$dX_t = f(X_t, t) \, dt + g(X_t, t) \circ dB_t. \tag{8.7}$$

First we form the predictor with an Euler-Maruyama step:

$$Z_{t+h} = X_t + f(X_t, t)h + g(X_t, t)(B_{t+h} - B_t)$$

and note that since the Euler-Maruyama step is consistent with the Itô interpretation, this Z_{t+h} is *not* a good (i.e., consistent) approximation of X_{t+h}. Undeterred, we modify our estimates of drift and diffusion with this predictor:

$$\bar{f} = \frac{1}{2}(f(X_t, t) + f(Z_{t+h}, t + h)), \quad \bar{g} = \frac{1}{2}(g(X_t, t) + g(Z_{t+h}, t + h)).$$

Then we use this modified estimate of drift and diffusion for the final update:

$$X_{t+h} = X_t + \bar{f} \, h + \bar{g} \, (B_{t+h} - B_t).$$

The Heun method is strongly consistent with the Stratonovich equation (8.7) and has strong order 1 when the Brownian motion is scalar, or when the

Brownian motion is multidimensional but the noise structure is commutative (Section 8.6.1) (Rümelin, 1982). In the deterministic case $g = 0$, the Heun method has second-order global error. This is an indication that the scheme is likely to perform substantially better than the Euler-Maruyama method and the Mil'shtein method, when the noise g is weak. When the noise is additive, so that the Itô and Stratonovich interpretation is the same, there is generally no reason to use the Euler-Maruyama method rather than the Heun method.

Example 8.4.2 *We consider again Brownian motion on the unit circle, now described with the Stratonovich equation*

$$d \begin{pmatrix} X_t \\ Y_t \end{pmatrix} = \begin{pmatrix} -Y_t \\ X_t \end{pmatrix} \circ dB_t.$$

The predictor step returns the point Z_{t+h}

$$Z_{t+h} = \begin{pmatrix} X_t - Y_t \Delta B \\ Y_t + X_t \Delta B \end{pmatrix}.$$

The diffusion term evaluated at Z_{t+h} is

$$g(Z_{t+h}) \Delta B = \begin{pmatrix} -Y_t - X_t \Delta B \\ X_t - Y_t \Delta B \end{pmatrix} \Delta B$$

so that the corrected update is

$$\begin{pmatrix} X_{t+h} \\ Y_{t+h} \end{pmatrix} = \begin{pmatrix} X_t - Y_t \Delta B - \frac{1}{2} X_t (\Delta B)^2 \\ Y_t + X_t \Delta B - \frac{1}{2} Y_t (\Delta B)^2 \end{pmatrix}.$$

Note that this exactly agrees with the Mil'shtein scheme for the corresponding Itô equation, which we examined in Example 8.4.1.

An advantage of the scheme is that it allows to assess the local error at each time step by comparing the drift f and noise intensity g evaluated at (X_t, t) and at $(Z_{t+h}, t+h)$. This can be used for adaptive time step control.

8.4.5 The Weak Order

In many situations, the individual sample path is not of interest, only statistics over many realizations. For example, we may use Monte Carlo to compute the distribution of X_T or some other "analytical" property of the model. Then, the strong error is an overly harsh measure of the performance of a scheme. In this section, we consider an alternative, the weak order, which measures the speed of convergence *in distribution* as the time step tends to 0.

To this end, it is practical to assume that the objective of the simulation study is to determine

$$\mathbf{E} k(X_T)$$

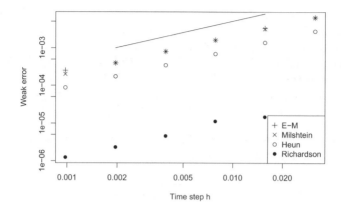

Figure 8.5 The weak error of the Euler-Maruyama, Mil'shtein, and Heun schemes for geometric Brownian motion, and of the Richardson extrapolation of the Heun approximation. Parameters and realizations as in Figure 8.3. For reference, a straight line with slope 1, indicating weak order 1.

for a given function k. This test function k could be a low-order polynomial if we are interested in the moments of X_T, or an indicator function of some set if we are interested in transition probabilities.

We say that the scheme has *weak order p* with respect to a family of test functions, if for each test function k in the class there exists a constant $C > 0$ such that

$$\left| \mathbf{E}k(X_T^{(h)}) - \mathbf{E}k(X_T) \right| \leq Ch^p$$

holds. Notice that the weak order is concerned only with the statistics of the solution; the actual sample paths may be different, and it is not required that $X_T^{(h)} \to X_T$ in any stronger sense than convergence in distribution.

Figure 8.5 shows results of the same experiment as in Figure 8.3. Here we have used the test function $k(x) = x$; the simulations are divided into 10 batches in order to reduce statistical fluctuations. Note that the results are consistent with a weak order of 1, for all schemes. Notice also that the Euler-Maruyama and Mi'lshtein schemes have roughly the same weak error, while the Heun method has lower error, for each time step.

Comparing with Figure 8.3, we notice that the weak order is never smaller than the strong order, for all schemes. This holds in general, as long as the test functions are locally Lipschitz continuous (e.g., smooth) and the solutions are bounded (Kloeden and Platen, 1999).

Figure 8.5 also includes the error obtained with the so-called Richardson extrapolation applied to the Heun method. This extrapolation, which is a general technique in numerical analysis, combines results obtained with two different time steps in order to extrapolate to the limit of vanishing time steps.

Let $\hat{k}(h)$ be the estimate of $\mathbf{E}k(X_T)$ obtained with a time step h; then the Richardson extrapolation combines $\hat{k}(h)$ and $\hat{k}(2h)$ to get

$$\hat{k}(0; h, 2h) = \frac{2^n \hat{k}(h) - \hat{k}(2h)}{2^n - 1}.$$

Here, $\hat{k}(0; h, 2h)$ indicates that we estimate $\hat{k}(0)$ using information obtained with step sizes h and $2h$. n is the weak order of the scheme; in this case, $n = 1$. We see that Richardson extrapolation yields a very substantial improvement. With the Mil'shtein scheme we obtain comparable results, which are left out of the plot in order to avoid clutter. The estimates obtained with the Euler-Maruyama method, however, need even more samples for the Richardson extrapolation to offer significant improvements.

8.4.6 A Bias/Variance Trade-off in Monte Carlo Methods

When estimating a quantity such as $\mathbf{E}k(X_T)$ through simulation, a trade-off arises between the accuracy of the individual sample path and the number of sample paths. Here, we illustrate this trade-off with the example of estimating the mean in geometric Brownian motion

$$dX_t = \lambda X_t \, dt + \sigma X_t \, dB_t, \quad X_0 = 1$$

using the Euler-Maruyama discretization

$$X_{t+h}^{(h)} = X_t^{(h)} + \lambda X_t^{(h)} \, h + \sigma X_t^{(h)} \, (B_{t+h} - B_t).$$

The purpose of the simulation experiment is to estimate $\mu = \mathbf{E}X_T$, knowing that the true result is $\mathbf{E}X_T = e^{\lambda T}$. We use the crude Monte Carlo estimator

$$\hat{\mu} = \frac{1}{N} \sum_{i=1}^{N} X_T^{(h,i)}$$

where N is the number of replicates (sample paths) and $X_T^{(h,i)}$, for $i = 1, \dots, N$, are independent samples obtained with the Euler-Maruyama method using a step size of h. Taking expectation in Euler-Maruyama method, we obtain, for each i

$$\mathbf{E}X_T^{(h,i)} = (1 + \lambda h)^{T/h} = e^{\lambda T}(1 - \frac{1}{2}T\lambda^2 h + o(h))$$

where the last equality follows from differentiating w.r.t. h and letting $h \to 0$. We see that the Euler-Maruyama method has a bias of $-e^{\lambda T}\frac{1}{2}T\lambda^2 h$, to lowest order in h. The variance of X_T, in turn, is known from the properties of geometric Brownian motion (Exercise 7.1):

$$\mathbf{V}X_T = (e^{\sigma^2 T} - 1)e^{(2\lambda + \sigma^2)T}.$$

When the time step is small, the variance of $X_T^{(h)}$ is roughly equal to that of X_T, and therefore we get the following mean-square error on the estimate $\hat{\mu}$:

$$\mathbf{E}|\hat{\mu} - \mu|^2 = |\mathbf{E}\hat{\mu} - \mu|^2 + \frac{1}{N}\mathbf{V}X_T$$

$$= \frac{1}{4}T^2\lambda^4 h^2 e^{2\lambda T} + \frac{1}{N}(e^{\sigma^2 T} - 1)e^{(2\lambda+\sigma^2)T}. \qquad (8.8)$$

To reduce the first term originating from the bias, we need small time steps h. To reduce the second term originating from statistical error, we need many sample paths N. With limited computational effort available, we need to choose between these two conflicting objectives. So assume that the computational effort E depends on number of replicates N and the step size h as $E = N/h$. We then face the problem

$$\min_{N,h} ah^2 + b/N \text{ subject to the constraint } E = N/h,$$

where a and b are the coefficients in mean square error (8.8):

$$a = \frac{1}{4}T^2\lambda^4 e^{2\lambda T}, \quad b = (e^{\sigma^2 T} - 1)e^{(2\lambda+\sigma^2)T}.$$

The solution to this optimization problem is found by substituting $N = Eh$ and differentiating w.r.t. h, yielding:

$$h = \left(\frac{b}{2a}\right)^{1/3} E^{-1/3}, \quad N = \left(\frac{b}{2a}\right)^{1/3} E^{2/3}.$$

Thus, if the available effort E increases with a factor 1000, then we should reduce the step size with a factor 10 and increase the number of samples with a factor 100, and this reduces the r.m.s. error with a factor 10. Moreover, the prefactor

$$\frac{b}{2a} = \frac{2(e^{\sigma^2 T} - 1)e^{\sigma^2 T}}{\lambda^4 T^2}$$

says, for example, that if the noise level σ is increased, then we should increase the step size h and the number N of samples in order to reduce the statistical uncertainty, even at the cost of greater error on the individual sample.

We were able to carry through this analysis because we know the properties of geometric Brownian motion, so that Monte Carlo estimation is unnecessary. In a real application, the qualitative conclusions still stands, even if one would have to assess the statistical error and the discretization error by numeric experiments.

8.4.7 Stability and Implicit Schemes

Recall the motivation for implicit schemes in the numerical solution of ordinary (deterministic) differential equations: Consider the first order equation

$$dX_t = -\lambda X_t \, dt$$

where $\lambda > 0$, and the explicit Euler scheme $X_{t+h}^{(h)} = (1-\lambda h)X_t^{(h)}$. We now ask how the error $X_t^{(h)} - X_t$ depends on $t \in \{0, h, 2h, \ldots\}$ for a given time step h. When $\lambda h > 1$, the numeric approximation $X_t^{(h)}$ becomes an oscillatory function of t, and when $\lambda h > 2$, the numerical approximation becomes unstable, so that $|X_t^{(h)}| \to \infty$ as $t \to \infty$. The conclusion is that the Euler scheme should never be used with time steps larger than $2/\lambda$; if the sign is important, then the bound on the time step is $h < 1/\lambda$.

Consider next the two-dimensional system

$$dX_t = \begin{bmatrix} -1 & 0 \\ 0 & -\lambda \end{bmatrix} X_t \, dt \tag{8.9}$$

where λ is large and positive. For stability, we must use a time step smaller than $2/\lambda$. The slow dynamics has a time scale of 1, so we need $\lambda/2$ time steps to resolve the slow dynamics. If λ is on the order of, say, 10^6, the Euler method is prohibitively inefficient.

This system is the simplest example of a so-called *stiff system*, i.e., one which contains both fast and slow dynamics (numerically large and small eigenvalues). Many real-world systems are stiff in this sense, such as a guitar string which has higher harmonics. Stiff models arise when discretizing partial differential equations, and we shall see this later when discussing transition probabilities (Section 9.11.5). In contrast to the trivial example (8.9), it may not be easy to separate the fast dynamics from the slow dynamics, so we need numerical methods which can handle stiff systems.

The simplest method for stiff systems is the implicit Euler method. For the general nonlinear system in $\mathbf{R}^n$

$$dX_t = f(X_t) \, dt$$

we evaluate the drift at $X_{t+h}^{(h)}$ and get

$$X_{t+h}^{(h)} - X_t^{(h)} = f(X_{t+h}^{(h)}) \, h.$$

This equation must be solved for $X_{t+h}^{(h)}$ at each time step, which is the downside of the scheme. The advantage is that the scheme has nice stability properties. To see this, consider the linear case $f(x) = Ax$ where x is an n-vector and A is an n-by-n matrix; we then get

$$(I - Ah)X_{t+h}^{(h)} = X_t^{(h)} \quad \text{or} \quad X_{t+h}^{(h)} = (I - Ah)^{-1}X_t^{(h)}$$

We see that the discrete-time system is stable whenever A is, regardless of $h > 0$. *Exercise: Verify this by finding the eigenvalues of $(I - Ah)^{-1}$, assuming you know the eigenvalues of A, and that $1/h$ is not an eigenvalue of A.* Even if we cannot reach quite such a strong conclusion for the general nonlinear system, the implicit Euler method still has much more favorable stability properties than the explicit scheme.

For stochastic differential equations, the motivation for implicit schemes is the same, but the situation is complicated by the stochastic integral and the need for consistency with the Itô interpretation. The most common solution is to update implicitly with respect to the drift but explicitly with respect to the noise. For the general Itô equation, we obtain the semi-implicit scheme:

$$X_{t+h}^{(h)} = X_t^{(h)} + f(X_{t+h}^{(h)})\, h + g(X_t^{(h)})(B_{t+h} - B_t),$$

which is solved for $X_{t+h}^{(h)}$ at each time step. This solves issues of stiffness that arise from the drift f, but not those that origin from g. A (trivial) example where stiffness arises from g is the two-dimensional system

$$dX_t = X_t\, dB_t, \quad dY_t = \epsilon Y_t\, dB_t.$$

Discussion of implicit schemes, that also address systems like this, can be found in (Kloeden and Platen, 1999; Higham, 2001; Burrage et al., 2004).

8.5 CONCLUSION

Few people in science and engineering are concerned with questions of existence and uniqueness, when working with ordinary differential equations. For stochastic differential equations, most standard models in the literature have a unique solution (in the strong sense). However, if you aim to build a novel model of a new system or phenomenon, you may inadvertently specify a model which fails to have a unique solution. It may be difficult to diagnose the problem; for example, numerical simulation may give unclear results, where fundamental flaws in the model can be obscured by discretization effects or confused with coding errors. Therefore, it is worthwhile to be familiar with theorems for existence and uniqueness and work within the confines of these. The theorems presented in this chapter are the most commonly used, but there are many different variants in the literature.

You would often reject a model where solutions are not unique. In contrast, there are situations where the model should capture the possibility of explosions, so that global solutions do not exist. Indeed, an objective of the analysis could be to examine if, when and how the solution explodes. In this exposition, however, we prefer to work with processes that are defined at all times, so we prefer to ensure against explosions. An explosion requires a superlinearity that takes the state away from the origin (Theorem 8.3.1).

Both the existence and uniqueness theorems employ a general principle that linear bounds on the dynamics imply exponential bounds on trajectories; a principle which is made precise by the Grönwall-Bellman lemma. In the case of uniqueness, the reasoning is that two nearby trajectories cannot diverge faster than exponentially from each other, if the model is locally Lipschitz continuous. In the case of existence, the reasoning is that a trajectory cannot diverge faster than exponentially from the origin, if the model satisfies a linear growth bound. These questions, and the analysis in this chapter that addresses them, foreshadow the more general stability analysis in Chapter 12.

Numerical simulation of sample paths is often a main method of model analysis. Unfortunately, high-performance numerical algorithms are not as readily available for stochastic differential equations as they are for deterministic ordinary differential equations. The algorithms are substantially more complicated than their deterministic counterparts, and although software libraries exist and are improving, they are not as developed. Also, stochastic simulation experiments often include extra elements such as handling of stopping times, which make it more complicated to design general-purpose simulation algorithms. In practice, many still rely on their own low-level implementations.

The Euler-Maruyama method is often preferred due to its simplicity, but its performance is not impressive, as indicated by the strong order 0.5. The Mil'shtein scheme is a useful extension which improves the strong order, and easy to implement when the noise is commutative and the derivatives ∇g are available. The Heun method for Stratonovich equations can lead to substantial improvements over the Euler-Maruyama method for the corresponding Itô equation. Performing the simulation in a Lamperti transformed domain can be recommended, when possible.

In many applications, the objective of the stochastic simulation is to investigate some statistical property of the solution. In this case, the weak order may be more relevant than the strong order. However, sensitivity studies are often most effectively performed on individual sample paths, to block the effect of the realization. In such situations, it is the strong order of the scheme which is more relevant.

It is good practice to assess the discretization error, e.g. repeating the simulation on a coarser grid. This is a form of sensitivity analysis, so it is most effective when done strongly, i.e., using the same realizations of Brownian motion, to block chance. Then, statistics collected with the different time steps can be combined, e.g. using Richardson extrapolation.

While the order of a scheme is an important characteristic, it should be kept in mind that our ultimate interest is the trade-off between effort and accuracy. This trade-off involves not just the order, but also the constant multiplier in the error term, stability bounds, and the number of replicates used to reduce statistical uncertainty.

In summary, numerical analysis of stochastic differential equations is a vast technical topic. Here, we have presented fundamental algorithms and discussed fundamental issues that serve as a useful starting point. Some further notes are given in the following.

8.6 NOTES AND REFERENCES

Itô (1950) proved existence using the Picard iteration, and again in (Itô, 1951b) under slightly weaker assumptions. Has'minskiĭ (1980) give even weaker assumptions to rule out explosions, using Lyapunov functions (see Section 12.9), and also presents a similar approach to guarantee uniqueness. When Maruyama (1955) introduced the Euler-Maruyama scheme, he also proved existence and uniqueness by showing that the solutions of the discretized equations converge in the limit of small time steps.

A standard reference for numerical solution of stochastic differential equations is (Kloeden and Platen, 1999), which also contains an extensive bibliography and further references. Many different schemes and techniques exist; useful overviews are found in (Iacus, 2008; Särkkä and Solin, 2019) and include leapfrog methods and approximations based on linearization. Some domains of applications have developed specialized algorithms that address issues of particular importance in that field.

Variance reduction techniques can improve the accuracy obtained with simulation experiments, which little extra effort (Kloeden and Platen, 1999; Milstein and Tretyakov, 2004; Iacus, 2008), extending the analysis in Section 8.4.6. Step size control is possible, although not used as widely as in the deterministic case (Lamba, 2003; Iacus, 2008). Results with different time steps can be combined in more sophisticated ways than the Richardson extrapolation; for example so-called multilevel Monte Carlo (Giles, 2008).

8.6.1 Commutative Noise

We have mentioned that the Mil'shtein scheme simplified in the case of commutative noise.

Definition 8.6.1 *We say that the two scalar noise terms in the Itô equation*

$$dX_t = f(X_t) \ dt + g_1(X_t) \ dB_t^{(1)} + g_2(X_t) \ dB_t^{(2)}.$$

commute if the differential operators L_1 and L_2 commute, where $L_i h = \nabla h \cdot g_i$ for a (smooth) function $h : \mathbf{X} \mapsto \mathbf{R}$.

Recall that two operators commute if $L_1 L_2 h = L_2 L_1 h$ for all V. Evaluating these expressions and simplifying, we see that the two noise terms commute when the identity

$$\nabla g_1 \ g_2 = \nabla g_2 \ g_1$$

holds on the entire state space. This implies, loosely, that we could apply the noise terms one at a time in the Euler-Maruyama scheme and the ordering of the noise terms would not matter. The noise is commutative, for example, when it is *diagonal*, i.e. the k'th Brownian motion only affects state variable number k and with an intensity which depends only on this state variable. In the case of commutative noise, the Mil'shtein scheme simplifies:

$$X_{t+h} = X_t + f\,h + \sum_{k=1}^{m} g_k \Delta B^{(k)} + \sum_{k=1}^{m} \nabla g_k\, g_k\, ((\Delta B^{(k)})^2 - h)$$

$$+ \sum_{1 \le k < l \le m} \nabla g_k\, g_l\, (\Delta B^{(k)})(\Delta B^{(l)}).$$

where $\Delta B^{(k)} = B_{t+h}^{(k)} - B_t^{(k)}$. Commutative noise simplifies also other numerical schemes (Kloeden and Platen, 1999) and also relates to the question if there exists an invariant manifold in state space (Section 9.11.1); there, we re-state the property of commuting noise in terms of a so-called Lie bracket.

8.7 EXERCISES

Exercise 8.3 Existence and uniqueness: Consider the double well system in Section 6.2.2 with $q > 0$. Show that the model satisfies the conditions in Theorem 8.3.2 for existence and uniqueness of a solution. Then do the same for the stochastic van der Pol oscillator in Section 6.2.4 with $\mu > 0$.

Exercise 8.4: Consider stochastic logistic growth as in Example 7.7.1 with initial condition $X_0 = x > 0$. Show that existence and uniqueness is guaranteed.

Exercise 8.5: Consider (again) the Cox-Ingersoll-Ross process given by the stochastic differential equation (7.5) with positive parameters λ, ξ, γ. Show that existence and uniqueness is guaranteed as long as $X_t > 0$. *Note:* A theorem in (Karatzas and Shreve, 1997) shows existence and uniqueness even if the process hits 0. Heuristically, if this happens, then the noise vanishes and the drift will reflect the process back into the domain $x > 0$.

Exercise 8.6 Non-commutative noise: Consider the scalar equation $dX_t = dB_t^{(1)} + X_t\, dB_t^{(2)}$. Show that the noise terms do not commute.

Exercise 8.7 The Brownian unicycle: Consider the Brownian unicycle, which is a process taking values in $\mathbf{R}^3$:

$$d\Theta_t = dB_t^{(1)}, \quad dX_t = \cos\Theta_t\, dB_t^{(2)}, \quad dY_t = \sin\Theta_t\, dB_t^{(2)}.$$

Here, Θ_2 is the current heading and (X_t, Y_t) is the position in the plane. Show that the noise terms do not commute.

Exercise 8.8: Consider geometric Brownian motion given by the Stratonovich equation

$$dX_t = rX_t\, dt + \sigma X_t \circ dB_t,$$

and its Heun discretization. Show that the update can be written

$$X_{t+h} = X_t(1 + (rh + \sigma \Delta B) + \frac{1}{2}(rh + \sigma \Delta B)^2);$$

i.e., the analytical solution

$$X_{t+h} = X_t \exp(rh + \sigma \Delta B)$$

written as a Taylor series in $rh + \sigma \Delta B$ and truncated to second order.

CHAPTER 9

Transition Probabilities and the Kolmogorov Equations

In the previous chapters, we have considered the sample paths of Itô diffusions, and how they can be determined (also numerically) from the sample paths of Brownian motion. In this chapter, we ask how the solution X_t is distributed at a given time t. In many cases, the distribution of X_t will have a probability density function, so our task is to determine this density.

As we will show, the probability density function of X_t evolves in time according to a partial differential equation; specifically, an advection-diffusion equation of the same type we considered in Chapter 2. At the end of this chapter, we will therefore have connected Itô's stochastic calculus, which centers on the sample path, with the notion of diffusive transport, which describes how probability spreads over the state space. We previewed this connection already in Chapter 2 when we considered particle tracking schemes to explore diffusive transport.

The key property that allows this connection, is that Itô diffusions are *Markov* processes: Given the present state, future states are independent of past states. This means that their statistics (e.g., their finite-dimensional distribution) can be determined from their transition probabilities.

The resulting partial differential equations are known as the Kolmogorov equations (although also under other names). They exist in two versions, a *forward* equation which described how probability is redistributed as time marches forward, and an adjoint *backward* equation which describes how the transition density depends on the initial state.

The Kolmogorov equations let us find the stationary distribution of the state (when it exists). The equations can only be solved analytically in special cases; in other situations, approximate or numerical methods can be used.

Biography: Andrei Andreyevich Markov (1856–1922)

A Russian mathematician who in 1906 introduced the class of stochastic processes that we today know as Markov chains. His purpose was to show that a weak law of large numbers could hold, even if the random variables were not independent. Markov lived and worked in St. Petersburg, and was influenced by Pafnuty Chebyshev, who taught him probability theory.

9.1 BROWNIAN MOTION IS A MARKOV PROCESS

The Markov property is loosely formulated as "given the present, the future is independent of the past". To make this precise, we use a filtration $\{\mathcal{F}_t : t \geq 0\}$ of the probability space $(\Omega, \mathcal{F}, \mathbf{P})$ to denote the information available at time $t \geq 0$ about the past and the present. Then, to make predictions about the future, we use an arbitrary test function $h : \mathbf{R}^n \mapsto \mathbf{R}$ and consider predictions of $h(X_t)$.

Definition 9.1.1 (Markov Process) *A process $\{X_t \in \mathbf{R}^n : t \geq 0\}$ is said to be a Markov process w.r.t. the filtration $\{\mathcal{F}_t\}$ if:*

1. *$\{X_t\}$ is adapted to $\{\mathcal{F}_t\}$, and*

2. *for any bounded and Borel-measurable test function $h : \mathbf{R}^n \mapsto \mathbf{R}$, and any $t \geq s \geq 0$, it holds almost surely that*

$$\mathbf{E}\{h(X_t)|\mathcal{F}_s\} = \mathbf{E}\{h(X_t)|X_s\}.$$

Since the test function is arbitrary, the definition requires that the conditional distribution of X_t given $\mathcal{F}_s$ is identical to the conditional distribution of X_t given X_s. In other words, the current state X_s is a sufficient statistic of the history $\mathcal{F}_s$ for the purpose of predicting the future state X_t. Note that the definition involves not just the process $\{X_t\}$, but also the filtration $\{\mathcal{F}_t\}$. If we just say that a stochastic process X_s is Markov without specifying the filtration, then we take the filtration to be the one generated by the process itself, $\mathcal{F}_t = \sigma(\{X_s : 0 \leq s \leq t\})$.

For a Markov process, the future state X_t is conditionally independent of any $\mathcal{F}_s$-measurable random variable, given the state X_s. This implies that the variables $X_{t_1}, \ldots, X_{t_n}$ have a dependence structure as depicted in Figure 9.1, for any set of time points $0 \leq t_1 < t_2 < \cdots < t_n$.

Theorem 9.1.1 *Let $\{B_t : t \geq 0\}$ be Brownian motion w.r.t. a filtration $\{\mathcal{F}_t : t \geq 0\}$. Then $\{B_t\}$ is a Markov process with respect to $\{\mathcal{F}_t\}$.*

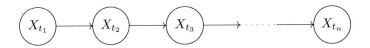

Figure 9.1 Probabilistic graphical model of a Markov process (w.r.t. its own filtration) evaluated at a set of time points $0 \leq t_1 < \cdots < t_n$. The Markov property implies, for example, that X_{t_3} and X_{t_1} are conditionally independent given X_{t_2}, so there is no arrow from X_{t_1} to X_{t_3}.

Proof: The case $s = t$ is trivial, so let $0 \leq s < t$. Write $B_t = B_s + X$ where $X = B_t - B_s$ is the increment; note that B_s is $\mathcal{F}_s$-measurable while the increment X is independent of $\mathcal{F}_s$ and Gaussian with expectation 0 and variance $t - s$. Let h be a test function as in Definition 9.1.1. We get:

$$\mathbf{E}\{h(B_t)|\mathcal{F}_s\} = \mathbf{E}\{h(B_s + X)|\mathcal{F}_s\}$$
$$= \int_{-\infty}^{+\infty} h(x + B_s) \frac{1}{\sqrt{2\pi(t-s)}} e^{-\frac{1}{2}\frac{x^2}{t-s}} \, dx.$$

The same calculation applies if we condition on B_s in stead of $\mathcal{F}_s$, since X and B_s are independent. The result follows. ■

Note that the proof does not rely on the increments being Gaussian, but just that they are independent, so the theorem holds for any process with independent increments.

Remark 9.1.1 (Markov Processes and State Space Systems) *The Markov property is tightly coupled to the notion of a state space model in system theory. But while it is universally agreed what the Markov property is, the term "state space model" is used slightly differently in different bodies of literature, in particular when there is noise and external inputs present. See (Kalman, 1963a) for one definition, relevant to deterministic systems with external inputs. Loosely, the state of a system is a set of variables which are sufficient statistics of the history of the system, so that any prediction about the future can be stated using only the current values of the state variables.*

9.2 DIFFUSIONS ARE MARKOV PROCESSES

Theorem 9.2.1 *Let $\{X_t \in \mathbf{R}^n : t \geq 0\}$ be the unique solution to the stochastic differential equation*

$$dX_t = f(X_t, t) \, dt + g(X_t, t) \, dB_t$$

where $\{B_t : t \geq 0\}$ is Brownian motion with respect to a filtered probability space $(\Omega, \mathcal{F}, \mathbf{P}, \{\mathcal{F}_t\})$, the initial condition X_0 is $\mathcal{F}_0$-measurable, and f and g satisfy the sufficient conditions for existence and uniqueness in Theorem 8.3.2. Then the process $\{X_t : t \geq 0\}$ is Markov with respect to $\{\mathcal{F}_t\}$ as well as with respect to its own filtration.

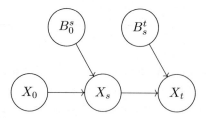

Figure 9.2 The Markov structure of solutions to stochastic differential equations. Here, B_0^s is shorthand for $\{B_u : 0 \le u \le s\}$ while B_s^t is shorthand for $\{B_u - B_s : s \le u \le t\}$.

Proof: We provide an outline; see e.g. (Øksendal, 2010) for a rigorous proof. The theorem states that the probabilistic graphical model in Figure 9.2 applies: Conditional on X_s, X_t is independent of X_0 and $\{B_u : 0 \le u \le u\}$. Therefore, Theorem 3.6.2 applies and gives the result.

The graphical model applies because existence and uniqueness holds: Define the state transition map $Y(t; s, x, \omega)$ for $0 \le s \le t$, $x \in \mathbf{R}^n$, and $\omega \in \Omega$, which returns the unique solution $X_t(\omega)$ to the stochastic differential equation with initial condition $X_s(\omega) = x$. Note that $Y(t; 0, x, \omega) = Y(t; s, Y(s; 0, x, \omega), \omega)$. For given t, x, s, the random variable $\omega \mapsto Y(t; s, x, \omega)$ depends only on the increments of the Brownian motion between times s and t, i.e., it is measurable w.r.t. the σ-algebra generated by $B_u - B_s$ for $u \in [s, t]$. This σ-algebra is included in $\mathcal{F}_t$ and independent of $\mathcal{F}_s$.

This state transition map allows us to construct the graphical model in Figure 9.2: First, X_0, B_0^s and B_s^t are all independent. Next, we have $X_s(\omega) = Y(s; 0, X_0(\omega), \omega)$ and $X_t(\omega) = Y(t; s, X_s(\omega), \omega)$. Thus, the graphical model applies. Theorem 3.6.2 then allows us to conclude that $\mathbf{E}\{h(X_t)|\mathcal{F}_s\}$ is X_s-measurable.

To see that $\{X_t\}$ is also Markov with respect to its own filtration $\{\mathcal{G}_t\}$, note that $\mathcal{G}_t \subset \mathcal{F}_t$. Since the extra information in $\mathcal{F}_s$ beyond X_s does not improve predictions of X_t, neither can the information in $\mathcal{G}_s$. ■

In fact, for an Itô diffusion $\{X_t\}$, the Markov property 9.1.1 holds also when then initial time s replaced with a Markov time τ and the terminal time is replaced by $\tau + t$ where $t > 0$ is deterministic. We say that Itô diffusions have the *strong Markov property*. See (Øksendal, 2010).

9.3 TRANSITION PROBABILITIES AND DENSITIES

Since diffusions are Markov processes, a key to their description is the transition probabilities

$$\mathbf{P}^{X_s = x}(X_t \in A) \text{ for } s \le t, \ x \in \mathbf{X}, \ A \subset \mathbf{X},$$

i.e., if the process starts with $X_s = x$, what is the probability that at a later time $t \geq s$, the state X_t will reside in a given Borel set A?

We will for the moment assume that the probability distribution of X_t admits a density [1] $p(s, x, t, y)$ so that

$$\mathbf{P}^{X_s = x}(X_t \in A) = \int_A p(s, x, t, y) \, dy.$$

This function p is called the *transition density*. To see the central role of transition probabilities and densities, note that the finite-dimensional distributions of a Markov process can be found from the transition probabilities. Consider Figure 9.1 and assume that $X_0 = x_0$ is fixed; then the joint density of $X_{t_1}, X_{t_2}, \ldots, X_{t_n}$, evaluated at a point $(x_1, x_2, \ldots, x_n) \in \mathbf{X}^n$ is

$$\prod_{i=1}^{n} p(t_{i-1}, x_{t_{i-1}}, t_i, x_{t_i}). \tag{9.1}$$

A similar expression holds even if the transition probabilities do not admit densities. Being a function of four variables, the transition density p is a quite complex object. To simplify, we may fix the initial condition (s, x) and get the density ϕ of X_t:

$$\phi(t, y) = p(s, x, t, y).$$

On the other hand, if we fix the terminal condition (t, y), we get

$$\psi(s, x) = p(s, x, t, y)$$

which determines the probability of ending near the target state y, seen as a function of the initial condition $X_s = x$. We can think of ψ as the likelihood function of the initial condition, in case the initial condition is unknown but the terminal position X_t has been measured. Note that $\psi(s, \cdot)$ does not necessarily integrate to one; $\psi(s, \cdot)$ is *not* a probability density of the initial state X_s.

When the dynamics (f, g) are time invariant, the transition densities p will depend on s and t only through the time lag $t - s$. It is however convenient to keep all four arguments.

In rare situations we can determine the transition probabilities analytically; we will see a few examples shortly. However, this is not possible in general. The transition probabilities can be determined numerically with ease in one spatial dimension, if the model is relatively well-behaved; with more effort also in two or three dimensions (see Section 9.11.5). Figure 9.3 shows the transition probabilities for a specific double well model (compare Section 6.2.2)

$$dX_t = 10X_t(1 - X_t^2) \, dt + dB_t.$$

---

[1] Examples where densities fail to exist include the case $g \equiv 0$ as well as - more interestingly - Brownian motion on the circle (page 152). See Section 9.11.1.

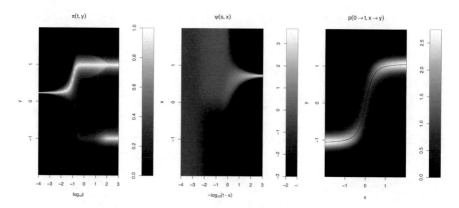

Figure 9.3 Transition probabilities for the double well model. Note the log-scale of the time axis. *Left panel:* The distribution of X_t, graphed using the credibilities $\pi(t, y)$. *Center panel:* The log-likelihood $\log \psi(s, x)$, truncated at -3. *Right panel:* The transition probability densities $p(0, x, t, y)$ as function of x and y for a time interval of $t = 0.1$; the black line is the state transition map for the noise-free model.

Both the p.d.f. $\phi(t, y)$, and the likelihood $\psi(s, x)$, are typically badly scaled which is a challenge for graphical visualization. The figure shows two ways to circumvent this: For the p.d.f. $\phi(t, y)$, we transform to the function

$$\pi(t, y) = \int_{\{x : \phi(t,x) \leq \phi(t,y)\}} \phi(t, x) \; dx.$$

We can use the function π to form credible regions for X_t, so we will call π the *credibility*. For each time t, higher p.d.f. $\phi(t, y)$ implies higher credibility $\pi(t, y)$, and because the credibilities are between 0 and 1 for all t, they are easier to visualize. For the likelihood $\psi(s, x)$, we log-transform and truncate. Both of these graphical techniques help highlighting the region around the maximum, for each time s or t.

For the forward view (Figure 9.3, left panel) we see that the distribution spreads out, then gravitates to the closest well ($y = 1$). Then, on a much slower time scale, the distribution equilibrates between the two wells, and eventually a stationary probability density is reached. For the backward view (same figure, center panel), we see that when s is near t, the likelihood effectively vanishes unless x is near y. Then, on an intermediate time scale, then likelihood is constant in the one well and effectively vanishes in the other well. On this time scale, the process mixes inside the well, so that all initial positions inside the well are roughly equally likely, but the process does not mix between wells, so initial positions in the other well are very unlikely. Then, eventually and on a much larger time scale, the process mixes between the two wells so that all initial conditions are equally likely.

Finally, we can fix the initial time s and the terminal time t, and show the transition density $p(s, x, t, y)$ as a function of x and y. Figure 9.3 (right panel) does this for the double well model for $t = 0.1$. A vertical line in the plot shows the probability density function of the terminal state, while a horizontal lines corresponds to the likelihood of the initial state. The white curve in the graph shows the state transition map for the corresponding deterministic model; specifically, a point $(x = X_0, y = X_t)$ on this line is the endpoints of a trajectory $\{X_s : 0 \le u \le t\}$ obtained with $B_u \equiv 0$. For a short transition time $t = 0.1$, where it is improbable that the noise pushes the trajectory from one well to another, the stochastic trajectories can be seen as small perturbations of the deterministic model, and the transition densities are concentrated near the deterministic state transition map.

Only in simple and very special situations, can we write down analytical expressions for the transition probabilities.

Example 9.3.1 *If X_t is Brownian motion in one dimension, then*

$$p(s, x, t, y) = \frac{1}{\sqrt{2\pi(t-s)}} \exp\left(-\frac{1}{2}\frac{(x-y)^2}{t-s}\right).$$

9.3.1 The Narrow-Sense Linear System

For the narrow-sense linear SDE in $\mathbf{R}^n$

$$dX_t = (AX_t + w_t)\, dt + G\, dB_t$$

where w_t is deterministic, we have previously established an integral expression for the solution (Exercise 7.4). For given initial condition $X_s = x$, X_t is Gaussian with a mean value $\mu(t)$ and a variance $\Sigma(t)$, which can be determined from their governing ordinary differential equations:

$$\frac{d\mu}{dt} = A\mu(t) + w_t, \quad \mu(s) = x$$

i.e.

$$\mu(t) = \exp(A(t-s))x + \int_s^t \exp(A(t-u))w_u\, du$$

For the variance $\Sigma(t)$, we found in Section 5.9 the differential Lyapunov equation

$$\frac{d}{dt}\Sigma(t) = A\Sigma(t) + \Sigma(t)A^\top + GG^\top$$

with the initial condition $\Sigma(0) = 0$ (see also Exercise 5.8). If the variance $\Sigma(t)$ is positive definite, then X_t admits a density (see Section 9.11.1 for a discussion of when this is the case). This is the well-known density of a multivariate Gaussian distribution, which evaluates at y to

$$p(s, x, t, y) = \frac{1}{(2\pi)^{d/2}|\Sigma|^{1/2}} \exp\left(-\frac{1}{2}(y-\mu)^\top\Sigma^{-1}(y-\mu)\right)$$

where $\mu = \mu(t)$ and $\Sigma = \Sigma(t)$ are as in the previous.

Biography: Andrei Nikolaevich Kolmogorov (1903–1987)

One of the most prominent mathematicians of the 20th century, he contributed to the foundations of probability theory and stochastic processes, to filtering and prediction, and to the description of turbulence. His approach to diffusion processes was to consider continuous-time, continuous-space Markov processes for which the transition probabilities are governed by advection-diffusion equations. Photo credit: CC BY-SA 2.0 de.

9.4 THE BACKWARD KOLMOGOROV EQUATION

In the examples given so far, we have found the transition probabilities using knowledge about the solution of the stochastic differential equation. This will not work for the majority of stochastic differential equation, where we can say very little about the solution. In stead, our objective of this chapter is to derive partial differential equations which govern these transition probabilities. We shall see that they are second order equations of the advection-diffusion type, known as the *Kolmogorov* equations.

We establish the Kolmogorov equations as follows: First, we find the *backward* equation which governs $\psi(s,x) = p(s,x,t,y)$, i.e., the transition densities as a function of the initial condition $X_s = x$ for fixed terminal condition $X_t = y$. Then, we obtain the forward equation governing the p.d.f. $\phi(y,t)$ using a duality argument. To follow this programme, we have to take into consideration that the transition probabilities can be concentrated on subsets of measure zero so that the densities do not exist. Therefore, consider again the Itô diffusion $\{X_t : t \geq 0\}$ from Theorem 9.2.1, stop the process at a fixed (deterministic) time t, and evaluate the function h at X_t. Here, h is a bounded C^2 test function on state space such that also ∇h is bounded. We now define the process $\{Y_s : 0 \leq s \leq t\}$ given by

$$Y_s = \mathbf{E}\{h(X_t) \mid \mathcal{F}_s\},$$

i.e., the expected terminal reward, conditional on the information obtained by observing the Brownian motion up to time s. The objective of introducing the test function h is that Y_s is always well defined, even if transition densities are not. To recover the transition densities, think of h as an approximation of the indicator function of a small set containing y; then Y_s is the probability of ending near y, conditional on $\mathcal{F}_s$.

First, notice that $\{Y_s\}$ must be a *Doob's* martingale w.r.t. $\{\mathcal{F}_s\}$ (Exercise 4.19). To repeat the argument and be explicit, for $0 \leq s \leq u \leq t$ we have

$$\begin{aligned}
\mathbf{E}\{Y_u \mid \mathcal{F}_s\} &= \mathbf{E}\{\mathbf{E}[h(X_t) \mid \mathcal{F}_u] \mid \mathcal{F}_s\} \\
&= \mathbf{E}\{h(X_t) \mid \mathcal{F}_s\} \\
&= Y_s.
\end{aligned}$$

Here we have used that the information available at time s is also available at time u, i.e., $\mathcal{F}_s \subset \mathcal{F}_u$, which allows us to use the Tower property of conditional expectations. *Exercise: Verify that $\{Y_s\}$ also possesses the other defining properties of a martingale (Definition 4.5.1 on page 78).*

Next, the Markov property of the process $\{X_s : s \geq 0\}$ implies that Y_s can only depend on $\mathcal{F}_s$ through X_s, i.e., Y_s is X_s-measurable. Therefore, by the Doob-Dynkin lemma, there must exist a function $k(s,x)$ such that

$$Y_s = k(s, X_s) \text{ where } k(s,x) = \mathbf{E}\{h(X_t)|X_s = x\}.$$

By existence and uniqueness, this conditional expectation must agree with the expectation of $h(X_t)$, if $\{X_t\}$ solves the stochastic differential equation with initial condition $X_s = x$. We write

$$\mathbf{E}^{X_s = x} h(X_t)$$

for this expectation. It can be shown (Øksendal, 2010) that k is smooth, so according to Itô's lemma, $\{Y_s\}$ is an Itô process satisfying

$$\begin{aligned}
dY_s &= \dot{k}\, ds + \nabla k\, dX_s + \frac{1}{2} dX_s^\top\, \mathbf{H}k\, dX_s \\
&= (\dot{k} + Lk)\, ds + \nabla k\, g\, dB_s.
\end{aligned}$$

Here, we have omitted the arguments (X_s, s), and we have introduced the differential operator L given by

$$(Lk)(s,x) = \nabla k(x,s) f(x,s) + \frac{1}{2} \mathrm{tr}\left[g(x,s)g^\top(x,s)\mathbf{H}k(s,x)\right]$$

defined for functions $k(s,x)$ which are twice differentiable in x. This is the operator that appears in Itô's lemma. Now, since $\{Y_s\}$ is a martingale, its drift term must vanish, so k must satisfy the partial differential equation

$$\frac{\partial k}{\partial s} + Lk = 0,$$

which is Kolmogorov's backward equation. Finally, assume that the densities $p(s,x,t,y)$ exist. Then we have

$$k(s,x) = \int_{\mathbf{X}} p(s,x,t,y) h(y)\, dy$$

and we therefore find that the densities satisfy the same linear partial differential equation as k. We summarize the findings:

Theorem 9.4.1 (Kolmogorov's Backward Equation) *Let* $h : \mathbf{R}^n \mapsto \mathbf{R}$ *be* C^2 *with bounded support; let* $t > 0$. *Then the function* $k(s, x)$, *given by*

$$k(s, x) = \mathbf{E}^{X_s = x} h(X_t) \tag{9.2}$$

for $0 \le s \le t$, *satisfies the* backward Kolmogorov equation

$$\frac{\partial k}{\partial s} + Lk = 0, \quad k(t, x) = h(x). \tag{9.3}$$

Moreover, a bounded solution to the backward Kolmogorov equation (9.3) has the characterization (9.2). Finally, assume that the transition probabilities admit a density $p(s, x, t, y)$. *Fix the terminal condition* (t, y) *arbitrarily and define* $\psi(s, x) = p(s, x, t, y)$. *Then*

$$\frac{\partial \psi}{\partial s} + L\psi = 0$$

for $0 \le s < t$.

Remark 9.4.1 *If the densities* $p(s, x, t, y)$ *do not exist in the classical sense, i.e., as continuous functions of* y, *then they can still be defined in a sense of distributions, and they still satisfy the Kolmogorov backward equation in a weak sense. The requirement that* h *has bounded support is a quick way to exclude examples such as* $dX_t = dB_t$, $h(x) = \exp(x^2)$, *where the expectations diverge (Exercise 7.22).*

Example 9.4.2 (Likelihood Estimation of the Initial Condition) *If we have observed* $y = X_t(\omega)$ *for some* $t > 0$ *and want to estimate the initial condition* $X_0 = x$, *then the likelihood function is*

$$\Lambda(x) = p(0, x, t, y)$$

assuming that the transition probabilities admit a density p. *To determine this likelihood, we solve the Kolmogorov backward equation* $\dot{k} + Lk = 0$ *for* $s \in [0, t]$, *with terminal conditional* $k(t, x) = \delta(x - y)$, *a Dirac delta. Then,* $\Lambda(x) = k(0, x)$. *Following the Maximum Likelihood paradigm for statistical estimation (Pawitan, 2001), we would estimate the initial condition as* $\hat{x} = \mathrm{Arg\,max}_x \Lambda(x)$ *and derive confidence regions etc. from* $\Lambda(\cdot)$. *More generally, we may have only an imprecise measurement of* X_t, *or indirect information about* X_t *obtained at time* t *or later. This information can be summarized in the function* $h(y)$, *which we view as a likelihood function of* $X_t = y$ *applicable at time* t. *To estimate the initial condition* $X_0 = x$, *we solve the Kolmogorov backward equation* $\dot{k} + Lk = 0$ *for* $s \in [0, t]$, *with terminal conditional* $k(t, y) = h(y)$. *Then, the likelihood function of* $X_0 = x$ *is* $\Lambda(x) = k(0, x)$. *Thus, Kolmogorov's backward equation pulls likelihoods backward in time.*

Example 9.4.3 *[What is the fair price of an option?] We consider* an option *which gives the owner the right (but not the obligation) to buy a certain stock* *at an "expiry" time t for a "strike" price of K. The fair price of such an option at time s < t is*

$$Y_s = e^{r(s-t)}\mathbf{E}\{(X_t - K)^+|\mathcal{F}_s\}$$

where X_t denotes the price of the stock at time t, r is the discount rate, $x^+ = \max\{0, x\}$, and $\mathcal{F}_s$ is the information available at time s. Here, we assume risk neutrality, i.e., the current value equals the discounted expected future value. We see that the fair price of the option can be found with a backward Kolmogorov equation, if a stochastic differential equation is posed for the price $\{X_t\}$ of the stock. In their seminal work, Black and Scholes (1973) assumed geometric Brownian motion:

$$dX_s = X_s(r\, ds + \sigma\, dB_s)$$

where $\{B_s\}$ is standard Brownian motion w.r.t. $\{\mathcal{F}_s\}$, so that

$$Y_s = e^{r(s-t)}k(s, X_s) \text{ where } \dot{k} + rxk' + \frac{1}{2}\sigma^2 x^2 k'' = 0.$$

The terminal condition is $k(t,x) = h(x) = (x-K)^+$. For this particular case, Black and Scholes (1973) established an analytical solution (Exercise 9.4).

9.5 THE FORWARD KOLMOGOROV EQUATION

We now turn to the forward equations, i.e., partial differential equations that govern the transition probabilities $\phi(t,y) = p(s,x,t,y)$ as functions of the end point t, y, for a given initial condition $X_s = x$. These equations will also govern the probability density of the state X_t as a function of time, when the initial condition is a random variable.

Theorem 9.5.1 *Under the same assumptions as in Theorem 9.2.1, assume additionally that the distribution of X_t admits a probability density for all $t > s$ and use $\phi(t,y)$ to denote this density at y. Then the forward Kolmogorov equation*

$$\dot{\phi} = -\nabla \cdot (f\phi) + \nabla \cdot \nabla(D\phi)$$

holds for all $t > s$ and for all $y \in \mathbf{R}^n$. Here, all functions are evaluated at t, y, and the diffusivity matrix is

$$D(y,t) = \frac{1}{2}g(y,t)g^\top(y,t)$$

Here, $\nabla \cdot \nabla(D\phi)$ can be written explicitly as $\sum_{i=1}^n \sum_{j=1}^n \frac{\partial^2}{\partial y_i \partial y_j}(D_{ij}\phi)$. Note that we here use y for the spatial coordinate, for a consistent notation where we use x for the initial position and y for the terminal position. If the forward equation is considered in isolation, and not in concert with the backward equation, then we usually use x for the spatial coordinate.

Proof: Let $Y_s = k(s, X_s) = \mathbf{E}\{h(X_t)|\mathcal{F}_s\}$ as in the previous section, then we can compute $\mathbf{E}Y_s$ by integration over state space:

$$\mathbf{E}Y_s = \int_{\mathbf{X}} \phi(s, x) k(s, x) \, dx$$

Since $\{Y_s\}$ is a martingale, this expectation is independent of time s. Differentiating with respect to time s, we obtain:

$$\int_{\mathbf{X}} \dot{\phi} k - \phi L k \, dx = 0 \tag{9.4}$$

where we have omitted arguments and used $\dot{k} = -Lk$. First note (omitting technicalities) that this equation may be written $\langle \dot{\phi}, k \rangle - \langle \phi, Lk \rangle = 0$, where $\langle \cdot, \cdot \rangle$ denotes the inner product defined through the integral. Then rewrite this as $\langle \dot{\phi}, k \rangle - \langle L^* \phi, k \rangle$, where L^* is the formal adjoint operator of L. Since this must hold for all k, we conclude that $\dot{\phi} = L^* \phi$. We now repeat this argument, filling in the specifics of the formal adjoint operator. Consider the term $\int_{\mathbf{X}} \phi Lk \, dx$:

$$\int_{\mathbf{X}} \phi Lk \, dx = \int_{\mathbf{X}} \phi \cdot (\nabla k \ f + \operatorname{tr}[D\mathbf{H}k]) \, dx$$

For the first term, we find

$$\int_{\mathbf{X}} \phi f \cdot \nabla k \, dx = -\int_{\mathbf{X}} k \nabla \cdot (f\phi) \, dx$$

using the divergence theorem on the vector field $k\phi f$, and the product rule $\nabla \cdot (k\phi f) = k\nabla \cdot (f\phi) + \phi f \cdot \nabla k$. Here we have used that the boundary term is zero at $|x| = \infty$: There can be no flow to infinity, since the process cannot escape to infinity in finite time. For the second term, we find

$$\int_{\mathbf{X}} \phi \operatorname{tr}[D\mathbf{H}k] \, dx = -\int_{\mathbf{X}} \nabla(\phi D) \cdot \nabla k \, dx$$

using the divergence theorem on the vector field $\phi D\nabla k$, and omitting boundary terms. Here, $\nabla(\phi D)$ is the vector field with elements

$$\nabla(\phi D)_i = \sum_j \frac{\partial(\phi D_{ij})}{\partial x_j}.$$

Repeating, now using the divergence theorem on the vector field $k\nabla(\phi D)$, we can summarize the steps in

$$\int_{\mathbf{X}} \phi \operatorname{tr}[D\mathbf{H}k] \, dx = -\int_{\mathbf{X}} \nabla(\phi D) \cdot \nabla k \, dx = \int_{\mathbf{X}} k\nabla \cdot \nabla(\phi D) \, dx.$$

Inserting this in equation (9.4), we obtain

$$\int_{\mathbf{X}} (\dot{\phi} + \nabla \cdot (f\phi) - \nabla \cdot \nabla(\phi D))k \ dx = 0$$

Since the test function h is arbitrary and s is arbitrary, k can be arbitrary, so we conclude that

$$\dot{\phi} + \nabla \cdot (f\phi) - \nabla \cdot \nabla(D\phi) = 0$$

which is the result we pursued. ■

9.6   DIFFERENT FORMS OF THE KOLMOGOROV EQUATIONS

We summarize the different forms in which the forward and backward Kolmogorov can be written. In the scalar case, we have:

The forward Kolmogorov equation: $\quad \dot{\phi} \ = -(f\phi)' + (D\phi)''$,
The backward Kolmogorov equation: $\quad -\dot{\psi} \ = \psi' f + D\psi''$

where the diffusivity is $D = \frac{1}{2}g^2$. In the multivariate case, the vector calculus is often a source of confusion. We have

$$\dot{\phi} = -\nabla \cdot (f\phi) + \nabla \cdot \nabla(D\phi), \quad -\dot{\psi} = \nabla\psi \cdot f + \text{tr}(\mathbf{D}\mathbf{H}\psi).$$

Here, $D = \frac{1}{2}gg^\top$ while $\mathbf{H}\psi$ is the Hessian of ψ, i.e.

$$(\mathbf{H}\psi)_{ij} = \frac{\partial^2 \psi}{\partial x_i \partial x_j}.$$

In turn, $\nabla \cdot \nabla(D\phi)$ is the divergence of the vector field $\nabla(D\phi)$, which is

$$(\nabla(D\phi))_i = \sum_j \frac{\partial(D_{ij}\phi)}{\partial x_j} \quad \text{so } \nabla \cdot \nabla(D\phi) = \sum_{i,j} \frac{\partial^2(D_{ij}\phi)}{\partial x_i \partial x_j}.$$

The equations can also be written in advection-diffusion form

$$\dot{\phi} = -\nabla \cdot (u\phi - D\nabla\phi), \quad -\dot{\psi} = \nabla\psi \cdot u + \nabla \cdot (D\nabla\psi).$$

Here, we have introduced the advective field

$$u = f - \nabla D \quad \text{where } (\nabla D)_i = \sum_j \frac{\partial D_{ij}}{\partial x_j}.$$

This formulation of the forward equation expresses how probability is redistributed in space while being conserved, so this is also termed the *conservative* form. Specifically, the local increase is balanced by the local net export, which is the divergence of the *flux* $J = u\phi - D\nabla\phi$. This flux J has an advective contribution $u\phi$ and a diffusive contribution $-D\nabla\phi$, as in Chapter 2.

It is convenient to also use a shorthand "operator" form

$$\dot{\phi} = L^*\phi, \quad -\dot{\psi} = L\psi$$

which emphasizes that the forward operator L^* and the backward operator L are formally adjoint. It allows us to use the compact notation for the solution (in the case where the dynamics are time invariant)

$$\phi_t = e^{L^*(t-s)}\phi_s, \quad \psi_s = e^{L(t-s)}\psi_t.$$

Here, $\{\exp(Lt) : t \geq 0\}$ can be accepted as a convenient notation for the operation of solving the backward Kolmogorov equation for a given terminal value. However, it can also be understood in the framework of functional analysis, and we briefly sketch this. As usual, we let $\mathcal{L}_\infty(\mathbf{X}, \mathbf{R})$ denote the space of bounded measurable real-valued functions on state space equipped with the supremum norm. For $t \geq 0$, define the linear operator $P_t : \mathcal{L}_\infty(\mathbf{X}, \mathbf{R}) \mapsto \mathcal{L}_\infty(\mathbf{X}, \mathbf{R})$ by

$$(P_t k)(x) = \mathbf{E}^x k(X_t)$$

for $t \geq 0$ and $k \in \mathcal{L}_\infty(\mathbf{X}, \mathbf{R})$. Here, we use the shorthand $\mathbf{E}^x$ for $\mathbf{E}^{X_0=x}$. Then it is easy to see that $\|P_t k\| \leq \|k\|$; e.g. if $|k(x)| \leq 1$ for all x, then also $|\mathbf{E}^x k(X_t)| \leq 1$ for all x. We say that P_t is a *contraction*, $\|P_t\| \leq 1$. The Markov property of $\{X_t\}$ and the law of total expectation gives $P_t(P_s k) = P_{t+s}k$, which implies that the family $\{P_t : t \geq 0\}$ of operators has the structure of a semigroup. This semigroup is right continuous in the sense that $P_t k \to k$ as $t \to 0$, at least when k is smooth and has bounded support. We say that $\{P_t\}$ forms a *weak Feller* semigroup and that diffusions are Feller processes. The semigroup is *generated* by L in the sense that $Lk = \lim_{t\to0}(P_t k - k)/t$, at least if k is smooth and has bounded support. We can therefore write $P_t = \exp(Lt)$. Next, since P_t is a bounded operator, it has an adjoint P_t^*, which acts on the dual space $\mathcal{L}_1(\mathbf{X}, \mathbf{R})$ (alternatively, on the space of finite measures on $\mathbf{X}$). This operator P_t^* propagates an initial distribution ϕ on state space forward in time, and the family $\{P_t^* : t \geq 0\}$ is a semigroup which is generated by L^*. This formalism is useful, not just because it is compact, but also because it highlights the parallels between the Kolmogorov equations for diffusions and for Markov chains on finite state spaces, where the generator L is a matrix and $P_t = \exp(Lt)$ is the matrix exponential.

9.7 DRIFT, NOISE INTENSITY, ADVECTION, AND DIFFUSION

The relationships $D = \frac{1}{2}gg^\top$ and $f = u + \nabla D$ between drift f, advection u, noise intensity g and diffusivity D, are important. The following is an elementary way to establish these relationships, i.e. without Itô calculus – for simplicity, in one dimension, and skipping technicalities. Start with an advection-diffusion equation governing the density $C(x,t)$ of some particles:

$$\dot{C} = -(uC - DC')' \text{ with no-flux boundary conditions at } |x| = \infty.$$

Choose a random particle; its position is X_t, so its expected position is

$$\mu(t) = \mathbf{E}X_t = \int_{-\infty}^{+\infty} x\, C(x,t)\, dx$$

and therefore the expected position changes with time as

$$\dot{\mu}(t) = \int_{-\infty}^{+\infty} x\, \dot{C}(x,t)\, dx = - \int_{-\infty}^{+\infty} x(uC - DC')'\, dx.$$

Integrate twice by parts to find

$$\dot{\mu}(t) = \int_{-\infty}^{+\infty} uC - DC'\, dx = \int_{-\infty}^{+\infty} (u + D')C\, dx = \mathbf{E}(u(X_t) + D'(X_t)).$$

In particular, if the particle is found at a position x_0 at time 0, i.e. the initial concentration is $C(x,0) = \delta(x - x_0)$, then initially the mean grows with rate $u(x_0) + D'(x_0)$. Note that this is exactly the drift:

$$f(x_0) = u(x_0) + D'(x_0).$$

Next, for the variance $\Sigma(t) = \mathbf{E}X_t^2 - (\mathbf{E}X_t)^2$, we have

$$\Sigma(t) = \int_{-\infty}^{+\infty} x^2\, C(x,t)\, dx - \mu^2(t)$$

and therefore

$$
\begin{aligned}
\dot{\Sigma}(t) &= \int_{-\infty}^{+\infty} -x^2\, (uC - DC')'\, dx - 2\mu(t)\dot{\mu}(t) \\
&= \int_{-\infty}^{+\infty} 2x\, (uC - DC')\, dx - 2\mu(t)\dot{\mu}(t) \\
&= \int_{-\infty}^{+\infty} 2x\, (u + D')C + 2DC\, dx - 2\mu(t)\dot{\mu}(t) \\
&= 2\mathbf{E}(f(X_t)X_t) + 2\mathbf{E}D(X_t) - 2\mathbf{E}X_t\, \mathbf{E}f(X_t) \\
&= 2\mathbf{Cov}(f(X_t), X_t) + 2\mathbf{E}D(X_t).
\end{aligned}
$$

Here, we have used integration by parts twice; the second time using $\int 2xDC'\, dx = -\int(2xD)'C\, dx = -\int(2D + 2xD')C\, dx$. The last step follows from the definition of covariance. Notice the conclusion: The variance changes with time due to two mechanisms. First, diffusion pumps variance into the system with rate $2\mathbf{E}D(X_t)$. Next, system dynamics amplify or dissipate variance with rate $2\mathbf{Cov}(f(X_t), X_t)$ - for example, if $f(x)$ is an increasing function of x, then system dynamics will amplify variance, whereas if $f(x)$ is decreasing function of x, then system dynamics dissipate variance.

Now let us focus on the initial growth of variance: If the concentration at time $t = 0$ is a Dirac delta at x_0, we find

$$\dot{\Sigma}(0) = 2D(x_0).$$

Notice that this agrees with the relationship between diffusivity and noise intensity, $D = g^2/2$, and the incremental variance in Itô processes.

The multivariate extensions of these formulas are

$$\dot{\mu}(t) = \mathbf{E}f(X_t), \quad \dot{\Sigma}(t) = \mathbf{Cov}(f(X_t), X_t) + \mathbf{Cov}(X_t, f(X_t)) + 2\mathbf{E}D(X_t)$$

where

$$f(x) = u(x) + \nabla D(x), \text{ still with } (\nabla D)_i = \sum_j \frac{\partial D_{ij}(x)}{\partial x_j}.$$

You should check that if the system is linear, $f(x) = Ax$, $g(x) = G$, then these formulas agree with what we found for linear systems; in particular the Lyapunov equation (5.21) for the variance.

9.8 STATIONARY DISTRIBUTIONS

Often, when the functions f, g do not depend on time, the forward Kolmogorov equation admits a stationary density $\rho(x)$, which by definition satisfies

$$L^*\rho = -\nabla \cdot (u\rho - D\nabla\rho) = 0. \tag{9.5}$$

If the initial condition X_0 is sampled from the stationary distribution, then X_t will also follow the stationary distribution, for any $t > 0$. Then, the Markov property assures that all statistics of the process are independent of time, and the process is stationary. If the stationary distribution has a finite variance, then the process will be weakly stationary in the sense of Section 5.4. Stationary distributions are as important for stochastic differential equations as equilibria are for ordinary differential equations; in many applications the main concern is the stationary distribution.

In general, there may not be a unique stationary density: There can be many stationary distributions, a single unique one, or none at all. For Brownian motion, the only non-negative stationary solutions to the forward Kolmogorov equation are the constant ones, which cannot be normalized to a probability density. Equilibrium points correspond to atomic stationary distributions which do not admits densities. For Brownian motion on the circle (7.6), any circle is invariant, so there is a family of stationary distributions and they do not all admit densities (see also Exercise 7.24). Despite these many possibilities, a common situation is that there is a unique stationary distribution which admits a density. In the following, we give a number of examples.

Example 9.8.1 (The General Scalar SDE) *Consider the scalar equation*

$$dX_t = f(X_t)\, dt + g(X_t)\, dB_t$$

with $g(x) > 0$. The stationary forward Kolmogorov equation is

$$-(\rho f)' + \frac{1}{2}(g^2 \rho)'' = 0$$

which we rewrite in advection-diffusion form:

$$-(\rho u - D\rho')' = 0$$

with $D(x) = \frac{1}{2}g^2(x)$, $u(x) = f(x) - D'(x)$. Integrate once to obtain

$$u\rho - D\rho' = j$$

where the integration constant j is the flux of probability:

$$\frac{d}{dt}\mathbf{P}(X_t > x) = j$$

for any $x \in \mathbf{R}$. If the process $\{X_t : t \geq 0\}$ is stationary, then this flux must equal 0. We elaborate on this point in Section 9.9. Proceeding, we find

$$\rho(x) = \frac{1}{Z}\exp\left(\int_{x_0}^x \frac{u(y)}{D(y)}\, dy\right) \tag{9.6}$$

where x_0 and Z are arbitrary. If Z can be chosen so that ρ integrates to 1, then this ρ is a stationary probability density function.

The stationary distribution can also be written in terms (f, g):

$$\rho(x) = \frac{2}{\bar{Z}g^2(x)}\exp\left(\int_{x_0}^x \frac{2f(y)}{g^2(y)}\, dy\right). \tag{9.7}$$

Exercise 9.1: Verify the form (9.7) for the stationary density.

Example 9.8.2 (Stochastic logistic Growth) *Consider the Itô equation (compare the introduction; page 6, and Example 7.7.1)*

$$dX_t = rX_t(1 - X_t/K)\, dt + \sigma X_t\, dB_t \text{ with } X_0 = x > 0,$$

where $r, K, \sigma > 0$. We expect that the solution will eventually fluctuate around the carrying capacity $K > 0$. If the stationary density exists, it is

$$\rho(x) = \frac{1}{Z\sigma^2 x^2}\exp\left(\int_K^x \frac{2ry(1 - y/K)}{\sigma^2 y^2}\, dy\right)$$

$$= \frac{1}{Z}(x/K)^{2r/\sigma^2 - 2}\exp\left(-\frac{2r}{\sigma^2}\frac{x}{K}\right)$$

where Z and $\bar{Z}$ are normalization constants. This is a Gamma distribution with scale parameter $\theta = \sigma^2 K/(2r)$ and shape parameter $k = 2r/\sigma^2 - 1$, so

$$\rho(x) = \frac{1}{\Gamma(k)\theta^k} x^{k-1} e^{-x/\theta}.$$

Recall that the Gamma distribution is only defined for $k > 0$, i.e., we require that $\sigma^2 < 2r$. If $k \leq 0$, the distribution cannot be normalized. The shape parameter k depends only on the non-dimensional noise level $\sigma/\sqrt{r}$ (compare Example 7.7.1). From the properties of the Gamma distribution, the stationary expectation is $\mathbf{E}X_t = \theta k = K(1 - \sigma^2/(2r))$. Thus, for low noise levels ($\sigma^2 \ll 2r$) we have $\mathbf{E}X_t \approx K$, but noise reduces the expectation (see Exercise 9.3 for an alternative derivation). As the noise is increased to $\sigma^2 = 2r$, the expectation approaches 0 and then the stationary distribution becomes an atom at the origin. At this point, the population collapses and goes extinct; we will return to this condition when discussing stability in Chapter 12.

Exercise 9.2 The Cox-Ingersoll-Ross Process: Consider the Cox-Ingersoll-Ross process

$$dX_t = \lambda(\xi - X_t)\, dt + \gamma\sqrt{X_t}\, dB_t$$

with $\lambda, \xi, \gamma > 0$ and for $x \geq 0$. Show that in stationarity, X_t is Gamma distributed with rate parameter $\omega = 2\lambda/\gamma^2$ and shape parameter $\nu = 2\lambda\xi/\gamma^2$, i.e., density

$$\rho(x) = \frac{\omega^\nu}{\Gamma(\nu)} x^{\nu-1} e^{-\omega x},$$

provided $\nu > 0$. Derive the mean and variance in stationarity and compare with the results of Exercise 8.5. *Note:* Also the transition probabilities are available in closed form; see Exercise 9.8.

Example 9.8.3 (The Linear SDE) *Consider the linear SDE in $\mathbf{R}^n$*

$$dX_t = AX_t\, dt + G\, dB_t.$$

Section 5.9 shows that the stationary distribution exists if A is exponentially stable; then it is Gaussian with mean 0 and variance Σ which is the unique solution to the algebraic Lyapunov equation

$$A\Sigma + \Sigma A^\top + GG^\top = 0.$$

It can be shown that this Gaussian distribution satisfies the forward Kolmogorov equation, but the method of Section 5.9 is less laborious.

Example 9.8.4 (Kinesis) *With the stochastic differential equation in $\mathbf{R}^n$*

$$dX_t = \sqrt{2D(X_t)}\, dB_t$$

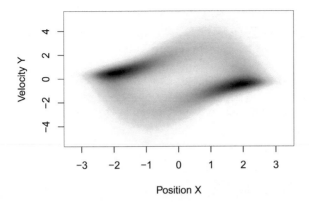

Figure 9.4 The stationary density for the van der Pol oscillator (6.9) with parameters $\mu = 1$, $\sigma = 1/2$. Compare with Figure 6.4.

where $\{B_t : t \geq 0\}$ is Brownian motion in $\mathbf{R}^n$ and $D : \mathbf{R}^n \to \mathbf{R}$ is smooth and non-negative, a candidate stationary distribution is

$$\rho(x) = \frac{1}{Z} \frac{1}{D(x)}$$

where Z is a normalization constant. If Z can be chosen so that ρ integrates to 1, then ρ is in fact a stationary distribution. To see this, note that the forward Kolmogorov equation is

$$\dot{\rho} = \nabla^2(D\,\rho).$$

Note that the process accumulates where the diffusivity is low. This $\{X_t : t \geq 0\}$ is a martingale, i.e., an unbiased random walk, but it is not pure Fickian diffusion, since pure Fickian diffusion has a uniform steady-state.

Example 9.8.5 (The van der Pol Oscillator) *For the system in Section 6.2.4, we compute the stationary density numerically using the methods in Section 9.11.5. The result is seen in Figure 9.4. We use the same parameters as in the simulation in Section 6.2.4, so the density can be compared visually with the trajectories in Figure 6.4.*

9.9 DETAILED BALANCE AND REVERSIBILITY

We have seen that a density ρ is stationary for the forward Kolmogorov equation, if and only if the flux $J := u\rho - D\nabla\rho$ is divergence free. A sufficient condition for this is that the flux J vanishes everywhere:

Definition 9.9.1 (Detailed Balance) *Let $\{X_t\}$ be a stationary Itô diffusion with advective flow field u and diffusivity D and stationary probability density ρ. We say that ρ (or $\{X_t\}$) satisfies detailed balance if $J := u\rho - D\nabla\rho = 0$ everywhere.*

Detailed balance is a first order equation in ρ and therefore easier than the second-order equation $\nabla \cdot (u\rho - D\nabla\rho) = 0$ that expresses stationarity. We used this when we identified the general expression for the stationary density of a scalar diffusion process (Example 9.8.1): In the scalar case, a divergence-free flux must be constant in space. If this constant flux is non-zero, then particles must be absorbed at the right boundary and re-emerge at the left boundary (or vice versa). Since we have not introduced such exotic boundary behavior, we conclude:

Proposition 9.9.1 *Let $\{X_t\}$ be a scalar stationary Itô diffusion. Then $\{X_t\}$ satisfies detailed balance.*

The corresponding result for Markov chains is that a stationary process on an acyclic graph must satisfy detailed balance (Grimmett and Stirzaker, 1992). In two dimensions or more, we cannot expect stationary distribution to satisfy detailed balance.

Detailed balance is important because it implies that the stationary process is *time reversible*; i.e., it has the same statistics as its time reversed version. This remarkable property of reversibility is central in equilibrium thermodynamics and statistical physics. To see why such a result should hold, note that detailed balance implies the net flow of probability over any surface is zero. This, in turn, means that the probability that a given particle moves from one region A to another region B in a specified time t, exactly equals the probability that this particle moves from B to A in the same time. More precisely, we have the following result:

Theorem 9.9.2 *Assume that $\{X_t\}$ is stationary and that its probability density ρ satisfies detailed balance. Let functions k, h be real-valued functions on state space such that $\mathbf{E}k^2(X_0) < \infty$, $\mathbf{E}h^2(X_0) < \infty$, and let $t \geq 0$. Then $\mathbf{E}h(X_0)k(X_t) = \mathbf{E}k(X_0)h(X_t)$.*

For example, if h and k are indicator functions of sets A and B, then the theorem states that transitions $A \rightarrow B$ happen as frequently as transitions $B \rightarrow A$. This, together with the Markov property, implies that the statistics of the process are preserved if we change the direction of time.

Proof: We conduct the proof under the additional assumption that $\mathbf{E}|Lh(X_t)|^2 < \infty$, $\mathbf{E}|Lk(X_t)|^2 < \infty$. Define the weighted inner product $\langle h, k \rangle_\rho = \int_{\mathbf{X}} h(x)\rho(x)k(x) \, dx$, then

$$\langle h, Lk \rangle_\rho = \int h\rho(\nabla k \cdot u + \nabla \cdot (D\nabla k)) \, dx \qquad \text{(The definition of } L\text{)}$$

$$= \int h\rho\nabla k \cdot u - \nabla(\rho h) \cdot D\nabla k \, dx \qquad \text{(Divergence theorem)}$$

$$= \int h\rho\nabla k \cdot u - \rho\nabla h \cdot D\nabla k - h\nabla\rho \cdot D\nabla k \, dx \qquad \text{(Product rule)}$$

$$= \int -\rho\nabla h \cdot D\nabla k \, dx. \qquad \text{(Detailed balance)}$$

Since this end result is symmetric in (h, k), we see that $\langle h, Lk \rangle_\rho = \langle Lh, k \rangle_\rho$. Thus, L is formally self-adjoint under the weighted inner product $\langle \cdot, \cdot \rangle_\rho$. Then also the semigroup $\exp(Lt)$ generated by L is self-adjoint, i.e.,

$$\langle h, e^{Lt}k \rangle_\rho = \langle e^{Lt}h, k \rangle_\rho.$$

Recall that the Kolmogorov's backward equation states that $(e^{Lt}k)(x) = \mathbf{E}^x k(X_t)$, so the probabilistic interpretation of the left hand side is

$$\langle h, e^{Lt}k \rangle_\rho = \mathbf{E}\left[h(X_0)\mathbf{E}\{k(X_t)|X_0\}\right] = \mathbf{E}h(X_0)k(X_t)$$

and with a similar interpretation of $\langle e^{Lt}h, k \rangle_\rho$ we conclude that $\mathbf{E}h(X_0)k(X_t) = \mathbf{E}k(X_0)h(X_t)$. ■

Example 9.9.1 (Gibbs' Canonical Distribution) *Consider diffusive motion in a potential given by*

$$dX_t = -\nabla U(X_t)\, dt + \sigma\, dB_t.$$

where $X_t \in \mathbf{R}^n$, $U : \mathbf{R}^n \mapsto \mathbf{R}$ is a smooth potential *and σ is a scalar constant; $\{B_t : t \geq 0\}$ is n-dimensional Brownian motion. The* Gibbs canonical distribution *is given by the density*

$$\rho(x) = \frac{1}{Z} \exp(-U(x)/D)$$

where $D = \sigma^2/2$ and $Z > 0$ is a constant. Then ρ is stationary under the forward Kolmogorov equation. To see this, compute the flux:

$$J = u\rho - D\nabla\rho = -\rho\nabla U - D(-\rho/D \cdot \nabla U) = 0$$

i.e., ρ satisfies detailed balance and is therefore stationary. If ρ is integrable, then we say that the potential is confining, *and we choose Z as the* partition function *so that ρ integrates to 1. Then, the stochastic differential equation admits a stationary solution $\{X_t\}$ which is time reversible. This may appear counter-intuitive, considering the drift $f(x) = -\nabla U(x)$: Say that $\{X_t\}$ is stationary and that we observe $X_t = x$, then*

$$\mathbf{E}\{X_{t+h}|X_t = x\} = \mathbf{E}\{X_{t-h}|X_t = x\},$$

and, by the Euler-Maruyama scheme, both of these expectations equal $x - h \cdot \nabla U(x) + o(h)$ for $h > 0$.

Example 9.9.2 (The Ornstein-Uhlenbeck Process) *Here, $\{X_t\}$ is given by*

$$dX_t = -\lambda X_t\, dt + \sigma\, dB_t$$

with $\lambda > 0$, $\sigma > 0$, and has as stationary distribution a Gaussian with mean 0 and variance $\sigma^2/(2\lambda)$. Since the process is scalar, this stationary distribution *must satisfy detailed balance, which can be confirmed by direct calculation: We have*

$$\rho(x) = s^{-1}\phi(x/s)$$

where $s = \sqrt{\sigma^2/(2\lambda)}$ is the stationary standard deviation and $\phi(\cdot)$, as usual, is the standard Gaussian probability density function. Therefore, the stationary flux at x is

$$J(x) = -\lambda x \rho(x) - \frac{1}{2}\sigma^2 \rho'(x).$$

The gradient is

$$\rho'(x) = s^{-2}\phi'(x/s) = -s^{-2}x\phi(x/s) = -s^{-1}x\rho(x)$$

so $J(x) = (-\lambda x + \sigma^2/(2s))\rho(x) = 0$, as claimed. Note that we could also shown that detailed balance holds by using Example 9.9.1, since the flow $x \mapsto -\lambda x$ is potential flow (or gradient flow), $-\lambda x = -U'(x)$ with $U(x) = \lambda x^2/2$. It follows that the stationary Ornstein-Uhlenbeck process $\{X_t\}$ is reversible.

For diffusions in higher dimensions, there is a generalized concept of detailed balance where we allow some state variables, e.g. velocities, to change sign as we reverse time. See (Gardiner, 1985).

9.10   CONCLUSION

We have seen that Itô diffusion processes, the solutions to stochastic differential equations, are Markov processes and their transition probabilities are governed by partial differential equations of the advection-diffusion type. The link is quite natural, bearing in mind our initial motivation of following a diffusing molecule in fluid flow, and is anticipated by using the name "Itô diffusion" for solutions of stochastic differential equations.

We have seen that there are two Kolmogorov equations, which both govern the transition densities: The forward equation, which describes how the probability density evolves as time marches forward, and the backward equation, which describe how the likelihood of the initial condition varies as the initial time is pulled backward. Importantly, the backward and forward equations are formally *adjoint*, so they contain the same information. Most students find it easier to develop intuition for the forward equation, but the backward equation is equally important: it governs expectations to the future which is key in stability analysis, performance evaluation, and dynamic optimization.

For mathematical models, it is well defined if a process is Markov or not, but for a real-world system, it is not trivial which variables constitute the state and can be assumed to have the Markov property. For simple physical systems, the answer is often given by the laws of physics. For a Newtonian particle moving in a potential, the position itself is not a Markov process: To predict future positions, we need also the current velocity; these two variables together form a state vector. In practical models, we may end up with both smaller and larger states spaces. The potential may evolve over time and we need one or more states to describe this. On the other hand, if the particle is small and embedded in a fluid, we may be uninterested in the small-scale velocity fluctuations, so we approximate the position with a Markov process and disregard the velocity.

For complex systems (a human body seen as a physiological system, an ocean seen as an ecosystem, or a nation seen as an economic system), it is less clear how many and which state variables are needed, and the answer typically requires some simplification which is justified by the limits in the intended use of the model.

One great advantage of the Markov property, and state space models in greater generality, is that a large number of analysis questions can be reduced to analysis on state space, rather than sample space. For Markov chains with a finite number of states, this means that computations are done in terms of vectors, representing probability distributions or functions on state space, and linear operators on these, i.e., matrices. For stochastic differential equations and diffusion processes, the main computational tool is partial differential equations which govern probability densities and functions on state space. Applied mathematicians have developed a large suite of methods for partial differential equations of advection-diffusion type - both analytical and numerical - and we can now put all this to work to study stochastic differential equations.

The solutions to Kolmogorov's equation can only be written up explicitly in special cases, even if these special cases are over-represented in applications due to their tractability. Approximate methods and numerical methods are important; in the notes following this chapter we will discuss numerics. There, we will also discuss boundaries and boundary conditions.

9.11   NOTES AND REFERENCES

We refer to the two key equations in this chapter as the "backward and forward Kolmogorov equations" due to work of Kolmogorov (1931), who called them the first and second differential equations. He considered only scalar diffusion processes, but also Markov processes on finite and countable state spaces. The forward equation was previously derived by Adrian Fokker in 1914 and by Max Planck in 1917, so it is also referred to as the Fokker-Planck equation. Kolmogorov (1937) considered time reversibility and found, among other results, that particles diffusing stationarily in potentials have time-reversible trajectories (Example 9.9.1).

9.11.1   Do the Transition Probabilities Admit Densities?

The transition probabilities may be degenerate in the sense that all probability is concentrated on a point, a line, or in general on a manifold in state space with (Lebesgue) measure 0; this manifold could evolve in time. Then, the transition probabilities do not admit densities. A trivial example is when $g(x) \equiv 0$; a more interesting one is Brownian motion on the n-dimensional sphere (Exercise 7.24): The sphere with radius $r > 0$, $S = \{\xi : |\xi| = r\}$, is invariant for all $r > 0$ so if the initial condition $X_0 = x$ is on the sphere, then $\mathbf{P}^{X_0=x}\{X_t \in S\} = 1$ for any $t > 0$, despite S having Lebesgue measure 0.

A *sufficient* condition for transition densities to exist is that $g(x)g^\top(x) > 0$ at each x; i.e., diffusion acts in all directions. But this condition is not necessary. An example is the system of two linear equations

$$dX_t = V_t \, dt, \quad dV_t = dB_t,$$

i.e., the position of a particle the velocity of which is Brownian motion. Here, diffusion is singular since it acts only in the v-direction. Nevertheless, (X_t, V_t) follows a bivariate Gaussian distributed for all $t > 0$ with non-singular covariance matrix, and so admits a density.

For linear time-invariant systems, the transition probabilities are Gaussian, and therefore a density will exist if and only if the covariance matrix is non-singular. We have the following result:

Theorem 9.11.1 *Consider the linear stochastic differential equation*

$$dX_t = AX_t \, dt + G \, dB_t, \quad X_0 = x \in \mathbf{R}^n.$$

Let $\Sigma(t)$ be the covariance matrix of X_t. Then the following are equivalent:

1. *There exists a time $t > 0$ such that $\Sigma(t) > 0$.*

2. *For all times $t > 0$ it holds that $\Sigma(t) > 0$.*

3. *For all left eigenvectors $p \neq 0$ of A, $pG \neq 0$.*

4. *The "controllability matrix" $[G, AG, A^2G, \dots, A^{n-1}G]$ has full row rank.*

If these hold, we say that the pair (A, G) is controllable.

See e.g. (Zhou et al., 1996) for the proof, the idea of which is as follows: If a left eigenvector $p \neq 0$ exists such that $pG = 0$, then the projection $Y_t = pX_t$ satisfies a deterministic differential equation, so that the covariance matrix $\Sigma(t)$ must be singular for all t. Also, in this case $p[G, AG, A^2G, \dots, A^{n-1}G] = 0$, so that the controllability cannot have full row rank. Finally, the image of the controllability matrix equals the image of $\{\exp(At)G : t \geq 0\}$ (due to the Taylor expansion of $\exp(At)$ and the Cayley–Hamilton theorem that matrix A satisfies its own characteristic equation). The integral form (5.20) for the covariance $\Sigma(t)$ then establishes the connection between full rank of the controllability matrix and of the covariance matrix.

The conclusion from this theorem is that the transition densities exist if and only if the pair (A, G) is controllable, which is a weaker requirement than the diffusivity $D = \frac{1}{2}GG^T$ being non-singular.

In the nonlinear case, with $X_t \in \mathbf{R}^n$ and $B_t \in \mathbf{R}^m$,

$$dX_t = f(X_t)\, dt + \sum_{i=1}^{m} g_i(X_t)\, dB_t^{(i)}$$

there is the complication that the different noise channels can interact to perturb the state in other directions than those spun by $\{g_i(X_t)\}$, even without drift. This phenomenon is also considered by the geometric approach to non-linear control theory, which in turn builds on differential geometry: Two noise channels g_1 and g_2 can perturb the state not only in the directions g_1 and g_2, but also in the direction $[g_1, g_2] = (\nabla g_1)g_2 - (\nabla g_2)g_1$; here, $[\cdot, \cdot]$ is termed the *Lie bracket*. This connection between control theory and stochastic analysis manifests itself in the so-called *Support Theorem* of Stroock and Varadhan, which equates the support the distribution of X_t with those states, which are reachable in the sense of control theory. This generalizes the result for linear systems which we stated in the previous. In the linear case, those directions in state space which are affected by noise is the image of the controllability matrix; a theorem due to Hörmander generalizes this to the nonlinear case through a Lie algebra, formed by iteratively taking Lie brackets. The conclusions is that transition densities exist if this Lie algebra has full rank for each x. So-called Malliavin calculus makes these statements precise; see (Rogers and Williams, 1994b; Pavliotis, 2014) and the example of the Brownian unicycle in Exercise 8.7.

9.11.2 Eigenfunctions, Mixing, and Ergodicity

Often, when a unique invariant density ρ exists, the stationary process $\{X_t\}$ *mixes* in the sense that $\mathbf{E}^x h(X_t) \to \mathbf{E}h(X_0) = \langle \rho, h \rangle$ as $t \to \infty$, for any function h such that the expectations exists; the convergence is often even exponential. This is related to the spectrum (eigenvalues) of the Kolmogorov operators L and L^*: In this situation, these operators have a discrete spectrum (i.e., countably many eigenvalues) with a unique eigenvalue at 0, and so that the remaining eigenvalues all have negative real parts, which are bounded away from 0. Then, we can think of L as an infinite-dimensional matrix and L^* as its transpose, and we can extract useful information from the eigenvalues and the eigenfunctions.

Exercise 9.5 goes through the example of the Ornstein-Uhlenbeck process, where the eigenvalues and eigenfunctions can be found analytically. Exercise 9.7 covers the example of biased random walks on the circle, leading to the so-called von Mises distribution, where the analysis is done numerically. In these exercises, and in most applications, we focus on the first few eigenfunctions of L^*, i.e., those with the numerically smallest eigenvalues. These slow modes, which correspond to large-scale fluctuations in state space, determine the transients and fluctuations of the stationary process. For example, for the Ornstein-Uhlenbeck process, the two slowest modes describe how the expectation and variance of X_t converge to stationary values.

A number of technicalities are necessary to complete this picture; this is outside our scope, but let us briefly mention them. First, we must require that the solutions of the stochastic differential equation do not diverge to infinity, but return to a "central part" of the state space, which in Markov terminology is positively recurrent. See Section 12.9. We must also require that there are no invariant subsets of state space, which is related to whether the transition densities exist, and holds e.g. if the diffusivity is bounded away from 0 (Section 9.11.1). Next, the operator L is unbounded because short waves are attenuated very quickly; we address by considering the so-called *resolvent* operator $R_\alpha = (\alpha - L)^{-1}$ for $\alpha > 0$ which has the same eigenfunctions as L, but bounded eigenvalues.

Now, assume that a unique stationary distribution exists and has density $\rho(\cdot)$, and consider a function $h(\cdot)$ on state space for which $\mathbf{E}|h(X_0)|^2 < \infty$. This defines a second-order stationary process $\{Y_t = h(X_t)\}$, i.e., the expectation $\mathbf{E}Y_t = \mu_Y$ is constant and the autocovariance $r_Y(\cdot)$ depends only on the time lag (Section 5.4):

$$\mathbf{E}(Y_t - \mu_Y)Y_s = \mathbf{E}(Y_{t-s} - \mu_Y)Y_0 = r_Y(t - s) \text{ for } 0 \leq s \leq t.$$

Exercise 9.6 demonstrates how to find the autocovariance function r_Y. Now assume exponential mixing, then the autocovariance function is bounded by a two-sided exponential:

$$|r_Y(t)| \leq Ce^{-\lambda|t|} \text{ for some } C, \lambda > 0$$

and so the variance spectrum $S_Y(\omega)$ is well defined (Section 5.5).

This has implications for the so-called *ergodic* properties of $\{Y_t\}$, i.e., whether we can estimate statistics of $\{Y_t\}$ based on a single realization of the stationary process $\{X_t\}$. To see this, consider the time average

$$\bar{h} = \frac{1}{T} \int_0^T Y_t \, dt.$$

Then $\bar{h}$ is a random variable with expectation $\mathbf{E}\bar{h} = \mu_Y$ and variance

$$\mathbf{V}\bar{h} = \frac{1}{T^2} \int_0^T \int_0^T r_Y(t - s) \, ds \, dt.$$

As $T \to \infty$, the exponential bounds on the autocovariance function imply that we can approximate the double integral:

$$\int_0^T \int_0^T r_Y(t - s) \, ds \, dt = T \, S_Y(0) + O(1).$$

Here, $S_Y(0) = \int_{-\infty}^{+\infty} r_Y(t) \, dt$ and $O(1)$ means a term which is bounded as $T \to \infty$. We conclude that the variance $\mathbf{V}\bar{h}$ vanishes as fast as $1/T$, as $T \to \infty$:

$$\mathbf{V}\bar{h} = \frac{1}{T} S_Y(0) + O(T^{-2})$$

The result which we have just outlined is contained in the ergodic theorem of von Neumann, which states that the time average $\bar{h}$ converges to the expectation $\mu_Y = \mathbf{E}h(X_t)$ in mean square. This allows us, for example, to estimate ensemble averages (i.e., expectations) from a single very long simulation.

9.11.3 Reflecting Boundaries

In some applications the diffusion process $\{X_t\}$ evolves inside a domain W and is *reflected* if it approaches the boundary ∂W. For example, consider a gas molecule in a metal container: The molecule diffuses in the interior of the container, but once it hits the metal, it bounces back into the interior. Such reflecting boundaries are most easily described in terms of the forward Kolmogorov equation governing the density ϕ, where we may impose a "no flux" boundary condition:

$$(u\phi - D\nabla\phi) \cdot n = 0 \text{ for } x \in \partial W. \tag{9.8}$$

Here, n is a normal vector to the boundary ∂W at x, directed outwards. This condition ensures that the flux of probability across the boundary is 0. In terms of the backward Kolmogorov equation governing the likelihood function ψ, the corresponding (or dual) boundary condition is

$$(D\nabla\psi)\, n = 0 \text{ for } x \in \partial W. \tag{9.9}$$

This can be shown by repeating the duality argument from Section 9.5 and including boundary terms, found by the divergence theorem:

$$\langle \phi, L\psi \rangle = \int_W \phi \left(\nabla\psi \cdot u + \nabla \cdot (D\nabla\psi) \right) \, dx$$
$$= \int_W \psi \left[-\nabla \cdot (u\phi - D\nabla\phi) \right] \, dx + \int_{\partial W} \psi(u\phi - D\nabla\phi) + \phi D\nabla\psi \, dn(x)$$
$$= \langle L^*\phi, \psi \rangle$$

provided the boundary terms vanish, which they do if ϕ satisfies the no-flux condition (9.8) and ψ satisfies the homogeneous Neumann condition (9.9).

Once we have established boundary conditions for the Kolmogorov equations, we can ask what the corresponding sample paths look like. Here, we outline this construction, skipping details. Consider first a scalar diffusion $dX_t = f(X_t) \, dt + g(X_t) \, dB_t$; we aim to impose a reflecting boundary at $x = 0$. To this end, add an extra drift term to obtain:

$$dX_t = [f(X_t) - U'(X_t/\epsilon)] \, dt + g(X_t) \, dB_t.$$

Here, $U(x)$ is a smooth potential defined for $x \neq 0$ such that $\lim_{x \to 0} U(x) = \infty$ and $U(x) = 0$ for $|x| > \epsilon$. The parameter ϵ controls the width of the "repulsion zone". For each $\epsilon > 0$ and each initial condition, there is a unique solution, and this solution never hits $x = 0$ (see Section 11.7). As $\epsilon \to 0$, the process converges to a limit which satisfies

$$dX_t = f(X_t) \, dt + dL_t + g(X_t) \, dB_t$$

where dL_t is a monotone process, which increases when the process hits the boundary from the right, and which is constant on time intervals during which the process does not hit the boundary. Think of L_t as the cumulated repulsion force that acts on the particle in the interval $[0, t]$; $dL_t \approx -U'(X_t/\epsilon) \, dt$. In terms of Euler-Maruyama simulation, a simple approach is to simulate sample paths using reflection as follows (for the case $X_t > 0$):

$$X_{t+h} = |X_t + f(X_t)h + g(X_t)(B_{t+h} - B_t)|.$$

In summary, it is easy to include reflecting boundaries in the Kolmogorov equations. Next, it is possible but more difficult to modify the Itô stochastic differential equation to include reflection at the boundary, while it is straightforward to simulate reflected sample paths.

Example 9.11.1 (Reflected Brownian Motion with Drift) *Consider Brok wnian motion with negative drift, starting at $X_0 > 0$, and reflected at $x = 0$:*

$$dX_t = -u \, dt + dL_t + \sigma \, dB_t$$

with $u > 0$; define $D = \frac{1}{2}\sigma^2$. The forward Kolmogorov equation is $\dot\phi = u\phi' + D\phi''$ with boundary condition $u\phi(0) + D\phi'(0) = 0$. The stationary p.d.f. is $\phi(x) = Z^{-1} \exp(-ux/D)$ with $Z = D/u$.

The analysis extends relatively straightforward to higher dimensions, provided that the diffusivity matrix $D(x)$ is nonsingular at the boundary point x and that the normal vector n is an eigenvector of $D(x)$. Without these assumptions, both modeling, analysis, and simulation must be done with care.

Example 9.11.2 (Reflection with a Position/Velocity System) *Consider the coupled equations*

$$dX_t = V_t \, dt, \quad dV_t = -\lambda V_t \, dt + s \, dB_t$$

and aim to reflect X_t at $x = 0$. If this is a model of a gas molecule in a container, then the most obvious way to model this reflection is by including a short-range repulsive potential, which modifies the equation governing $\{V_t\}$ to

$$dV_t = -[\lambda V_t + U'(X_t/\epsilon)] \, dt + s \, dB_t.$$

When $\epsilon \to 0$, this corresponds to the particle colliding elastically with the boundary at $x = 0$. At this collision, the particle instantaneously changes its velocity from V_t to $-V_t$. Since this implies that $\{V_t\}$ is discontinuous, the process $\{(X_t, V_t)\}$ is an Itô diffusion for $\epsilon > 0$ but the limit as $\epsilon \to 0$ is not an Itô process. However, it is still a Markov process, the transition probabilities of which are governed by the forward Kolmogorov equation with the boundary condition

$$\phi(0, -v, t) = \phi(0, v, t).$$

9.11.4 Girsanov's Theorem

We now ask how the law of an Itô diffusion $\{X_t\}$ changes when we change the drift term in the governing stochastic differential equation. The answer is given by a theorem due to Girsanov, which we now derive in a simplified version. Let $T > 0$ and consider a scalar Itô diffusion $\{X_t : 0 \leq t \leq T\}$ given by the driftless Itô equation

$$dX_t = g(X_t, t)\, dB_t, \quad X_0 = x,$$

where $g \neq 0$; $\{B_t\}$ is Brownian motion w.r.t. $\{\mathcal{F}_t\}$ and $\mathbf{P}$. We aim to compare the law of this $\{X_t\}$ with that arising from the same equation with drift, but with $\{B_t\}$ replaced by a new process $\{W_t\}$:

$$dX_t = f(X_t, t)\, dt + g(X_t, t)\, dW_t, \quad X_0 = x. \tag{9.10}$$

We do not change the random variables X_t as functions on sample space, but we change their distribution by changing the probability measure. Specifically, we aim to construct a new probability measure $\mathbf{Q}$ on the same sample space, so that $\{W_t\}$ defined by (9.10) is Brownian motion w.r.t. $\{\mathcal{F}_t\}$ and $\mathbf{Q}$. As we will see, this can be done with a process $\{L_t\}$ which we call a likelihood:

$$L_t = \exp\left(\int_0^t \frac{f(X_s, s)}{g(X_s, s)}\, dB_s - \frac{1}{2} \int_0^t \frac{f^2(X_s, s)}{g^2(X_s, s)}\, ds \right).$$

This process can be written as an Itô integral (compare Exercise 7.25) so it is a local martingale; we assume that it is in fact a martingale. We can then define a probability measure $\mathbf{Q}$ on $(\Omega, \mathcal{F})$ by

$$\mathbf{Q}(A) = \int_A L_T(\omega)\, d\mathbf{P}(\omega) \text{ for } A \in \mathcal{F}. \tag{9.11}$$

Note that $\mathbf{Q}(\Omega) = 1$ since $\{L_t\}$ is a martingale. This probability measure $\mathbf{Q}$ has the same null sets as $\mathbf{P}$, and the random variable L_T can be thought of as the density of $\mathbf{Q}$ w.r.t. $\mathbf{P}$; we call L_T the Radon-Nikodyn derivative of $\mathbf{Q}$ w.r.t. $\mathbf{P}$, $L_T = d\mathbf{Q}/d\mathbf{P}$. For (conditional) expectation of a random variable Z w.r.t. $\mathbf{Q}$, we write

$$\mathbf{Q}Z = \mathbf{E}\{L_T Z\}, \quad \mathbf{Q}\{Z|\mathcal{G}\} = \mathbf{E}\{L_T Z|\mathcal{G}\},$$

while $\mathbf{E}$ still denotes expectation w.r.t. $\mathbf{P}$.

Now, the process $\{W_t\}$ given by (9.10) is a martingale w.r.t. $\mathbf{Q}$: Let $0 \leq s \leq t \leq T$, then the martingale property of $\{L_t\}$ and the tower property of conditional expectations show that $\mathbf{Q}\{W_t|\mathcal{F}_s\} = \mathbf{E}\{W_t L_t|\mathcal{F}_s\}$. In turn, Itô's lemma shows that $\{W_t L_t\}$ is an Itô integral w.r.t. $\{B_t\}$ (which we assume is a martingale). Therefore, $\{W_t\}$ is a $\mathbf{Q}$-martingale.

Next, the quadratic variation of $\{W_t\}$ is $[W]_t = t$ almost surely w.r.t. $\mathbf{P}$ and therefore also w.r.t. $\mathbf{Q}$, so Lévy's characterization (Exercise 4.8) tells us that

$\{W_t\}$ is Brownian motion on $(\Omega, \mathcal{F}, \{\mathcal{F}_t\}, \mathbf{Q})$. This is Girsanov's theorem. We have therefore succeeded in changing the measure to $\mathbf{Q}$ so as to include drift: The process $\{X_t\}$ satisfies the Itô equation (9.10) where $\{W_t\}$ is Brownian motion w.r.t. $\mathbf{Q}$.

Note that L_T can be computed from the sample path of $\{X_t\}$:

$$L_t = \exp\left(\int_0^t \frac{f(X_s, s)}{g^2(X_s, s)}\, dX_s - \frac{1}{2}\int_0^t \frac{f^2(X_s, s)}{g^2(X_s, s)}\, ds\right). \tag{9.12}$$

Girsanov's theorem appears in many different and more general forms; see (Øksendal, 2010). We now sketch a number of applications of this result:

Model Analysis: Exercise 9.14 shows how to compute the hitting time distributions for Brownian motion with drift, by exploiting that we know these distributions for Brownian motion without drift (Section 4.3) and combining with Girsanov's theorem. So Girsanov's theorem lets us solve a "tricky" analysis problem by transforming it to a simpler model, where the answer is available, and then map the result back to the original.

Likelihood Estimation of the Drift: If the drift function f is unknown but assumed to belong to some family Φ, and we have observed a state trajectory $\{X_t : 0 \leq t \leq T\}$, then we can estimate f by maximizing the likelihood $d\mathbf{Q}/d\mathbf{P}$ w.r.t. f over Φ using (9.12). If the family Φ is a linear space, then this maximization problem is a weighted least squares problem. Exercise 9.15 gives an example.

Importance Sampling: The context is here to use Monte Carlo to estimate the probability $\mathbf{Q}(A)$ of some rare event A; e.g., a transition in a double well system. Direct Monte Carlo is ineffective, because it would take many realizations to observe the event A. Then, we can simulate trajectories from a process with a different drift, chosen to make the event A more probable, and estimating $\mathbf{Q}(A)$ using weights as in (9.11). See, for example, (Milstein and Tretyakov, 2004).

Mathematical Finance: Here, we operate with two probability measures as explained in Section 3.9.1: The "real-world" measure which governs observed dynamics in the market, and the "risk-free" measure which governs the pricing of assets. Girsanov's theorem allows us to convert between these two measures.

9.11.5 Numerical Computation of Transition Probabilities

The transition probabilities can only be found analytically in special cases. In up to three dimensions, say, we may compute the transition probabilities numerically by discretizing the Kolmogorov equations. There are many software packages for numerical analysis of PDEs, ranging from commercial industrial-grade packages (e.g., Comsol, Ansys, Fluent) to community-driven open source projects (ReacTrans for R; FEniCS for python). Many packages and methods are designed for problems in fluid mechanics or thermodynamics, but can be applied to stochastic differential equations due to the similarity of the governing equations. Yet, for the newcomer, progress can be slow because these software packages typically have steep learning curves and are designed for different applications. While the vast topic of numerical analysis of partial differential equations largely is outside our scope, we provide here simple numerical methods for simple problems, focusing on the one-dimensional case. The methods are implemented in the toolbox SDEtools; in particular the functions fvade for scalar problems and fvade2d for the two-dimensional case.

It is convenient to first keep time continuous while space is discretized. This is called the *method of lines*. Thus, the forward Kolmogorov equation governing the evolution of the probability density function ϕ is discretized as a system of ordinary differential equations:

$$\dot{\phi} = L^*\phi \quad \mapsto \quad \dot{\bar{\phi}} = \bar{\phi}G. \tag{9.13}$$

From now on we focus on the operator L^* and its discretization G. We therefore consider the time t fixed and omit it, in order to obtain a simpler notation. The continuous object $\phi(\cdot)$, which is a smooth function defined on state space, is replaced by a vector $\bar{\phi} \in \mathbf{R}^N$. The linear partial differential operator L^*, in turn, is replaced by a matrix $G \in \mathbf{R}^{N \times N}$. The corresponding discretization of the backward Kolmogorov equation is

$$-\dot{\psi} = L\psi \quad \mapsto \quad -\dot{\bar{\psi}} = G\bar{\psi}. \tag{9.14}$$

Note that we view the discretization of the probability density function $\phi(\cdot)$ as a row vector $\bar{\phi}$ while the discretization of the likelihood function $\psi(\cdot)$ is a column vector $\bar{\psi}$. This convention reflects that ϕ and ψ are dual objects, as are $\bar{\phi}$ and $\bar{\psi}$. Since L^* and L are adjoint operators, it is the *same* matrix G that appears in the discrete version of two equations; the backward equation is obtained by multiplying $\bar{\psi}$ on G from the right, while the forward equation is obtained by multiplying $\bar{\phi}$ on G from the left.

In the following we focus on a discretization technique which ensures that G is the generator of a continuous-time Markov chain with state space $\{1, \ldots, N\}$. This process approximates the original diffusion process, and the equations in (9.13) and (9.14) are the forward and backward Kolmogorov equations for the original diffusion process and for its approximating Markov chain, respectively.

The discretization method is a finite volume method which uses a second order central scheme for the diffusion and a first order upwind scheme for the advection. We present the scalar case; see (Versteeg and Malalasekera, 1995) for an elaborate treatment and generalizations to higher dimensions. We truncate the real axis so that the computational domain is $[a, b]$ and partition the domain into N grid cells $\{I_i : i = 1, \ldots, N\}$. We let x_i be the center point in each grid cell and we use $x_{i-1/2}$ and $x_{i+1/2}$ for the left and right interfaces; thus $x_i = (x_{i-1/2} + x_{i+1/2})/2$ and $|I_i| = x_{i+1/2} - x_{i-1/2}$. Now, consider the forward Kolmogorov equation in advection diffusion form:

$$\dot{\phi} = L^*\phi = -\nabla \cdot (u\phi - D\nabla).$$

The discretized version considers only the total probability in each grid cell, so that the i'th element of $\bar{\phi}$ is

$$\bar{\phi}_i \approx \int_{x_{i-1/2}}^{x_{i+1/2}} \phi(x) \; dx.$$

The diffusive fluxes between cells i and $i+1$ is approximated by evaluating the diffusivity $D(x)$ at the interface, and assume that the concentration profile is linear over the two cells i and $i + 1$. We get

$$J_D(x_{i+1/2}) = -D(x_{i+1/2})\phi'(x_{i+1/2}) \approx -D(x_{i+1/2})\frac{\bar{\phi}_{i+1}/|I_{i+1}| - \bar{\phi}_i/|I_i|}{x_{i+1} - x_i}.$$

Now assume that the state is in cell i so that $\bar{\phi}_i = 1$, $\bar{\phi}_{i+1} = 0$. For the approximating Markov chain, this diffusive flux corresponds to a jump rate $G^D_{i(i+1)}$, which must therefore equal

$$G^D_{i(i+1)} = \frac{D(x_{i+1/2})}{|I_i|(x_{i+1} - x_i)}.$$

Similarly, diffusion gives rise to a jump rate from cell $i + 1$ to cell i, which is

$$G^D_{(i+1)i} = \frac{D(x_{i+1/2})}{|I_{i+1}|(x_{i+1} - x_i)}.$$

The diffusion operator $\phi \mapsto \nabla \cdot (D\nabla\phi)$ is therefore discretized as a matrix G^D which is tridiagonal and has these elements in the first off-diagonals.

Next, we approximate the advective flux between cells I_i and I_{i+1} as

$$u(x_{i+1/2})\phi(x_{i+1/2}) \approx \begin{cases} u(x_{i+1/2})\bar{\phi}_i/|I_i| & \text{when } u(x_{i+1/2}) > 0, \\ u(x_{i+1/2})\bar{\phi}_{i+1}/|I_{i+1}| & \text{when } u(x_{i+1/2}) < 0. \end{cases}$$

Note that we approximate the probability density at the interface with the average density in the cell where the flow comes from. This is the *upwind* principle. These fluxes correspond to jump rates:

$$G^A_{i(i+1)} = \begin{cases} u(x_{i+1/2})/|I_i| & \text{when } u(x_{i+1/2}) > 0, \\ 0 & \text{else,} \end{cases}$$

and

$$G^A_{(i+1)i} = \begin{cases} -u(x_{i+1/2})/|I_{i+1}| & \text{when } u(x_{i+1/2}) < 0, \\ 0 & \text{else.} \end{cases}$$

Here, G^A is the matrix representation of the advective operator $\phi \mapsto -\nabla \cdot (u\phi)$. To take both advection and diffusion into account, we simply add the two generators, i.e., $G = G^A + G^D$. So far we haven't discussed the diagonal, but we know that it is negative and ensures that row sums are 0:

$$G_{ii} = -G_{i(i+1)} - G_{i(i-1)}.$$

We have derived these formulas assuming that the interface is interior, so it remains to specify what to do at the left and right boundaries. There are three common situations:

1. *No-flux Boundaries.* Compare Section 9.11.3. We set $J_{1/2} = J_{N+1/2} = 0$, so that a particle reflects when hitting the boundary as in Section 9.11.3. In this case, the particle can only leave cell 1 to go to cell 2, and the rate with which this happens is as before.

2. *Periodic Boundaries.* We identify the left boundary point $x = x_{1-1/2}$ with the right one, $x = x_{N+1/2}$. Thus, we set $J(x_{1/2}) = J(x_{N+1/2})$ so what moves out of the left exterior boundary enters through the right. We compute this flux according to the same formulas as for interior cells. This leads to jump rates between cells 1 and N as in the previous. Specifically, the diffusive jump rate from cell 1 to N is

$$G_{1N}^D = \frac{D(x_{1/2})}{|I_1|(x_1 - x_{1/2} + x_{N+1/2} - x_N)}.$$

and the advective jump rate from cell 1 to N is,

$$G_{1N}^A = -u(x_{1/2})/|I_1| \text{ if } u(x_{1/2}) < 0 \text{ and } 0 \text{ else.}$$

with a similar expression for G_{N1}^A.

3. *Absorbing Boundaries.* We compute the fluxes $J_{1/2}$ and $J_{N+1/2}$ using the same formulas as for interior cells, and assume that there are two absorbing boundary cells, one to the left of cell 1 and one to the right of cell N. Here, an absorbing cell is one which is never left, so that the corresponding row in the generator G is all zeros. These boundary cells can be included, in which case we keep track of the probability of absorption at each boundary, or omitted, in which case we only keep track of the probability of the interior, which is then not conserved. In the terminology of partial differential equations, absorbing boundaries correspond to a homogeneous Dirichlet condition for the forward Kolmogorov equation and a (possible inhomogeneous) Dirichlet condition for the backward Kolmogorov equation. We discuss problems that lead to absorption further in Chapter 11.

9.11.5.1   *Solution of the Discretized Equations*

With the matrix G in hand, which discretizes the operators L^* and L, it remains to solve the ODE systems (9.13) and (9.14).

For time invariant systems with not too many grid cells (say, up to a few hundred), we can use the matrix exponential. For the forward equation:

$$\phi_t = \phi_0 e^{Gt}.$$

When G changes with time, the matrix exponential does not apply (except when the matrices $\{G_t : t \geq 0\}$ commute). Then, and for large computational grids, time stepping can be used. Standard methods apply, for example, Runge-Kutta schemes as implemented in ode for R. However, the structure of the equations should be preserved, i.e., they are sparse, stable (all eigenvalues have non-positive real part), positive ($\exp(Gt)$ has all non-negative elements; often even positive for $t > 0$), and conservative (row sums of G are zero, except with absorbing boundaries). Moreover, the equations are typically stiff, i.e., some eigenvalues have very negative real parts, corresponding to short spatial fluctuations disappearing fast. This suggests implicit methods, such as the implicit Euler method:

$$\phi_t(I - Gt) = \phi_0$$

which is solved for ϕ_t. Solvers may exploit that $I - Gt$ is sparse; often even tridiagonal. The implicit Euler method is unconditionally stable, conserves probability, and is positive in the sense that all elements of ϕ_t are non-negative whenever those of ϕ_0 are.

A common task is to identify the stationary distribution, i.e., find ϕ such that $\phi G = 0$ while $\phi \mathbf{e} = 1$ where $\mathbf{e}$ is a column vector of all ones. We expand the system with an extra scalar unknown, δ, and solve the system

$$[\phi \; \delta] \begin{bmatrix} G & \mathbf{e} \\ \mathbf{e}^\top & 0 \end{bmatrix} = [\mathbf{0} \; 1]$$

for $[\phi \; \delta]$. If the diffusivity is bounded away from 0, G is the generator of an *irreducible* Markov chain (i.e., any state can be reached from any other state), and then this system is non-singular and uniquely defines ϕ.

The eigenvalues of G gives useful information about how fast the distribution converges to the stationary one (compare Section 9.11.2). To compute eigenvalues except for small grids, it is worthwhile to use a sparse eigenvalue solver (e.g. RSpectra for R), and only compute the slowest modes, as it saves computations and the fast modes are sensitive to the discretization. See Exercises 9.5 and 9.7 for examples.

9.12 EXERCISES

Exercise 9.3 Stationary Moments in Logistic Growth: Sometimes we can find the stationary moments without finding the entire stationary distribution. To do this, note that if $\{X_t\}$ is stationary, then $\mathbf{E}Lh(X_t) = 0$ for any test function h such that the expectation exists. Now, consider stochastic logistic growth (Example 9.8.2) and assume stationarity.

1. Take as test function $h(x) = \log x$. Show that $\mathbf{E}X_t = K(1 - \sigma^2/(2r))$.

2. Take as test function $h(x) = x$. Show that $\mathbf{E}X_t^2 = K\mathbf{E}X_t$ so that $\mathbf{V}X_t = K^2(1 - \Sigma)\Sigma$ where $\Sigma = \sigma^2/(2r)$.

3. Show that, for given K, the variance is no greater than $K^2/4$, and that this is attained with $\sigma^2 = r$. Explain why maximum variance is obtained with intermediate noise intensity.

Exercise 9.4 Black-Scholes Option Pricing: As in example 9.4.3, let the price of a stock evolve according to $dX_s = X_s(r\ ds + \sigma\ dB_s)$ and let

$$k(s, x) = \mathbf{E}^{X_s = x}(X_t - K)^+$$

be the undiscounted fair price at time s of an option to buy the stock at time t for a price of K, if the stock price is $X_s = x$. Show that

$$k(0, x) = e^{rt} x \Phi(d + \sigma\sqrt{t}) - K\Phi(d)$$

where

$$d = \frac{\log(x/K) + (r - \frac{1}{2}\sigma^2)t}{\sigma\sqrt{t}},$$

so that the discounted price is

$$e^{-rt} k(0, x) = x\Phi(d + \sigma\sqrt{t}) - e^{-rt} K\Phi(d).$$

Hint: Either use the backward Kolmogorov equation - in that case, symbolic software like `Maple` is useful - or use that the transition probabilities of $\{X_t\}$ are known. Finally, graph the discounted fair price as a function of $t \in [0, T]$ and $x \in [0, 2K]$, for parameters $r = 0.05$, $\sigma = 0.02$, $K = 1$, $T = 10$.

Exercise 9.5 Eigenmodes in the O-U Process: Consider the (rescaled) Ornstein-Uhlenbeck process

$$dX_t = -X_t\ dt + \sqrt{2}\ dB_t.$$

1. Write up the forward and backward Kolmogorov operators.

2. Verify that the functions

$$H_0(x) = 1, \quad H_1(x) = x, \quad H_2(x) = x^2 - 1, \quad H_3(x) = x^3 - 3x$$

are eigenfunctions for the backward Kolmogorov operators and compute the corresponding eigenvalues. *Note:* These are the first four of the (probabilistic) Hermite polynomials.

3. For the case of H_1 and H_2, show that these eigenmodes are consistent with what we know from Section 7.4.1 about the evolution of the mean $\mathbf{E}^x X_t$ and the variance $\mathbf{V}^x X_t$ for the Ornstein-Uhlenbeck process.

4. Show that if a stationary density ρ for an Itô diffusion satisfying detailed balance and ψ is an eigenfunction of the backward Kolmogorov operator, then $\phi = \rho\psi$ is an eigenfunction of the forward Kolmogorov operator. *Hint:* In Section 9.9, we showed that in this case, L is self-adjoint under the inner product $\langle h, k \rangle_\rho = \int h\rho k\ dx$.

5. Compute and plot the corresponding eigenfunctions for the forward Kolmogorov operator, for the Ornstein-Uhlenbeck process.

6. Compare the analytical results with a numerical computation of eigenvalues and -vectors, discretizing the interval $x \in [-4, 4]$ with e.g. 800 grid cells and imposing reflection at the boundaries.

Exercise 9.6 The Autocovariance Function: Let $\{X_t\}$ be a stationary solution to a stochastic differential equation with generator L. Let ρ be the stationary density, and let $h(\cdot)$ be a scalar function on state space such that $\mathbf{E}h^2(X_t) < \infty$. Show that

$$\mathbf{E}[h(X_0)h(X_t)] = \int_{\mathbf{X}} \rho(x)h(x)[e^{Lt}h](x) \; dx.$$

This is the autocovariance function of $\{h(X_t)\}$ if $\mathbf{E}h(X_0) = 0$; otherwise, we first remove the bias. Next, verify that this formula is consistent with what we already know about the Ornstein-Uhlenbeck process $(Lh = -\lambda x h' + D h'')$ when $h(x) = x$. Finally, how does this autocovariance function look if h is a (real) eigenfunction of L? *Note:* The next exercise includes a numerical analysis of this formula.

Exercise 9.7 Random Directions and the von Mises Distribution: Consider the Itô stochastic differential equation

$$dX_t = -\sin X_t \; dt + \sigma \; dB_t,$$

which can be viewed as a random walk on the circle which is biased towards $X_t = 2n\pi$ for $n \in \mathbf{N}$. This is a popular model for random reorientations, when there is a preferred direction.

1. Write up the forward Kolmogorov equation and show that a stationary solution is the so-called *von Mises* (or Tikhonov) distribution $\rho(x) = Z^{-1} \exp(\kappa \cos x)$, where $\kappa = 2/\sigma^2$. Here, Z is a normalization constant so that ρ integrates to 1 over an interval of length 2π; i.e., we consider the state X_t an angle which is only well defined up to adding a multiple of 2π.

2. Take $\sigma = 1$. Simulate the process $\{X_t\}$ starting at $x = 0$ over the time interval $t \in [0, 100]$. Plot the trajectory and the histogram of the state. Compare the histogram with the stationary distribution.

3. Discretize the generator on $x \in [-\pi, \pi]$ using periodic boundary conditions. Determine the stationary distribution numerically from the generator and compare it, graphically, with the results from the previous question. *Note:* Unless the spatial grid is very fine, some numerical diffusion stemming from the discretization will affect the numerical solution.

4. Estimate the autocovariance function of $\{\sin X_t\}$ from the time series (using a built-in routine in your favorite software environment) and plot it. Use a sufficient large number of lags until you can see how long it takes for the process to decorrelate.

5. Compute the aucovariance function numerically from the following formula (compare Exercise 9.6):

$$\mathbf{E}[(h(X_0) - \mu)h(X_t)] = \int_{\mathbf{X}} \rho(x)(h(x) - \mu)[e^{Lt}h](x)\ dx.$$

Here, we take $h = \sin$ and define $\mu = \mathbf{E}h(X_0) = \int_{\mathbf{X}} \rho(x)h(x)\ dx$. Add this autocovariance to the empirical plot from the previous.

6. Compute the slowest 3 eigenmodes of L from the numerical discretization. Add to the plot of the autocovariance an exponentially decaying function $e^{-\lambda t}\mathbf{V}\sin X_0$ where λ is the largest non-zero eigenvalue of L. Comment on the agreement. Then, in a different plot, plot the slowest 3 eigenfunctions of L as well as of L^* and describe their role.

Exercise 9.8 The Cox-Ingersoll-Ross Process: Consider (again) the Cox-Ingersoll-Ross process, given by the Itô equation

$$dY_t = \lambda(\xi - Y_t)\ dt + \gamma\sqrt{Y_t}\ dW_t$$

where $\{W_t\}$ is Brownian motion.

1. Write up the forward Kolmogorov equation for the process.

2. Using the interpretation that X_t can be written as the sum-of-squares of Ornstein-Uhlenbeck processes (Section 7.5.2) for certain parameter combinations, argue verbally that the transition distribution of X_t is a re-scaled non-central chi-squared distribution.

3. Make the argument precise and find the transition densities in CIR process. Specifically, show that given $Y_0 = y_0$, Y_t has density at y

$$\mathbf{P}(Y_t \in dy|Y_0 = y_0) = ce^{(-2cy+\nu)/2}\left(\frac{2cy}{\nu}\right)^{n/4-1/2}I_{n/2-1}(\sqrt{2c\nu y})\ dy$$

where I is a modified Bessel function of the first kind, and

$$c = \frac{2\lambda}{\gamma^2(1 - e^{-\lambda t})}, \quad \nu = 2cy_0e^{-\lambda t}.$$

Exercise 9.9 Monte Carlo Solution of Advection-Diffusion Equations: Consider the advection-diffusion equation governing $C(x,t)$ with $x \in \mathbf{R}^d$

$$\dot{C} = -\nabla \cdot (uC - D\nabla C)$$

where $u = u(x,t)$ and $D = D(x,t)$. The initial condition $C(\cdot, 0)$ is given. Consider next the following Monte Carlo method for solving this equation: Sample the trajectory of a single particle, i.e., the stochastic process $\{X_t : t \geq 0\}$ for $i = 1, \ldots, N\}$. The initial condition X_0 is sampled from $C(\cdot, 0)$, and the process $\{X_t\}$ satisfies

$$dX_t = f(X_t) \, dt + g(X_t) \, dB_t$$

where $\{B_t\}$ are independent Brownian motions. Find f and g such that the probability density of each X_t equals $C(\cdot, t)$. Then write the Euler-Maruyama scheme for each of these SDEs.

Exercise 9.10 The Density of a Time-Changed Process: Let $\{X_t\}$ be a stationary Itô diffusion satisfying $dX_t = f(X_t) \, dt + g(X_t) \, dB_t$ with density $\phi(\cdot)$. Let $\{Y_u : u \geq 0\}$ be a time-changed process as in Section 7.7, $Y_u = X_{T_u}$ where $dT_u = h^{-1}(Y_u) \, du$; we assume that h is bounded away from 0 and above. Show that an un-normalized stationary density of Y_u is $\phi(y) \, h(y)$.

Exercise 9.11 Detailed Balance in Linear Systems: Consider the stationary distribution of the narrow-sense linear system $dX_t = AX_t \, dt + G \, dB_t$ in n dimensions. Show that this distribution satisfies detailed balance iff $A\Sigma = \Sigma A^\top$, where Σ is the stationary variance. Assume $\Sigma > 0$. Give an example of a linear system where this is satisfied, and one where it is not.

Exercise 9.12 The Backward Equation for Linear Systems: Consider again the linear system $dX_t = AX_t \, dt + G \, dB_t$.

1. Write up the generator L for this process - first, for a general function $h(x)$ on state space, and next, for a constant-quadratic function $h(x) = x^\top Q x + q$, where $Q = Q^\top$.

2. Consider the backward Kolmogorov $\dot{k} + Lk = 0$ with terminal condition $k(t,x) = h(x)$ where again $h(x) = x^\top Q x + q$. Guess that the solution is also constant-quadratic, $k(t,x) = x^\top P_t x + p_t$, and write up ordinary differential equations for the coefficients P_t and p_t

The point to notice from this exercise is that for linear systems, the family of constant-quadratic functions is closed under the backward operator, so that evaluating expectations of quadratic functions can be done with matrix algebra (or analysis) in stead of partial differential equations.

Exercise 9.13 Densities of the Brownian Unicycle: Consider again the unicycle in Exercise 8.7.

1. Confirm that at each point $z = (\theta, x, y)$ in state space, the three vectors $g_1(z)$, $g_2(z)$ and $[g_1, g_2](z)$ together span $\mathbf{R}^3$. Here, $[g_1, g_2] = (\nabla g_1) g_2 - (\nabla g_2) g_1$ is the Lie bracket.

2. This tells us that the transition densities exist. Conduct a numerical experiment, simulate 1000 sample paths starting at $z = 0$ on the time interval $[0, 0.1]$ with a time step of 0.001, to verify this.

Exercise 9.14 Hitting Times for Brownian Motion with Drift: Let $X_t = B_t$ where $\{B_t\}$ is Brownian motion under $\mathbf{P}$ and define $\tau = \inf\{t : X_t \geq 1\}$; the p.d.f. of τ is given by Theorem 4.3.3. Now pose an alternative model, viz. $X_t = ut + W_t$ where $\{W_t\}$ is Brownian motion w.r.t. $\mathbf{Q}$. We pursue the distribution of τ under $\mathbf{Q}$; i.e., the hitting times distribution for Brownian motion with drift. Proceed as follows:

1. Let ω be a sample path such that $\tau = t$. Find the Radon-Nikodyn derivative $d\mathbf{Q}/d\mathbf{P}(\omega)$.

2. Verify that the p.d.f. of τ under $\mathbf{Q}$ is

$$\mathbf{Q}\{\tau \in [t, t+dt]\} = t^{-3/2}\phi(1/\sqrt{t})e^{u-u^2 t/2} \ dt.$$

3. Verify the result with stochastic simulation.

4. Verify, for example using numerical integration with a couple of values for u, that $\mathbf{Q}\{\tau < \infty\} = 1$ when $u > 0$, and that in this case, $\mathbf{Q}\tau = 1/u$. Then verify that $\mathbf{Q}\{\tau < \infty\} = \exp(1/(2u))$ when $u < 0$. (We will establish this last result in Exercise 11.7 using different methods.)

Exercise 9.15 Likelihood Estimation Using Girsanov's Theorem: We perform a simulation-reestimation experiment, using the biased random walk on the circle of in Exercise 9.7, but now with two parameters:

$$dX_t = a_s \sin(\theta_s - X_t) \ dt + \sigma \ dB_t.$$

1. Simulate a sample path from the model. Take $a_s = 1$, $\theta_s = 0$, $\sigma = 1$. Use a time step of 0.1 and simulate up to time $T = 100$.

2. We now reestimate a_s and θ_s from the time series. Using Girsanov's theorem, write down the likelihood function $d\mathbf{Q}/d\mathbf{P}(\omega) = L(a, \theta)$, where the reference $\mathbf{P}$-model is given by $dX_t = \sigma \ dB_t$ and the alternative $\mathbf{Q}$-model is given by $f(x) = a\sin(\theta - x)$.

3. Show that, for a given value of θ, the likelihood is maximized w.r.t. a by

$$\hat{a}(\theta) = \left(\int_0^T \sin^2(\theta - X_t) \ dt\right)^{-1} \int_0^T \sin(\theta - X_t) \ dX_t$$

and that the log-likelihood there is

$$\max_a \log L(a, \theta) = \frac{1}{2\sigma^2} \frac{\left(\int_0^T \sin(\theta - X_t) \ dX_t\right)^2}{\int_0^T \sin^2(\theta - X_t) \ dt}.$$

4. For 1000 values of θ in the interval $[0, \pi]$, tabulate and plot the profile log-likelihood $\max_a \log L(a, \theta)$. Identify the maximizing argument $\hat{\theta}$ and the corresponding estimate $\hat{a}(\hat{\theta})$. Compare with the true values.

III

Applications

Appendices

CHAPTER 10

State Estimation

This chapter considers the problem of inference, or estimation, in stochastic differential equations based on observed time series. There are many different variants of this problem; the most fundamental is that of *real-time state estimation:* If we have taken a sequence of measurements on the system at different points in time, how should we estimate the state of the system at the time of the last measurement? The problem of state estimation arises in many applications, for example, in decision and control: If we have to decide on an action that affects the future dynamics, we need to first estimate the current state.

In this chapter, we focus on a *recursive* solution to this problem, where we process the measurements one at a time. This solution combines a *time update*, which involves Kolmogorov's forward equation, with a *data update*, which involves Bayes' rule. These recursions are easy to write up but, in general, difficult to implement. In the special case of linear stochastic differential equations with Gaussian measurement errors, the recursions simplify to algebraic equations for the conditional mean and variances; this is the celebrated *Kalman filter*. For the nonlinear systems in one or two dimensions, brute-force numerical solution is feasible.

We will also see different but related versions of the estimation problem: In forecasting or hindcasting, we aim to estimate future or past values of the state. Also, we consider estimation of unknown parameters in the model rather than states, using likelihood methods. These problems build on and expand the theory of state estimation.

The approach of recursive filtering has many advantages, conceptually and computationally. We also present an alternative formulation, which casts the problem as a statistical *mixed-effects* problem. This approach has become feasible in recent years thanks to powerful computational methods for mixed-effects problems.

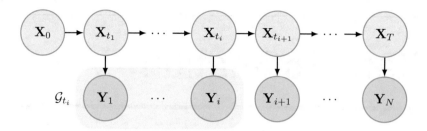

Figure 10.1 Probabilistic graphical model of the random variables in the filtering problem. The measurement equation specifies the conditional distribution of Y_i given X_{t_i}, and the stochastic differential equation specifies the conditional distribution of $X_{t_{i+1}}$ given X_{t_i}. The shaded region contains the measurements in $\mathcal{G}_{t_i}$, i.e., those which are used for the state estimate ψ_i and the state prediction ϕ_{i+1}

10.1 RECURSIVE FILTERING

We consider the following problem (see Figure 10.1): We have a diffusion process $\{X_t : t \geq 0\}$ governed by a general, possibly non-linear and time-varying, Itô equation

$$dX_t = f(X_t, t) \, dt + g(X_t, t) \, dB_t. \tag{10.1}$$

We envision an observer who, at certain points of time $0 \leq t_1 < \cdots < t_N = T$ takes measurements $Y_1, Y_2, \ldots, Y_N$. These are random variables; we assume that each measurement Y_i depends on the state X_{t_i} of the system at the time of the measurement, but is (conditionally) independent of everything else. The information available to this observer at time t is measurements taken no later than time t, i.e., the σ-algebra

$$\mathcal{G}_t = \sigma(\{Y_i : t_i \leq t\})$$

Note that $\{\mathcal{G}_t : t \geq 0\}$ is a filtration. The problem of state estimation is to determine the conditional distribution of X_t given $\mathcal{G}_t$.

A graphical model of the situation is seen in Figure 10.1. The graphic illustrates that our problem is one of inference in a Hidden Markov Model (Zucchini and MacDonald, 2009): The state process $\{X_t : t \geq 0\}$, sub-sampled at the times $\{t_0 = 0, t_1, \ldots, t_N = T\}$, constitute a discrete-time Markov process $\{X_{t_i} : i = 0, \ldots, N\}$. The states are unobserved or "hidden", but we aim to infer them indirectly from measurements $\{Y_i : i = 1, \ldots, N\}$.

State estimation in stochastic differential equations, using this Hidden Markov Model approach, involves the following steps:

1. Specification of the model: The stochastic differential equation that propagates the state, and the conditional distribution of observations given the state.

2. A recursive filter, which runs forward in time and processes one measurement at a time. This filter can run *on-line* in real time, i.e., process measurements as they become available, although it is probably most often used in off-line studies. This filter consists of a *time update*, which addresses the link between state $X_{t_{i-1}}$ and state X_{t_i} in Figure 10.1, and a *data update*, which addresses the link between state X_{t_i} and observation Y_i.

3. This filter yields *state estimates*, such as $\mathbf{E}\{X_{t_i}|\mathcal{G}_{t_i}\}$, and *state one-step predictions* $\mathbf{E}\{X_{t_{i+1}}|\mathcal{G}_{t_i}\}$. Along with these conditional means come variances and, more generally, the entire conditional distributions.

4. With the filter in hand, we are able to evaluate the *likelihood* of any unknown underlying parameters in the stochastic differential equation, or in the measurement process, and therefore estimate these parameters.

5. A second recursion improves the state estimates to include also information based on future observations, i.e., form conditional expectations such as $\mathbf{E}\{X_{t_i}|\mathcal{G}_T\}$. This is the *smoothing* filter. It is most useful in off-line studies where past values of the state are *hindcasted*.

6. Finally, one may be interested in not just the marginal conditional distribution of X_{t_i} given $\mathcal{G}_T$, but in the entire joint distribution of $\{X_t : 0 \le t \le T\}$ given $\mathcal{G}_T$. It is possible to draw samples from this distribution, i.e., to simulate *typical tracks*, and to identify the mode of the distribution, i.e., the *most probable track*.

In the following we address these steps one at a time. To make the presentation more specific, we will use a running example, namely state estimation in the Cox-Ingersoll-Ross process. Specifically, we consider the process

$$dX_t = (1 - X_t)\, dt + \sqrt{X_t}\, dB_t. \tag{10.2}$$

where we interpret X_t as the abundance of bacteria. Figure 10.2 displays the simulation of the process. In the following sections, we assume that this trajectory is unobserved but aim to estimate it from available data, which consists of bacteria counts in samples.

10.2 OBSERVATION MODELS AND THE STATE LIKELIHOOD

We now focus on the observations Y_i. These are random variables; the model specifies their distribution by means of the conditional density of Y_i given X_{t_i}, i.e., $f_{Y_i|X_{t_i}}(y|x)$. Ultimately, we aim to apply Bayes' rule to find the conditional distribution of X_{t_i} given the measurement Y_i. For example, if we take a single measurement $Y_i(\omega) = y_i$, and have no other information, then we find the conditional density of X_{t_i} as

$$f_{X_{t_i}|Y_i}(x|y_i) = \frac{1}{c} f_{X_{t_i}}(x) f_{Y_i|X_{t_i}}(y_i|x)$$

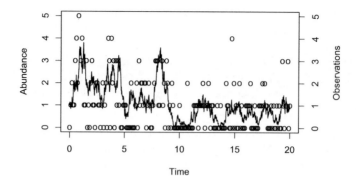

Figure 10.2 *Solid line:* Simulation of the Cox-Ingersoll-Ross process (10.2). We assume that this trajectory is not available to the observer, who only has access to the observed time series $\{Y_i\}$ (*open circles*). The objective is to estimate the state trajectory based on the measurements.

where $f_{X_{t_i}}(x)$ is the prior density of the state X_{t_i} at the point x. Here c is the normalization constant $f_{Y_i}(y_i) = \int f_{X_{t_i}}(x) f_{Y_i|X_{t_i}}(y_i|x)\, dx$. This precise formulation assumes that the joint distribution of X_{t_i} and Y_i is continuous; similar expressions exist when e.g. Y_i is discrete.

We see from this example that information in the measurement Y_i about the state X_{t_i} is quantified by the function $f_{Y_i|X_{t_i}}(y_i|x)$. We introduce the term *state likelihood function*, and the symbol $l_i(x)$, for this function:

$$l_i(x) = f_{Y_i|X_{t_i}}(y_i|x)$$

Note that to the observer, $y_i = Y_i(\omega)$ is a known quantity at time t_i and thereafter, so $l_i(x)$ is known, for each value of the state x, at time t_i. In other words, $l_i(x)$ is a $\mathcal{G}_{t_i}$-measurable random variable $\omega \mapsto f_{Y_i|X_{t_i}}(Y_i(\omega)|x)$, for each x.

Let us give a few specific examples of how $l_i(x)$ can arise. A common situation is that the measurements Y_i are noisy observations of some function of the state X_{t_i}. For example, we may have

$$Y_i = c(X_i) + s(X_i)\xi_i$$

where the function $c(x)$ specifies the quantity which is measured, and where $\{\xi_i\}$ are Gaussian measurement errors with mean 0 and variance 1, which are independent of each other and of the Brownian motion $\{B_t : t \geq 0\}$. In this situation, the state likelihood functions are

$$l_i(x) = \frac{1}{\sqrt{2\pi}s(x)} \exp\left(-\frac{1}{2}\frac{(y_i - c(x))^2}{s(x)^2}\right).$$

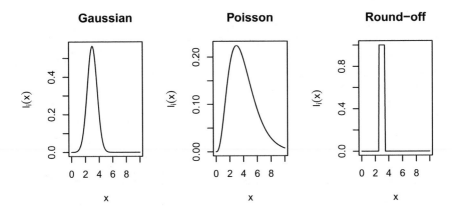

Figure 10.3 Three examples of state likelihood functions: To the left, an observation $Y = 3$ with $Y \sim N(X, 1/2)$. In the middle, an observation $Y = 3$ with $Y \sim \text{Poisson}(X)$. To the right, an observation $Y = 3$ with Y being X rounded to the nearest integer. Note that these are likelihood functions of the state X and *not* probability density functions of X.

Note that this is *not* a probability density function of X_{t_i}. For example, it does not in general integrate to 1 – although it does when $c(x) \equiv x$ and $s(x)$ is a constant function of x.

As another example, the random variable Y_i given X_{t_i} may be conditionally Poisson distributed with a mean which depends on X_{t_i}, i.e.,

$$l_i(x) = \frac{(\mu(x))^{y_i}}{y_i!} e^{-\mu(x)}$$

where the function $\mu : \mathbf{X} \mapsto \mathbf{R}$ describes how the conditional mean of Y_i depends on X_{t_i}. As yet another example, we can have observed whether X_{t_i} is in some region A of state space or not:

$$l_i(x) = \begin{cases} \mathbf{1}(x \in A) & \text{if } y_i = 1 \\ \mathbf{1}(x \notin A) & \text{if } y_i = 0 \end{cases}$$

Regardless of the specific way measurements Y_i are taken, we see that this information is condensed to a state likelihood function $l_i(x)$, which describes the information about the state X_{t_i} which is acquired at time t_i. In agreement with Figure 10.1 we also assume likelihood independence, i.e., the joint likelihood of all unobserved states X_{t_i} is the product of the l_i's.

For our example of the Cox-Ingersoll-Ross process (10.2), we assume that the state is measured at regular time intervals $t_i = ih$ with a sampling interval $h = 0.1$. At each sampling time, a random variable Y_i is observed where

$$Y_i | X_{t_i} \sim \text{Poisson}(X_{t_i}).$$

The interpretation is that we count the number of organisms in a volume; we have rescaled the process so that this volume is 1. Figure 10.2 shows the simulated data set $\{Y_i\}$, overlaid with the true states $\{X_t\}$. Note that the individual observation contains very little information about the state; in order to estimate the states, it is important to take the state dynamics into account so that the state X_{t_i} is estimated not just from the single observation Y_i, but from all available observations, giving most weight to measurements taken at nearby times.

10.3 THE RECURSIONS: TIME UPDATE AND DATA UPDATE

We now combine the *process model*, the Itô stochastic differential equation (10.1) governing X_t, and the *observation model*, the state likelihood functions $l_i(\cdot)$, to obtain estimates of the state X_{t_i} based on the information $\mathcal{G}_{t_i}$. The approach is to process or "filter" the measurements one at a time. This recursive approach to estimation allows us to run the filter on-line, i.e., do the processing in real time as the measurements are taken.

To this end, we introduce two conditional distributions of the state X_{t_i}, which differ in how many observations are available to estimate of X_{t_i}. First,

$$\text{the predicted distribution } \phi_i(x)$$

is the p.d.f. of X_{t_i} given $\mathcal{G}_{t_{i-1}}$, i.e., all measurements taken strictly prior to time t_i. Similarly,

$$\text{the estimated distribution } \psi_i(x)$$

is the p.d.f. of X_{t_i} given $\mathcal{G}_{t_i}$, i.e., all measurements taken no later than time t_i.

The principle is to tie the predictions $\{\phi_i\}$ and the estimates $\{\psi_i\}$ together in a recursion. At each step in the recursion, we perform a *time update* and a *data update*.

The Time Update: This step connects the estimate ψ_i and the prediction ϕ_{i+1}. Notice that these are both based on observations available at time t_i. In the time interval $[t_i, t_{i+1}]$, the process $\{X_t\}$ evolves according to the stochastic differential equation. Using the Markov property of the process $\{X_t\}$, we let $\rho(x,t)$ be the p.d.f. of X_t, conditional on $\mathcal{G}_{t_i}$, and evaluated at point x. Then, for $t \geq t_i$, ρ is governed by the forward Kolmogorov equation

$$\dot{\rho} = -\nabla \cdot (u\rho - D\nabla\rho), \tag{10.3}$$

where we have used the advection-diffusion form of the forward Kolmogorov equation, with (as always) $D(x,t) = \frac{1}{2}g(x,t)g^\top(x,t)$ and $u(x,t) = f(x,t) - \nabla D(x,t)$. We solve this equation for $t \in [t_i, t_{i+1}]$ subject to the initial condition $\rho(x,t_i) = \psi_i(x)$. Then, we find the next state prediction as $\phi_{i+1}(x) = \rho(x,t_{i+1})$. This is the *time update*, which changes the time index of the estimated state without changing the information on which the estimation is based.

In the important special case of a time-invariant stochastic differential equation, i.e., when the drift term f and noise term g in (10.1) do not depend on time t, we can write

$$\phi_{i+1} = e^{L^*(t_{i+1}-t_i)}\psi_i$$

where L^* is the forward Kolmogorov operator.

The data update: At the time t_{i+1} the new observation $Y_{i+1}(\omega) = y_{i+1}$ becomes available to the observer. This causes us to modify the distribution of the state, using Bayes' rule (3.4):

$$\psi_{i+1}(x) = \frac{1}{c_{i+1}}\phi_{i+1}(x)l_{i+1}(x). \tag{10.4}$$

Here, c_{i+1} is the normalization constant

$$c_{i+1} = \int_{\mathbf{X}} \phi_{i+1}(x)l_{i+1}(x) \, dx$$

which is the probability density of the next observation Y_{i+1}, conditional on current information $\mathcal{G}_{t_i}$, and evaluated at the actual measurement y_{i+1}.

We summarize the algorithm:

1. Start at $t_0 = 0$ with $\psi_0(\cdot)$. Set $i = 0$.

2. *Time update:* Solve the forward Kolmogorov equation (10.3) on $t \in [t_i, t_{i+1}]$ with initial condition $\rho(x, t_i) = \psi_i(x)$. Advance time to t_{i+1} and set $\phi_{i+1}(x)$ equal to the solution $\rho(x, t_{i+1})$.

3. *Data update:* Compute $\psi_{i+1}(\cdot)$ from Bayes' rule (10.4).

4. Advance $i := i + 1$ and go to step 2.

Figure 10.4 illustrates the time update and the data update for the Cox-Ingersoll-Ross model. Here, we have discretized the state space and solved the equations numerically. The core of the code, written in R, is listed in listing 10.1.

```
P <- expm(G*dt)  # G discretizes the generator L

for(i in 2:length(tm))
{
    phi[i,] <- psi[i-1,] %*% P          # Time update
    psi[i,] <- phi[i,] * ltab[,i]       # Data update
    psi[i,] <- psi[i,] / sum(psi[i,])   # Normalization
}
```

Listing 10.1: R code for implementing the recursive filter. Simplified extract from the function HMMfilterSDE in the SDEtools package.

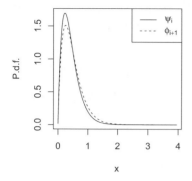 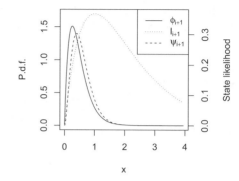

Figure 10.4 The time update (left panel) and the data update (right panel) for the Cox-Ingersoll-Ross process, illustrated for a single time step where the estimated state is low but a positive measurement $Y_i = 1$ arrives.

In the left panel, we see the time update from a time point t_i to t_{i+1}. In this illustration, the estimated distribution ψ_i is concentrated at lower states than the stationary mean $x = 1$, so the time update shifts the distribution slightly to the right towards the stationary distribution, and also widens it. The time step 0.1 is small relative to the decorrelation time 1 of the process, so the effects of the time update is slight, but nevertheless still important. To the right we see the effect of the data update. At time t_{i+1}, a measurement $Y_{i+1} = 1$ is made available. Since Y_{i+1} is conditionally Poisson distributed with mean $X_{t_{i+1}}$, the maximum of the state likelihood is obtained at $x = 1$. This is larger than the mean in the predicted distribution ϕ_{i+1}, so the data update shifts the distribution even further to the right.

Figure 10.5 displays the estimated state, defined as the mean in the estimated distribution ψ_i, as a function of time. Included is also the true state $\{X_t\}$ as well as lower and upper confidence limits on X_{t_i}, derived as 16.6% and 83.3% percentiles in the distribution ψ_i, respectively. Notice that the estimated state follows the true state reasonably accurately, and in particular that the true state is within the confidence intervals most of the time. Notice also that the estimated state appears to lag a bit behind the true state. This is because the estimated state is based on *past* measurements.

While this recursive Bayes algorithm in principle solves the filtering problem, it remains to be described how in practice to solve the forward Kolmogorov equation, and how to do the Bayes update. When the state space has low dimension (up to two or three, say, depending on our stamina) we can solve the equations directly using numerical discretization such as a finite volume method, as we described in Section 9.11.5. This is the approach we use in our example with the Cox-Ingersoll-Ross process. Depending on the specifics of the model, there may be other techniques for solving these equations:

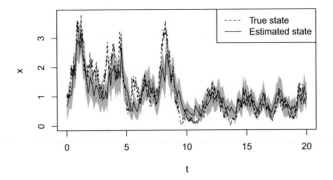

Figure 10.5 Estimated (solid) and true (dashed) state trajectory for the Cox-Ingersoll-Ross process. The estimate is the mean in the estimated distribution ψ_i. Included is also 67% confidence intervals, derived from the estimated distribution ψ_i.

1. The case of linear stochastic differential equations with linear observations and Gaussian noise. Here, the conditional distributions of the state given the observation is also Gaussian. The time update reduces to linear ordinary differential equations in the mean and covariance matrix, and the data update reduces to matrix algebra. This leads to the *Kalman filter* which we describe in Section 10.7.

2. In other situations we can approximate the solutions with Gaussians. This holds if the system is close to linear, i.e., if the non-linearities are weak at the scale given by the variance of the state estimate. The resulting algorithm is the *Extended Kalman filter* or variants, e.g., the unscented Kalman filter, or filters which include higher-order corrections.

3. In some situations it is convenient to use Monte Carlo to estimate the solutions. This leads to the *particle filter*.

Also hybrid techniques are possible (Simon, 2006).

With the state estimate ψ_i in hand, it is (in principle) straightforward to predict future values of the state, also beyond the one-step predictions. We can do this by solving the forward Kolmogorov equation for $t \geq t_i$ with initial condition ψ_i; in the time-invariant case, the conditional p.d.f. of X_t given $\mathcal{G}_{t_i}$ is $\exp(L^*(t - t_i))\psi_i$ for $t \geq t_i$. Alternatively, we can draw a random sample X_{t_i} from ψ_i and simulate a trajectory starting at X_{t_i}.

10.4 THE SMOOTHING FILTER

We now consider the offline situation, where we want to hindcast the process; i.e., we pursue estimates of X_{t_i} conditional on all measurements $\mathcal{G}_T$. We obtain these with the means of the so-called *smoothing filter*, which improves on the densities ϕ_i and ψ_i by including also information from observations taken after time t_i.

With the state estimate $\psi_i(\cdot)$ in hand, the way to include future measurements is the following: We consider ψ_i the *prior density* of X_{t_i} while we let $\pi_i(\cdot)$ denote the *posterior*, which conditions also on all information obtained after time t_i. To derive the posterior $\pi_i(\cdot)$ from the prior $\psi_i(\cdot)$, we apply Bayes' rule. This requires the likelihood $\mu_i(\cdot)$ of all future measurements, seen as a function of the realized value of X_{t_i}:

$$\mu_i(x) = f_{Y_{i+1}, Y_{i+2}, \ldots, Y_N | X_{t_i}}(y_{i+1}, y_{i+2}, \ldots, y_N, x).$$

To compute these likelihood functions, we perform a recursion over time. To this end, let $\lambda_i(x)$ denote the likelihood of all measurements taken at t_i or later

$$\lambda_i(x) = f_{Y_i, Y_{i+1}, \ldots, Y_N | X_{t_i}}(y_i, y_{i+1}, \ldots, y_N, x).$$

First, if we have computed $\mu_i(x)$, then we can compute also $\lambda_i(x)$ by including the likelihood of the measurement taken at time t_{i-1}:

$$\lambda_i(x) = \mu_i(x) \cdot l_i(x).$$

This is the analogy of the data update in the predictive filter. Next, with $\lambda_i(\cdot)$ in hand we need to compute $\mu_{i-1}(\cdot)$. Now, by the properties of the conditional expectation we have

$$\mu_{i-1}(x) = \mathbf{E}\{\lambda_i(X_{t_i}) \mid X_{t_{i-1}} = x\}.$$

This means that we can find μ_{i-1} by solving the backward Kolmogorov equation

$$-\dot{h} = u \cdot \nabla h + \nabla \cdot (D\nabla h)$$

for $t \le t_i$ together with the terminal condition $h(x, t_i) = \lambda_i(x)$. The interpretation of this function h is in fact the likelihood of measurements taken at time t_i or later, viewing $X_t = x$ as the initial condition. Then

$$\mu_{i-1}(x) = h(x, t_{i-1})$$

This is the analogy of the time update in the predictive filter. In the important special case of a time-invariant stochastic differential equation, we have

$$\mu_{i-1}(x) = e^{L(t_i - t_{i-1})}\lambda_i(x)$$

These two steps, the time update and the data update, are iterated backward in time, starting with

$$\lambda_N(x) = l_N(x)$$

Having completed the recursion, we find the posterior distribution

$$\pi_i(x) = \frac{1}{k_i}\psi_i(x)\mu_i(x)$$

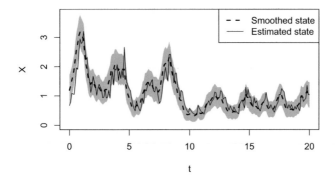

Figure 10.6 Estimated (dashed) and smoothed (solid) state trajectory for the Cox-Ingersoll-Ross process. Included is also 67% confidence intervals, derived from the smoothed distribution π_i.

which is the conditional distribution of X_{t_i} given all measurements, past, present, and future. Here k_i is a normalization constant ensuring that π_i integrates to 1. (Of course we could equally well have used $\phi_i(x)\lambda_i(x)$ after normalization).

Figure 10.6 displays the smoothed estimate for the Cox-Ingersoll-Ross process. For comparison, we have also included the estimate from figure 10.5. Comparing the two figures, we see that in this example the uncertainty only reduces slightly when including also future observations. On the other hand, the smoothed estimate does not display the lagging that the estimate has.

10.5 SAMPLING TYPICAL TRACKS

The smoothing filter provides us with the conditional distribution of X_{t_i} given all measurements. These distributions are marginal in time; for example, they do not specify the joint conditional distribution of X_{t_i} and $X_{t_{i+1}}$.

Ultimately, we would like to know the conditional law of the stochastic process $\{X_t : t \in [0, T]\}$ given all measurements, but on the other hand this is too complicated an object to operate with. The notable exception to this rule is the linear-Gaussian case, where the conditional law is given by the estimate $\mu_{t|T}$ and the autocovariance structure of the estimation error. For general, non-linear models, we have to be satisfied with the ability to sample "typical" tracks $(X_0, X_{t_1}, \ldots, X_T)$ from the posterior distribution. Such simulated trajectories are often very useful for communicating the results of the filtering problem, in particular to non-specialists. They can also be used to make Monte Carlo estimates of statistics that are otherwise difficult to compute, for example the distribution of the maximum $\max\{X_0, X_{t_1}, \ldots, X_T\}$, conditional on measurements.

An algorithm for this is as follows:

1. First, sample ξ_T from $\pi_N(\cdot)$.

2. Next, for each $i \in \{N - 1, N - 2, \ldots, 0\}$, do

   (a) Compute the conditional distribution of ξ_{t_i} given measurements *and* $X_{t_{i+1}} = \xi_{t_{i+1}}$. Unnormalized, this distribution has density at x

   $$\psi_i(x) \cdot p(t_i, x, t_{i+1}, \xi_{t_{i+1}})$$

   where p are the transition probabilities, which are most conveniently found by solving Kolmogorov's backward equation governing $p(s, x, t_{i+1}, \xi_{t_{i+1}})$ as a terminal value problem, with the terminal condition $p(t_{i+1}, x, t_{i+1}, \xi_{t_{i+1}}) = \delta(x - \xi_{t_{i+1}})$.

   (b) Sample ξ_{t_i} from this distribution.

In additional to the marginal distribution of the state at each time, and to the sampled trajectories, it is useful to compute the mode of the joint distribution of $X_0, X_{t_1}, \ldots, X_T$ conditional on measurements. This is called the *most probable track*. Maximizing the posterior distribution over all possible trajectories is a (deterministic) dynamic optimization problem which can be solved with dynamic programming, and the resulting algorithm is known as the Viterbi algorithm (Zucchini and MacDonald, 2009).

10.6 LIKELIHOOD INFERENCE

If the underlying model of system dynamics, or the measurement process, includes unknown parameters, then we can use the filter to estimate these parameters. In fact, maximum likelihood estimation of unknown parameters corresponds to "tuning" the predictive filter in the sense of adjusting parameters to make the predictive filter perform optimally.

To make this precise, assume that the parameters f and g in the stochastic differential equation (10.1) depend on some unknown parameter vector θ. This parameter may also enter in the state likelihood functions $l_i(x)$; for example, controlling the variance of measurement errors. Now, the likelihood function $\Lambda(\theta)$ is the joint probability density function of the observation variables $Y_1, \ldots, Y_N$, evaluated at the measured values $y_1, \ldots, y_N$, and for the given parameter θ:

$$\Lambda(\theta) = f_{Y_1, \ldots, Y_N}(y_1, \ldots, y_N; \theta).$$

This joint p.d.f. can be written in terms of the conditional densities, using the general result $f_{X,Y}(x, y) = f_X(x) f_{Y|X}(x, y)$. First, we single out the first measurement:

$$\Lambda(\theta) = f_{Y_1}(y_1; \theta) f_{Y_2, \ldots, Y_N | Y_1}(y_1, \ldots, y_N; \theta).$$

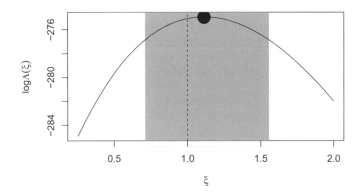

Figure 10.7 The log-likelihood $\log \Lambda(\xi)$ of the Cox-Ingersoll-Ross model. The curve displays the log-likelihood $\log \Lambda(\xi)$. The vertical line indicates the "true" value used to generate the simulated data set. The point indicates the maximum, corresponding to the maximum likelihood estimate. The gray region indicates a 95% confidence interval.

Next, in the second term we single out Y_2:

$$\Lambda(\theta) = f_{Y_1}(y_1; \theta) f_{Y_2|Y_1}(y_1, y_2; \theta) f_{Y_3,\ldots,Y_N|Y_1,Y_2}(y_1, \ldots, y_N; \theta)$$

and continuing this recursion we get

$$\Lambda(\theta) = \prod_{i=1}^{N} f_{Y_i|Y_1,\ldots,Y_{i-1}}(y_1, \ldots, y_i; \theta).$$

The terms in this product are exactly the normalization constants c_{i+1} in the data update of the filter (10.4). Thus, the likelihood of the unknown parameter can be written

$$\Lambda(\theta) = \prod_{i=0}^{N-1} c_{i+1}(\theta) \tag{10.5}$$

where we have stressed that the normalization constants depend on θ because the predictions and/or the state likelihood functions $l_i(x)$ do. Thus, maximizing the likelihood function corresponds to tuning the predictive filter so that it predicts the next measurement optimally in an average sense made precise by this product.

Figure 10.7 displays the likelihood function for the Cox-Ingersoll-Ross model. Here, we have rewritten the model (10.2) as

$$dX_t = (\xi - X_t)\, dt + \sqrt{X_t}\, dB_t$$

and assumed that the mean parameter ξ is unknown. Based on the same data set as in Figure 10.2, we run the filter with different values of this parameter ξ, and compute the likelihood (10.5) for each value of ξ. The result is seen in Figure 10.7. We find the maximum likelihood estimate

$$\hat{\xi} = \text{Arg} \max_{\xi} \Lambda(\xi)$$

and compute an approximate 95% confidence interval as

$$\{\xi : 2 \log \Lambda(\xi) \geq 2 \log \Lambda(\hat{\xi}) - \chi_1^2(0.95)\}.$$

Here, $\chi_1^2(0.95)$ indicates the 95% quantile in the χ^2-distribution with 1 degree of freedom. This way of generating confidence intervals is standard in statistics (Pawitan, 2001). We obtain the estimate $\hat{\xi} = 1.11$ and the confidence interval $[0.71, 1.56]$. We see that the "true" value, $\xi = 1$, is well within the confidence interval. The confidence interval is fairly wide, reflecting that the time series is quite short for this purpose. This indicates that the data set is not informative enough to estimate ξ with high fidelity; the more positive formulation is that the filter performs reasonably well even if it is based on an inaccurate parameter ξ.

This technique is only feasible for models with few states – so that we can implement the filter – and few parameters, so that optimization is feasible. From a modeling perspective, it is appealing that we can formulate the model based on domain knowledge. A disadvantage is that we can, in general, not say much about the structure of the likelihood function and the properties of the estimator. For example, is the likelihood function concave so that there is a unique maximum which we can find with standard optimization algorithms? There is a large specialized literature on alternative approaches to parameter estimation which lead to simpler algorithms and/or analysis. An important word of caution is that the Maximum Likelihood estimator may not be very robust. Specifically, when the actual data-generating system does not belong to the class of models considered, the estimated model may not be the one we would think of as being "closest" to the actual system. Since model families are often gross simplifications of true data generating systems, it is important to make a careful model validation after estimating parameters.

10.7   THE KALMAN FILTER

We now apply the general theory to the multivariate linear system

$$dX_t = (AX_t + u_t)\, dt + G\, dB_t \tag{10.6}$$

where $\{u_t\}$ is a deterministic function of time; A and G are matrices. At times $\{t_i : i = 1, \ldots, N\}$ we take the measurements

$$Y_i = CX_i + DV_i$$

Biography: Rudolf Emil Kalman (1930–2016)
Born in Hungary, his family emigrated to the United
States in 1943. A pioneer of "Modern Control Theory",
he developed the "Kalman" filter around 1960. Other
important contributions are the "Kalman axioms" for
dynamic systems, realization theory for linear systems,
and the recognition that classical and analytical mechanics would be useful in systems theory (Lyapunov
stability, Hamilton-Jacobi formalism for calculus of
variations). Photo credit: National Academy of Engineering and Alexander Benz.

where C and D are matrices. $\{V_i\}$ is a sequence of Gaussian variables with
mean 0 and unit variance matrix, independent of each other and of the Brow-
nian motion $\{B_t : t \geq 0\}$. As before, we let $\mathcal{G}_t$ be the information available at
time t, i.e., the σ-algebra generated by Y_i for $t_i \leq t$.

To perform the time update, assume that conditionally on $\mathcal{G}_{t_i}$, X_{t_i} is
Gaussian with mean $\mu_{t_i|t_i} = \mathbf{E}\{X_{t_i}|\mathcal{G}_{t_i}\}$ and variance $\Sigma_{t_i|t_i} = \mathbf{V}\{X_{t_i}|\mathcal{G}_{t_i}\}$.
Then conditional on the same information, the distribution of X_t remains
Gaussian for $t \in [t_i, t_{i+1}]$ (since the transition probabilities for a linear system
are Gaussian) with a conditional mean given by the vector ordinary differential
equation

$$\frac{d}{dt}\mu_{t|t_i} = A\mu_{t|t_i} + u_t$$

and a conditional variance given by the matrix ODE

$$\frac{d}{dt}\Sigma_{t|t_i} = A\Sigma_{t|t_i} + \Sigma_{t|t_i}A^\top + GG^\top.$$

By advancing these ODE's to time t_{i+1}, we have completed the time update.
Recall that Exercise 5.8 discussed the numerical solution of the variance equation.

For the data update, we use a standard result about conditioning in multivariate Gaussians (Exercise 3.20). To this end, note first that conditional on
$\mathcal{G}_{t_i}$, $X_{t_{i+1}}$ and Y_{i+1} are joint normal with mean

$$\mathbf{E}\left\{\begin{pmatrix} X_{t_{i+1}} \\ Y_{i+1} \end{pmatrix} \Big| \mathcal{G}_{t_i}\right\} = \begin{pmatrix} \mu_{t_{i+1}|t_i} \\ C\mu_{t_{i+1}|t_i} \end{pmatrix}$$

and covariance matrix

$$\mathbf{V}\left\{\begin{pmatrix} X_{t_{i+1}} \\ Y_{i+1} \end{pmatrix} \Big| \mathcal{G}_{t_i}\right\} = \begin{bmatrix} \Sigma_{t_{i+1}|t_i} & \Sigma_{t_{i+1}|t_i}C^\top \\ C\Sigma_{t_{i+1}|t_i} & C\Sigma_{t_{i+1}|t_i}C^\top + DD^\top \end{bmatrix}.$$

To ensure that the conditional distribution of Y_{i+1} is regular, we assume that $DD^\top > 0$, i.e., all measurements are subject to noise. We can now summarize the Kalman filter for linear systems with discrete-time measurements:

1. Time update: Advance time from t_i to t_{i+1} by solving the equations

$$
\frac{d}{dt}\mu_{t|t_i} = A\mu_{t|t_i} + u_t
$$
$$
\frac{d}{dt}\Sigma_{t|t_i} = A\Sigma_{t|t_i} + \Sigma_{t|t_i}A^\top + GG^\top
$$

for $t \in [t_i, t_{i+1}]$ with initial conditions $\mu_{t_i|t_i}$ and $\Sigma_{t_i|t_i}$.

2. Data update: Compute the so-called *Kalman gain*, i.e., the matrix

$$
K_{i+1} = \Sigma_{t_{i+1}|t_i} C^\top (C\Sigma_{t_{i+1}|t_i} C^\top + DD^\top)^{-1}
$$

and then include information at time t_{i+1} as follows:

$$
\mu_{t_{i+1}|t_{i+1}} = \mu_{t_{i+1}|t_i} + K_{i+1}(y_{i+1} - C\mu_{t_{i+1}|t_i})
$$
$$
\Sigma_{t_{i+1}|t_{i+1}} = \Sigma_{t_{i+1}|t_i} - K_{i+1}C\Sigma_{t_{i+1}|t_i}.
$$

Remark 10.7.1 *When the sampling interval $h = t_{i+1} - t_i$ is constant, the time update can be done more efficiently as follows: First, before the iteration starts, compute the matrix exponential $\exp(Ah)$ and solve the Lyapunov matrix ODE*

$$
\frac{d}{dt}S(t) = AS(t) + S(t)A^\top + GG^\top
$$

with initial condition $S(0) = 0$, for $t \in [0, h]$. See Exercise 5.8 for how to do this. Then

$$
\Sigma_{t_{i+1}|t_i} = e^{Ah}\Sigma_{t_i|t_i}e^{A^\top h} + S(h).
$$

To see this, use the general rule $\mathbf{V}X_{t_{i+1}} = \mathbf{EV}\{X_{t_{i+1}} \mid X_{t_i}\} + \mathbf{VE}\{X_{t_{i+1}} \mid X_{t_i}\}$ where expectation and variance are conditional on $\mathcal{G}_{t_i}$. Thus we can advance the variance matrix without solving a matrix ODE at each time step; matrix multiplication suffices. Depending on the driving term u_t, the same may be possible for the mean value. For example, if A is invertible and u_t is constant and equal to u over each time step (t_i, t_{i+1}), then

$$
\mu_{t_{i+1}|t_i} = e^{Ah}\mu_{t_i|t_i} + A^{-1}(e^{Ah} - I)u
$$

so also the vector ODE for $\mu_{t|t_i}$ needs not be solved numerically; rather the computations reduce to matrix algebra.

The theory of Kalman filtering for linear systems is quite complete and addresses many other issues than deriving the basic algorithm as we have just done. Just to give a few examples, there are matrix algebraic formulations of the smoothing step, conditions under which the forward filter converges to a steady state, the special case of periodic systems has been investigated, as have the robustness of the filter to various types of model errors.

10.7.1 Fast Sampling and Continuous-Time Filtering

So far we have discussed the Kalman filter with measurements that are taken at discrete points of time. In the early days of Kalman filtering, the filter would sometimes be implemented with analog electric circuits, so truly operating in continuous time. Nowadays, even if the filter runs in discrete time on digital hardware, the sampling rate may be so fast that the filter effectively runs in continuous time. In other situations we have not yet determined the sampling time, but start by examining the continuous-time filter and later choose the sampling frequency to be "fast enough". For these reasons we now investigate the Kalman filter when the measurements are available in continuous time.

In this situation, we need also a model of continuous-time measurement errors. A simple and useful model is that the measurements are subject to additive measurement noise, i.e.,

$$Z_t = CX_t + Dw_t$$

and we wish to estimate X_t based on observations of Z_s for $s \leq t$. Here, $\{w_t\}$ is the measurement noise signal, and the model must specify the statistical properties of this process. The simplest structure is that $\{w_t\}$ is white noise of unit spectral density. To make this model fit into our general framework, we replace it with its integrated version

$$dY_t = Z_t \, dt = CX_t \, dt + D \, dW_t \tag{10.7}$$

where $\{W_t\}$ is Brownian motion and we have allowed the informal equation $w_t \, dt = dW_t$; i.e., Brownian motion is integrated white noise.

To estimate X_t based on $\mathcal{G}_t = \sigma(\{Y_s : 0 \leq s \leq t\})$ we consider first a discretized version, obtained by sampling Y_t at regular intervals nh for $n \in \mathbf{N}$ where $h > 0$ is the sampling time. We solve this discretized problem with the material of the previous section. Then we let the sampling time h tend to 0. To this end, assume that we at time $t = nh$ have access to the measurements

$$Y_h, Y_{2h}, \ldots, Y_{nh}$$

and based on this information, X_t has conditional mean $\mu_{t|t}$ and variance $\Sigma_{t|t}$. Performing the time update as in the previous, we get

$$\mu_{t+h|t} = \mu_{t|t} + A\mu_{t|t}h + u_t h + o(h)$$

and

$$\Sigma_{t+h|t} = \Sigma_{t|t} + \left(A\Sigma_{t|t} + \Sigma_{t|t}A^\top + GG^\top \right) h + o(h).$$

We now turn to the data update. At time $t + h$ the new information is Y_{t+h}, but since we already know Y_t, we can also say that the new information is

$$\Delta Y_t = (Y_{t+h} - Y_t) = CX_t h + D(W_{t+h} - W_t) + o(h).$$

This agrees with the form in the previous section except for the covariance of the measurement error $(W_{t+h} - W_t)$ which is $h \cdot I$ rather than I. Making the obvious rescaling, and neglecting higher order terms, the data update is given by a Kalman gain

$$K_t = \Sigma_{t+h|t} C^\top h (C \Sigma_{t+h|t} C^\top h^2 + DD^\top h)^{-1}.$$

We now make the assumption that $DD^\top > 0$, so that all measurements are noisy, which implies that the Kalman gain has a well-defined limit as $h \searrow 0$:

$$K_t = \Sigma_{t|t} C^\top (DD^\top)^{-1} + O(h).$$

With this Kalman gain, the data update for the mean becomes

$$\mu_{t+h|t+h} = \mu_{t+h|t} + K_t (\Delta Y_n - C h \mu_{t+h|t} + o(h))$$

and for the variance

$$\Sigma_{t+h|t+h} = \Sigma_{t+h|t} - K_t C^\top h \Sigma_{t+h|t} + o(h).$$

Combining with the time update, we get

$$\mu_{t+h|t+h} = \mu_{t|t} + A \mu_{t|t} h + u_t h + K_t (\Delta Y_n - C h \mu_{t|t} + o(h))$$

and

$$\Sigma_{t+h|t+h} = \Sigma_{t|t} + \left(A \Sigma_{t|t} + \Sigma_{t|t} A^\top + GG^\top - K_n C^\top \Sigma_{t|t} \right) h + o(h).$$

Letting the time step h tend to zero, we can summarize the analysis:

Theorem 10.7.1 *Consider the filtering problem consisting of the state equation (10.6) and the continuous-time observation equation (10.7) with $DD^\top > 0$. In the limit $h \searrow 0$, the state estimate satisfies the Itô stochastic differential equation*

$$d\mu_{t|t} = (A \mu_{t|t} + u_t) \, dt + K_t (dY_t - C \mu_{t|t} dt).$$

Here, the Kalman gain is

$$K_t = \Sigma_{t|t} C^\top (DD^\top)^{-1}.$$

The variance of the estimation error satisfies the so-called Riccati equation, an ordinary differential matrix equation

$$\frac{d}{dt} \Sigma_{t|t} = A \Sigma_{t|t} + \Sigma_{t|t} A^\top + GG^\top - \Sigma_{t|t} C^\top (DD^\top)^{-1} C^\top \Sigma_{t|t}.$$

The technique of replacing measurements with their integrated version is useful also in other situations. For example, it allows simulation experiments that compare filter performance for different choices of the sample time: By fixing the Brownian motions $\{B_t\}$ and $\{W_t\}$ and discretizing with different time steps, we can obtain simulations that correspond to the same realization but different sample times. It also highlights that when the sampling is fast, the important property of the measurement noise is its spectral density, i.e., $DD^\top h$: Accuracy can be improved by decreasing the variance $DD^\top$ or the sampling interval h. The technique can be applied also when the state dynamics are governed by non-linear stochastic differential equations, and with other types of measurement noise. For example, for the Poisson measurements in Figure 10.2, we can consider a continuous-time limit where we measure more and more frequently, but count bacteria in smaller and smaller volumes. In the limit, the information becomes a train of Dirac deltas indicating the arrival times of individual bacteria. Phrased differently, we observe a Poisson process in real time.

10.7.2 The Stationary Filter

If the system parameters A, G, C, D are constant in time, then we can characterize the asymptotic behavior of the Kalman gain K_t, the variance $\Sigma_{t|t}$, and the estimation error $\tilde{X}_t := X_t - \mu_{t|t}$. For simplicity, we assume that the pair (A, C) is observable:

Definition 10.7.1 *Consider the matrix pair (C, A) where $A \in \mathbf{R}^{n \times n}$ and $C \in \mathbf{R}^{m \times n}$. The pair is said to be observable, if the following two equivalent conditions hold:*

1. *All right eigenvectors $v \neq 0$ of A satisfy $Cv \neq 0$.*

2. *The so-called observability matrix*

$$
\begin{bmatrix}
C \\
CA \\
CA^2 \\
\vdots \\
CA^{n-1}
\end{bmatrix}
$$

is injective (i.e., has full column rank).

The observability condition says that all eigenmodes of the system are visible in the output Y_t; this means that in the absence of noise we would be able to reconstruct the state perfectly using continuous-time measurements Y_t over any interval $t \in [0, T]$. When noise is present, it is sufficient to prevent the variance from growing beyond bounds. In fact, under this condition, it can be shown that the estimation variance $\Sigma_{t|t} = \mathbf{E}\tilde{X}_t\tilde{X}_t^\top$ converges to an asymptotic value Σ, which solves the algebraic Riccati equation (ARE)

$$
A\Sigma + \Sigma A^\top + GG^\top - \Sigma C^\top (DD^\top)^{-1} C\Sigma = 0.
$$

Next, we further assume that the pair (A, G) is controllable. Recall the definition in Theorem 9.11.1 and note that (A, G) is controllable iff $(G^\top, A^\top)$ is observable; these are dual properties. Controllability implies that all dynamics of the system are excited by the noise $\{B_t\}$. Since also all measurements are noisy $(DD^\top > 0)$, the variance $\Sigma_{t|t}$ of the estimation error will be positive definite for any $t > 0$, and also the limiting variance $\Sigma = \lim_{t\to\infty} \Sigma_{t|t}$ is positive definite.

Theorem 10.7.2 *Consider the filtering problem (10.6), (10.7) with $DD^\top > 0$, (C, A) observable and (A, G) controllable. Then the steady-state variance on the estimation error Σ exists and is positive definite. Moreover, it is the maximal solution to the algebraic Riccati equation in P*

$$AP + PA^\top + GG^\top - PC^\top(DD^\top)^{-1}C^\top P = 0 \qquad (10.8)$$

i.e., any P that satisfies this equation has $P \le \Sigma$. Finally, Σ is the unique stabilizing solution, i.e., the only solution P to this equation such that $A - PC(DD^\top)^{-1}$ is asymptotically stable.

The estimation error $\tilde{X}_t = X_t - \mu_{t|t}$ satisfies the stochastic differential equation

$$d\tilde{X}_t = (A - KC)\tilde{X}_t \ dt + G \ dB_t + K \ dW_t$$

and has a stationary distribution with mean 0 and variance Σ.

Note that the dynamics of the estimation error, i.e., its autocorrelation function and its decorrelation time, depends on $A - KC$. In particular, the eigenvalues of this matrix contains useful information on how fast the error decorrelates.

10.7.3 Sampling Typical Tracks

We now consider the problem of sampling typical tracks $\{X_t : 0 \le t \le T\}$ in the linear Kalman filter, conditional on all measurements $\{Y_{t_i} : i = 1, \dots, N\}$. Of course, one can use the general approach described in Section 10.5. However, the linear structure allows an algorithm that is simpler in the sense that it requires almost no extra implementation, just a re-use of the existing code.

To show this, recall the general technique for sampling in conditional Gaussian distributions (Exercise 4.12): If we can simulate all variables in the model, and compute conditional expectations, then we can combine these two to obtain conditional simulations. This general result for linear-Gaussian models applies to the Kalman filter to yield the following algorithm:

1. Construct the smoothed estimates $\hat{X}_{t|T}$ using the smoothing filter.

2. Simulate a random state trajectory $\bar{X}_t$ and measurements $\bar{Y}_i$ from the model.

3. Construct smoothed estimates $\hat{\bar{X}}_{t|T}$ of the simulated state trajectory $\{\bar{X}_t\}$ by passing the simulated measurements $\bar{Y}_i$ through the smoothing filter.

4. Construct a sample as $X_t = \hat{X}_{t|T} + \bar{X}_t - \hat{\bar{X}}_{t|T}$.

To reiterate the discussion from Section 10.5, these tracks can be used for visualization and for communicating the results. They can also be used for Monte Carlo computation of statistics that do not follow readily from the Gaussian distributions, such as the conditional probability that the state ever enters a given critical region.

10.8 ESTIMATING STATES AND PARAMETERS AS A MIXED-EFFECTS MODEL

An entirely different approach to state estimation is to formulate the model as a general statistical model where the unobserved states $\{X_{t_i}\}$ are considered random effects (i.e., unobserved random variables; latent variables), and use general numerical methods for inference in this model. One benefit of this approach is that it becomes a minor extension to estimate states and system parameters in one sweep. This approach has become feasible in recent years thanks to the availability of powerful software for such mixed-effects models. Here, we use the R package Template Model Builder (TMB) by (Kristensen et al., 2016) which combines numerical optimization of the likelihood function with the so-called Laplace approximation.

Let us illustrate this approach with the same example of state estimation in the Cox-Ingersoll-Ross process (10.2), now written as

$$dX_t = \lambda(\xi - X_t)\, dt + \gamma\sqrt{X_t}\, dB_t.$$

Let $\theta = (\lambda, \xi, \log\gamma)$ be the system parameter vector. Let

$$\phi(x_{t_0}, x_{t_1}, \ldots, x_T, y_1, \ldots, y_N; \theta)$$

denote the joint probability density of all states and observations, for a given set of system parameters θ. This can be written as

$$\phi = \psi_0(x_{t_0}) \cdot \prod_{i=1}^{N} p(t_{i-1}, x_{t_{i-1}}, t_i, x_{t_i}) \cdot l_i(x_{t_i}) \tag{10.9}$$

where we have omitted the arguments of ϕ and the dependence on parameters θ.

Our first task is to estimate the system parameters θ. For a given set of observations $y_1, \ldots, y_N$, the likelihood of θ is obtained by integrating out the unobserved states $\{X_{t_i}\}$, using the law of total probability:

$$\Lambda(\theta) = \int_{\mathbf{X}} \cdots \int_{\mathbf{X}} \phi(x_{t_0}, x_{t_1}, \ldots, x_T, y_1, \ldots, y_N; \theta)\, dx_{t_0} \cdots dx_{t_T} \tag{10.10}$$

and the maximum likelihood estimate of the system parameters is

$$\hat{\theta} = \arg\max_{\theta} \Lambda(\theta).$$

Factbox: [The Laplace approximation] Starting in one dimension, our objective is to approximate an integral

$$I = \int_{-\infty}^{+\infty} f(x) \, dx$$

where f is C^2, takes positive values, and has a unique maximum. The idea is to approximate $f(x)$ with a Gaussian bell $\hat{f}(x)$, which resembles the original function $f(x)$ around the maximum point, since this region presumably has the greatest contribution to the integral. We choose $\hat{f}$ such that the mode $\hat{x}$ of f and $\hat{f}$ coincide, so that $f(\hat{x}) = \hat{f}(\hat{x})$, and so that the curvatures agree, $f''(\hat{x}) = \hat{f}''(\hat{x})$. See Figure 10.8 where $\log f(x) = -\frac{1}{2}(x-1)^2 + \frac{1}{6}(x-1)^3 - \frac{1}{16}(x-1)^4$. Thus

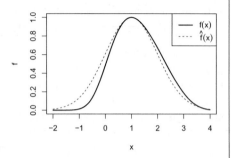

Figure 10.8 The function $f(x)$ and its approximation $\hat{f}(x)$.

$$\log \hat{f}(x) = \log f(\hat{x}) - \frac{1}{2}H \cdot (x - \hat{x})^2$$

where

$$\hat{x} = \arg\max_{x} f(x), \quad H = -\frac{d^2(\log f)}{dx^2}(\hat{x}).$$

Now, we know the area under Gaussian bells. Specifically,

$$\int_{-\infty}^{+\infty} (H/(2\pi))^{1/2} \exp(-\frac{1}{2}H \cdot (x - \hat{x})^2) \, dx = 1$$

since the integrand is the probability density of a Gaussian variable with mean $\hat{x}$ and variance H^{-1}. Therefore

$$I \approx \int_{-\infty}^{+\infty} \hat{f}(x) \, dx = (2\pi/H)^{1/2} f(\hat{x}).$$

This is the Laplace approximation of the integral I. It replaces the operation of integration with two simpler operations: First maximization of $\log f$ and next computation of the double derivative of $\log f$. In higher dimensions, we get

$$I \approx f(\hat{x}) \cdot \left| \frac{1}{2\pi} H \right|^{-1/2}$$

where $\hat{x}$ is the maximum point, H is the Hessian of $\log f$ evaluated at $\hat{x}$, and $|\cdot|$ denotes determinant.

We pursue a direct computation of this estimate through numerical optimization. This requires also a numerical method for evaluating the likelihood function, in particular integrating out the unobserved states. This is a challenging problem, since the integral in (10.10) can be over a very high-dimensional space. The method we employ is the *Laplace approximation*; see the fact box. In this approximation, we first estimate the unobserved states $\{X_{t_i}\}$, given the data $\{Y_i\}$ and a guess on parameters θ, using the modes in the conditional distribution as estimates $\{\hat{X}_{t_i}(\theta) : i = 0, \ldots, N\}$. Note that the mode can be found by optimizing ϕ numerically over the states $\{x_{t_i}\}$. Next, we approximate the integral in the likelihood function (10.10) with the integral of the approximating Gaussian bell; this corresponds to approximating the posterior distribution of the states with a multivariate Gaussian. The integral of this Gaussian can be found from the Hessian matrix of the log-density evaluated at the mode. This means that the computationally intractable task of integration in a very high-dimensional space is simplified to maximization, computation of derivatives, and matrix algebra. This yields a numerical approximation to the likelihood function, which we can maximize using standard methods for numerical optimization, to find the maximum likelihood estimate $\hat{\theta}$ and the corresponding state estimates $\{\hat{X}_{t_i}(\hat{\theta})\}$. These steps have all been implemented in TMB, so our only task is to specify the joint density of states and observations, i.e., the function ϕ in (10.9).

For our running example of estimation in the CIR process, the transition densities are available in closed form (Exercise 9.8). However, since this is not always the case, we have used a technique that does not require them: The model estimates the state at time points $0 = t_0 < t_1 < \cdots < t_n = T$, which includes those points in time where observations are taken, but also additional time points inserted between the times of observations. The mesh $t_i - t_{i-1}$ is sufficiently fine that we can approximate the transition probabilities on this finer grid with the Euler-Maruyama scheme. The states are estimated in the Lamperti domain, i.e., the latent variables are $\sqrt{X_{t_i}}$, for improved performance of the Euler-Maruyama scheme, and for consistency.

Table 10.1 shows the results of the parameter estimation. The true values of parameters are within the confidence limits, and the estimate of the parameter ξ agrees reasonably well with what we found in Section 10.6, as does the confidence interval.

Figure 10.9 shows the estimated states from TMB and, for comparison, the estimates from the HMM method employed in the previous Sections. The two approaches, TMB and HMM, give fairly consistent results. TMB makes use of a Gaussian approximation of the conditional distribution of the states, so there is no distinction between expectation, mode and median in TMB (in the Lamperti domain). This is in contrast to the HMM method, where we find the full posterior distribution (discretized) and can distinguish between these statistics. Here, we show the mean; for this example, the mode and median generally lie below the mean and closer to the TMB estimate.

TABLE 10.1 Parameter estimation in the Cox-Ingersoll-Ross model. For each parameter, the table displays the "true" value used in the simulation, the maximum likelihood estimate, and the standard deviation on that estimate as derived from the Fisher information.

Parameter	True value	Estimate ± s.d.
λ	1.0	1.2 ± 0.6
ξ	1.0	1.1 ± 0.2
$\log \gamma$	0.0	0.0 ± 0.3

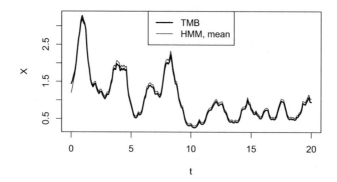

Figure 10.9 Comparison of state estimates with Template Model Builder (thick solid line) and the Hidden Markov Model method (thin solid line). For the HMM method, we show the mean in the smoothed distribution.

10.9 CONCLUSION

An important area of application of stochastic differential equations is statistical analysis of time series. Here, mechanistic understanding of system dynamics is represented in the model function (f, g), while the measurement uncertainty is taken into account so that data is processed optimally.

We have emphasized the "recursive filtering" approach to this analysis. Discretizing the diffusion processes to a discrete Markov model, corresponding to numerical solution of Kolmogorov's equations, means that we approximate the filtering problem for the diffusion process with a Hidden Markov Model. The resulting numerical algorithms are very direct translations of the mathematical formulas. This "brute force" approach is instructive, but only feasible for low-dimensional problems: One dimension is straightforward, two dimensions require a bit of work, three dimensions are challenging both for the modeller and for the computer, and higher dimensions are prohibitively laborious.

State estimation with very high-dimensional state spaces is only feasible in the linear paradigm; the Kalman filter can be implemented routinely with hundreds of states and, with some attention to implementation, even much higher number of dimensions. The Kalman filter is restricted to linear problems, or to problems that can be approximated with linear ones as in the *extended* Kalman filter or similar. In applications, it is a great advantage that linear-Gaussian models are tractable and easy to implement, so it is often a good idea to start with a filter based on a linear model, even if the original model contains nonlinearities.

These recursive filtering approaches to estimation in diffusion models have been well established since the 1960's. The alternative approach, that of treating the problem as a general mixed-effect model, is more recent, and it is not yet entirely clear which algorithms are applicable or most efficient in which situations. Markov Chain methods apply in principle to any problem, but they may be very tricky to get to work in practice, and often lead to excessive computing times. Numerical optimization of likelihood functions, using the Laplace approximation, is powerful but limited to situations where the posterior distributions are well approximated with Gaussian ones. This could imply that this approach is useful for the same problems as extended Kalman filters, in which case the choice between the two methods is primarily a matter of the time it takes for the modeler to implement the model and for the computer to conduct the computations.

Estimation in diffusion processes is a large and active research area, and there are several established text-books on the matter, and a steadily increasing number of journal papers both on methods and applications. This chapter serves only as a first introduction to the topic.

10.10   NOTES AND REFERENCES

Kalman published the filter that now bears his name in (Kalman, 1960) for the discrete-time case. Kalman and Bucy (1961) together published the continuous-time version after discovering that they had been working in parallel. This work differed from earlier work by Wiener by exploiting a state space formalism, that encompassed also non-stationary time series. Stratonovich published similar ideas, in a non-linear setting, around the same time in the Soviet literature.

The theory of Kalman filtering is now covered in numerous textbooks, e.g. (Harvey, 1989), and is a fundamental technique in time series analysis (Madsen, 2007). There are several extensions of the Kalman filter to non-linear systems. Many of these are covered in (Simon, 2006); a classical reference is (Jazwinski, 1970).

10.11 EXERCISES

Exercise 10.1 Kalman Filtering in the OU-Process: Let $dX_t = -\lambda X_t\, dt + \sigma\, dB_t$, and consider continuous-time measurements, $dY_t = cX_t\, dt + d\, dW_t$. Write up the algebraic Riccati equation (10.8) governing the stationary filter and identify the maximal solution. For each of the parameters in the model, check if the stationary variance increases or decreases with that parameter. Next, do the same for the Kalman gain K, and finally for eigenvalue $A - KC$ which governs the estimation error. Assume that c, d and σ are all positive. Investigate the limits $d \to 0$ and $d \to \infty$ (possibly numerically, taking $\lambda = \pm 1$, $\sigma = 1$, $c = 1$).

Exercise 10.2 Filtering in a Linear Mass-Spring-Damper System: Consider an object with position Q_t and velocity V_t which evolve according to the system

$$dQ_t = V_t\, dt, \quad dV_t = [-cV_t - kQ_t]\, dt + \sigma\, dB_t$$

and say that we measure the position with regular intervals $t_i = ih$ for $i \in \mathbf{N}$, where $h > 0$ is the sample interval:

$$Y_i = Q_{ih} + \frac{s}{h}(W_{ih} - W_{(i-1)h}).$$

Here, $\{B_t\}$ and $\{W_t\}$ are independent Brownian motions.

1. Simulate the system and the measurements over the time interval $t \in [0, 100]$. Take parameters $k = 1$, $c = 0.1$, $\sigma = 0.1$, $s = 1$, $h = 0.1$. Plot the true position and the measured position in one plot, and the true velocity in another.

2. Implement the Kalman filter. Plot the true position and the estimated position (with confidence intervals) in one plot, and the true velocity and estimated velocty (with confidence intervals) in another.

3. Compute the variance-covariance matrix of the estimation error. Compare with the matrix $\Sigma_{T|T}$ as computed by the Kalman filter.

4. Repeat with a different sample interval, e.g., $h = 0.2$ or $h = 0.05$. Verify that the estimation error has (approximately) the same variance.

Exercise 10.3 The Kalman Filter as an Optimal Luenberger Observer: An alternative to the stationary Kalman filter for continuous-time measurements is the so-called Luenberger observer, where we use the same filter equation as in Theorem 10.7.2, which we rewrite as

$$d\hat{X}_t = (A - KC)\hat{X}_t\, dt + u_t\, dt + K(dY_t - C\hat{X}_t\, dt),$$

but now we view the gain K as a design parameter.

1. Show that the estimation error $\tilde{X}_t = X_t - \hat{X}_t$ satisfies

$$d\tilde{X}_t = (A - KC)\tilde{X}_t + G \ dB_t + KCD \ dW_t,$$

provided the system dynamics (10.6) and the measurement equation (10.7) hold.

2. Pose the algebraic Lyapunov equation that governs the steady-state variance Σ of $\tilde{X}_t$, for a given value of K.

3. For the optimal Luenberger filter, we aim to minimize $\mathrm{tr}(\Sigma Q)$ w.r.t. Σ and K subject to the constraint from the last question. Here, $Q > 0$ is an arbitrary weight matrix. Show that at the optimum, K must have the form from Theorem 10.7.1, and Σ must satisfy the algebraic Riccati equation (10.8).

Note: In many applications, the noise characteristics G and D are poorly known, and then these parameters can be viewed as a convenient way of parametrizing Luenberger observers. This alternative view adds to our understanding of the problem and its solution, but allows us also to add additional design objectives and constraints.

Exercise 10.4 A Diffusion Bridge: Consider the double well system $dX_t = X_t(1 - X_t^2) \ dt + \sigma \ dB_t$ with $\sigma = 1/2$, where we have observed $X_0 = -1$ and $X_T = +1$ with $T = 100$. Compute numerically the conditional c.d.f. of X_t for $t = 0, 1, 2, \ldots, 100$ and plot it as a pseudocolor image. Include the conditional expectation plus/minus the conditional standard deviation.

CHAPTER 11

Expectations to the Future

In this chapter, we consider problems that involve computing expectations to the future, as functions of the current state. One example of such problems is the backward Kolmogorov equation; here, we extend this result by replacing the terminal time T in the backward Kolmogorov equation with a random stopping time. For example, we consider diffusions that evolve in a bounded domain, and stop the process when the process reaches the boundary. We then ask where and when we expect the process to exit. We will see that these expectations to the future are governed by boundary value problems, which involve the backward Kolmogorov operator.

Such "exit problems" appear in many different applications: In molecular dynamics, we may ask when a molecule reaches a surface where it may undergo a reaction, or when two molecules meet, or when the vibrations of one molecule are so violent that the molecule breaks apart. In control systems, we may ask when the controlled system reaches its target so that the mission can be terminated, or conversely, if we should be concerned that it may leave a safe operating domain. In finance, we may have decided to sell a stock when it reaches a target price, and ask when that happens.

We also consider the present-day value of future rewards, when their value is discounted with a rate that depends on the state trajectory between now and the time the reward is given. Discounting is standard in finance, where a euro is worth more today than in a year from now, but we do not yet know exactly how much more. Discounting also appears in evolutionary ecology, when the risk of dying needs to be included when computing the Darwinian fitness of animals. Discounting adds an extra term to the backward Kolmogorov equation, and results in the so-called Feynman-Kac formula.

In summary, this chapter connects expectations to the future with partial differential equations that involve the backward Kolmogorov operator.

11.1 DYNKIN'S FORMULA

We start by establishing a seemingly simple lemma, due to Dynkin, which is pivotal in the rest of the chapter. We consider an Itô diffusion $\{X_t\}$ in $\mathbf{R}^n$ given by the Itô equation $dX_t = f(X_t)\, dt + g(X_t)\, dB_t$, where f and g satisfy the usual conditions for existence and uniqueness of a solution (Theorem 8.3.2), and where $\{B_t\}$ is Brownian motion on a filtered probability space $(\Omega, \mathcal{F}, \{\mathcal{F}_t\}, \mathbf{P})$. We let L be the backward Kolmogorov operator given by $Lh = \nabla h \cdot f + \operatorname{tr}[gg^\top \mathbf{H}h]/2$.

Theorem 11.1.1 (Dynkin's Formula) *Let $h \in C_0^2(\mathbf{R}^n)$ and let τ be a stopping time such that $\mathbf{E}^x \tau < \infty$. Then*

$$\mathbf{E}^x h(X_\tau) = h(x) + \mathbf{E}^x \int_0^\tau Lh(X_s)\, ds.$$

Recall that the superscript in $\mathbf{E}^x$ means that the initial condition is $X_0 = x$. Note that τ is finite w.p. 1 (since $\mathbf{E}^x \tau < \infty$) so $h(X_\tau)$ is well defined w.p. 1. Dynkin's formula can be seen as a stochastic version of the fundamental theorem of calculus, combined with the chain rule: In the deterministic case we would have $h(x(t)) = h(x(0)) + \int_0^t h'(x(s))\, x'(s)\, ds$ for $0 < t$, whenever the involved derivatives exist.

Proof: Define $Z_t = h(X_t) - \int_0^t Lh(X_s)\, ds$, then $\{Z_t : t \geq 0\}$ is an Itô process which satisfies

$$dZ_t = Lh(X_t)\, dt + \frac{\partial h}{\partial x} g\, dB_t - Lh(X_t)\, dt = \frac{\partial h}{\partial x} g\, dB_t$$

Since $h \in C_0^2$, $\frac{\partial h}{\partial x} g$ is bounded, so the Itô integral $\{Z_t\}$ is a martingale (Theorem 6.3.2 on page 124). Let $T > 0$; then it follows that $\mathbf{E}^x Z_{\tau \wedge T} = Z_0 = h(x)$ (Lemma 4.5.1 on page 80), which can be restated as

$$\mathbf{E}^x h(X_{\tau \wedge T}) = h(x) + \mathbf{E}^x \int_0^{\tau \wedge T} Lh(X_s)ds. \tag{11.1}$$

Thus, Dynkin's formula holds when τ is bounded. Now let $T \to \infty$. Since h is bounded (say, $|h(x)| \leq K$), we have

$$|\mathbf{E}^x h(X_\tau) - \mathbf{E}^x h(X_{\tau \wedge T})| \leq \mathbf{E}^x |h(X_\tau) - h(X_{\tau \wedge T})| \leq 2K \mathbf{P}^x \{\tau > T\}$$

which vanishes as $T \to \infty$. Similarly, Lh is bounded (say, $|Lh(x)| \leq C$), so

$$\mathbf{E}^x \left| \int_0^\tau Lh(X_s)\, ds - \int_0^{\tau \wedge T} Lh(X_s)\, ds \right| \leq \mathbf{E}^x \int_{\tau \wedge T}^\tau C\, ds \leq C\mathbf{E}^x (\tau - (\tau \wedge T))$$

which also vanishes as $T \to \infty$, because $\mathbf{E}^x \tau \leq \infty$. Letting $T \to \infty$ in (11.1), we therefore obtain Dynkin's formula. ∎

The requirement that h has bounded support is quite harsh and can be relaxed; it serves to ensure that the Itô integral $\{Z_t\}$ in the proof is a martingale, and to assist with the limit $T \to \infty$, and this can also be obtained with other assumptions. However, see Exercise 11.6.

Biography: Eugene Borisovich Dynkin (1924–2014)

Dynkin was a Soviet mathematician who worked in Moscow until he emigrated to the USA and Cornell University in 1976. His early contributions were within Lie algebra. Having Kolmogorov as his Ph.D. advisor, he is considered one of the founders of modern Markov theory, which he laid out in his 1959 textbook in Russian and later in (Dynkin, 1965). Photograph reprinted with the permission of the Institute of Mathematical Statistics.

11.2 EXPECTED EXIT TIMES FROM BOUNDED DOMAINS

In the following sections, we consider a diffusion $\{X_t : t \geq 0\}$ on $\mathbf{R}^n$ which starts inside or at the boundary of a domain $\Omega \subset \mathbf{R}^n$. This domain is a bounded, open, and connected subset of the state space; do not confuse it with the sample space Ω! We investigate when (and, in a later section, where) the process first reaches the boundary of Ω. Throughout the section, we let τ denote the time of first exit, $\tau = \inf\{t \geq 0 : X_t \notin \Omega\}$.

Theorem 11.2.1 *Let* $h : \bar{\Omega} \mapsto \mathbf{R}$ *be* C^2 *and such that*

$$Lh(x) + 1 = 0 \text{ for } x \in \Omega, \text{ and } h(x) = 0 \text{ for } x \in \partial\Omega. \qquad (11.2)$$

Then $h(x) = \mathbf{E}^x\tau$.

Proof: We first show that τ has finite expectation: Let $x \in \bar{\Omega}$ and $T > 0$ be arbitrary, then $\tau \wedge T = \min(\tau, T)$ is bounded and therefore Dynkin's formula gives

$$\mathbf{E}^x h(X_{\tau \wedge T}) = h(x) - \mathbf{E}^x(\tau \wedge T).$$

Since the domain is bounded and h is continuous, the left side is bounded. This implies that $\mathbf{E}^x(\tau \wedge T)$ is a bounded function of T, and since it also monotone, it must converge as $T \to \infty$. This implies that $\mathbf{E}^x\tau < \infty$ and therefore Dynkin's formula gives

$$0 = \mathbf{E}^x h(X_\tau) = h(x) + \mathbf{E}^x \int_0^\tau Lh(X_s) \, ds = h(x) - \mathbf{E}^x\tau$$

as claimed. ∎

11.2.1 Exit Time of Brownian Motion with Drift on the Line

Consider Brownian motion with drift, $dX_t = u \, dt + \sigma \, dB_t$, on the domain $\Omega = (0, l)$, where l, u and σ are positive constants. For given initial condition $X_0 = x$, what is the expected time to exit of the domain?

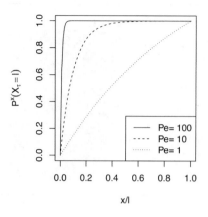 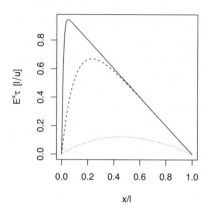

Figure 11.1 Exit behavior for Brownian motion with drift, for various Péclet numbers. *Left panel:* The probability of exiting to the right, as a function of the dimensionless initial position x/l. *Right panel:* The expected time to exit, as a function x/l, and measured in units of the advective time scale l/u.

The expected time of exit, $h(x) = \mathbf{E}^x \tau$, is governed by the equation

$$uh' + Dh'' + 1 = 0 \text{ for } x \in [0, l],$$

and the boundary condition $h(0) = h(l) = 0$. Here, the diffusivity is $D = \sigma^2/2$, as usual. We find the general solution

$$h(x) = -\frac{x}{u} - \frac{c_1 D}{u} \exp(-xu/D) + c_2$$

where we determine c_1 and c_2 from the boundary conditions:

$$h(x) = \frac{l}{u} \left(1 - \frac{x}{l} - \frac{\exp(-\frac{x}{l}\text{Pe}) - \exp(-\text{Pe})}{1 - \exp(-\text{Pe})} \right).$$

Here, we use the Péclet number $\text{Pe} = ul/D$ to write the solution in terms of Pe, the non-dimensional position x/l, and the advective time scale l/u.

This solution is shown in Figure 11.1 (right panel). The first term, $l/u \, (1 - x/l)$, is the time it takes a particle to travel from position x to position l when moving with constant speed u. This straight line is visible in the figure and approximates the solution for high Péclet numbers. The second term involving exponentials is a correction, effective in the diffusive boundary layer, which takes into account that the process may exit quickly to the left rather than traversing the domain and exiting to the right. Relative to pure advection, diffusion lowers the time to exit, in particular near the left boundary.

11.2.2 Exit Time from a Sphere, and the Diffusive Time Scale

How long time does it take for pure diffusion in $\mathbf{R}^n$ to travel a given distance? To answer this question, let $X_t = \sigma B_t$ where B_t is Brownian motion in $\mathbf{R}^n$, and let the domain be the R-sphere, i.e., $\Omega = \{x \in \mathbf{R}^n : |x| < R\}$. The expected time of exit, $h(x) = \mathbf{E}^x \tau$, is governed by the boundary value problem

$$D\nabla^2 h(x) + 1 = 0 \text{ for } |x| < R, \quad h(x) = 0 \text{ for } |x| = R$$

with $D = \sigma^2/2$. The solution to this boundary value problem is

$$h(x) = \frac{R^2 - |x|^2}{2nD}.$$

To see this, recall that $\nabla^2 |x|^2 = 2n$. In particular, when starting at $X_0 = x = 0$:

$$\mathbf{E}^0 \tau = h(0) = R^2/2nD.$$

This is the "diffusive time scale" corresponding to the length scale R, i.e., the expected time it takes to move a distance R from the starting point. Note that $\mathbf{E}^0 |X_t|^2 = 2nDt$ (Sections 2.1.3 and 4.3), and with $t = \mathbf{E}^0 \tau = R^2/2nD$ we get $\mathbf{E}^0 |X_t|^2 = R^2$ which is a noteworthy consistency. The scaling $\mathbf{E}^0 \tau \sim R^2/D$ also follows from a dimensional analysis.

11.2.3 Exit Times in the Ornstein-Uhlenbeck Process

We consider the scalar Ornstein-Uhlenbeck process

$$dX_t = -X_t \, dt + \sqrt{2} \, dB_t.$$

The stationary distribution is a standard Gaussian $N(0, 1)$, and the decorrelation time is 1, so both time and space has been rescaled. We aim to find the expected time until the process leaves the domain $\Omega = (-l, l)$ where $l > 0$. This time $\mathbf{E}^x \tau = h(x)$ is governed by the equation

$$h''(x) - xh'(x) + 1 = 0$$

with boundary conditions $h(l) = h(-l) = 0$. To solve this equation, we first reduce it to a first order equation. Define $k = h'$, then

$$k'(x) - xk(x) + 1 = 0$$

and using symmetry $k(0) = 0$, we find

$$k(x) = -\int_0^x e^{x^2/2 - y^2/2} \, dy = -\sqrt{2\pi} e^{x^2/2} \left(\Phi(x) - \frac{1}{2} \right)$$

where $\Phi(x)$ is the c.d.f. of a standard Gaussian random variable. We therefore get

$$h(x) = h(0) - \sqrt{2\pi} \int_0^x e^{y^2/2} \left(\Phi(y) - \frac{1}{2} \right) \, dy.$$

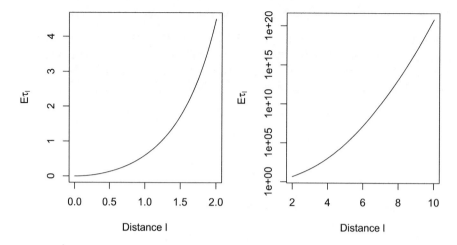

Figure 11.2 Expected time for an Ornstein-Uhlenbeck process to exit the domain $[-l, l]$, as a function of the domain size l. *Left panel:* The expected time for low values of l. *Right panel:* The expected time for higher values of l. Note the log scale. In this panel, the approximation (11.3) would be indistinguishable from the full expression.

We now use the boundary condition $h(l) = 0$, and focus on the expected time to exit, given that we start in the center:

$$h(0) = \mathbf{E}^0 \tau = \sqrt{2\pi} \int_0^l e^{y^2/2} \left(\Phi(y) - \frac{1}{2} \right) \, dy.$$

Figure 11.2 shows this expected exit time as a function of the threshold l, computed numerically. In the left panel, we see the result for low values of l where the expected exit time is comparable to the decorrelation time. In the right panel, we see that as the domain grows, the expected time to exit grows exceedingly fast. In fact, for large values of l, the approximation

$$\mathbf{E}^0 \tau = h(0) \approx \sqrt{\frac{\pi}{2}} \frac{1}{l} e^{l^2/2} \tag{11.3}$$

is useful. This approximation follows from the asymptotic expansion $\int_0^l \exp(y^2/2) \, dy = \exp(l^2/2) \cdot (l^{-1} + O(l^{-3}))$, using that $1 - \Phi(l) \approx l^{-1}\phi(l)$ (Exercise 2.7), which implies that we can approximate $\Phi(y) \approx 1$ when evaluating the integral in the expression for $h(0)$.

Exercise 11.1: Repeat the derivation for the dimensional model $dX_t = -\lambda X_t \, dt + \sigma \, dB_t$, to show that the expression in that case is

$$\mathbf{E}^0 \tau \approx \frac{\sigma \sqrt{\pi}}{2l\lambda^{3/2}} e^{l^2 \lambda/\sigma^2}.$$

This result has implications for the reaction rate of chemical and biological processes, as demonstrated by Kramers (1940), who used a somewhat different but essentially equivalent approach. Consider, for example, a complex molecule such as a protein that vibrates due to thermal noise and can break apart when deflections become too large. We can think of the potential $U(x) = \lambda x^2/2$ as a measure of the energy in the system when it is in a given state x. Then, $E_a := U(l) = \lambda l^2/2$ is an "activation" energy at the threshold $x = l$, while the average energy, i.e., the temperature, is $T := \mathbf{E}U(X_t) = \sigma^2/4$. Formulated in terms of these quantities, the expected lifetime is

$$\mathbf{E}\tau = \sqrt{\frac{T}{E_a}} \frac{\sqrt{\pi/2}}{\lambda} e^{E_a/4T}$$

and thus, the rate with which the molecules desintegrate, is proportional to

$$\sqrt{\frac{E_a}{T}} \lambda e^{-E_a/4T}.$$

Thus this model is consistent with the (modified) Arrhenius equation which quantifies how chemical processes run faster at higher temperatures.

11.2.4 Regular Diffusions Exit Bounded Domains in Finite Time

Recall that we assumed that the domain Ω is bounded, but even so, it does not hold in general that the expected time to exit is finite. For example, Brownian motion on the sphere never exits a larger, enclosing sphere. A useful result is the following:

Theorem 11.2.2 *Let the diffusion $\{X_t : t \geq 0\}$ be regular in the sense that there exists a $d > 0$ such that*

$$\frac{1}{2} g(x) g^\top(x) > dI$$

for all $x \in \Omega$. As before, let $\tau = \inf\{t \geq 0 : X_t \notin \Omega\}$ be the time of first exit. Then $\mathbf{E}^x \tau < \infty$ for all $x \in \Omega$.

One reason the result is useful is that it allows us to apply Dynkin's lemma, which requires that $\mathbf{E}^x \tau < \infty$. The proof of the theorem establishes a bound on $\mathbf{E}^x \tau$, using an inequality version of the result (11.2):

Lemma 11.2.3 *Let $h : \bar{\Omega} \mapsto \mathbf{R}$ be a smooth function such that*

$$h(x) \geq 0, \quad Lh(x) + 1 \leq 0 \text{ for } x \in \Omega.$$

Let $\tau = \inf\{t \geq 0 : X_t \notin \Omega\}$. Then $\mathbf{E}^x \tau \leq h(x)$ for $x \in \Omega$.

Proof: Let $T > 0$, then Dynkin's formula gives

$$0 \leq \mathbf{E}^x h(X_{\tau \wedge T}) = h(x) + \int_0^{\tau \wedge T} Lh(X_t) \, dt \leq h(x) - \mathbf{E}^x (\tau \wedge T),$$

i.e., $\mathbf{E}^x (\tau \wedge T) \leq h(x)$. Letting $T \to \infty$, we obtain the conclusion. ■

Aiming to find such a bounding function h, we take a worst-case view and first consider the scalar version:

Exercise 11.2: Consider a biased random walk $dX_t = -u \, dt + \sigma \, dB_t$ on $(0, R)$ which is reflected at $x = 0$. Let $\tau = \inf\{t \geq 0 : X_t \geq R\}$ be the time of first exit at $x = R$. Show that

$$\mathbf{E}^x \tau = \frac{x - R}{u} + \frac{D}{u^2} \left(e^{uR/D} - e^{ux/D} \right)$$

for $x \in (0, R)$.

This scalar result allows us to guess a bounding function h, with which we can prove Theorem 11.2.2:

Proof: It suffices to consider the case where the domain Ω is the R-sphere for some $R > 0$, i.e., $\Omega = \{x \in \mathbf{R}^n : |x| < R\}$. First, we bound the drift and the diffusion: Let d be as in the theorem and let $u > 0$ be such that $|f(x)| \leq u$ for $|x| \leq R$. Then define

$$\psi(r) = \exp(r) - r, \quad h(x) = \frac{d}{u^2} \left(\psi(uR/d) - \psi(u|x|/d) \right).$$

Note that

$$\psi'(r) = e^r - 1, \quad \psi''(r) = e^r,$$

so that ψ is convex and increasing for $r \geq 0$. We have

$$\nabla h(x) = -\frac{1}{u} \psi'(u|x|/d) \, \nabla |x| = -\frac{1}{u} \psi'(u|x|/d) \frac{x}{|x|}.$$

and

$$\mathbf{H}h(x) = -\frac{1}{d} \psi''(u|x|/d) \frac{xx^\top}{|x|^2} - \frac{1}{u} \psi'(u|x|/d) \mathbf{H}|x| \leq -\frac{1}{d} \psi''(u|x|/d) \frac{xx^\top}{|x|^2} \leq 0.$$

We now bound Lh, using the bounds on drift and diffusivity:

$$\begin{aligned} Lh(x) &= \nabla h \cdot f + \mathrm{tr} D \mathbf{H} h \\ &\leq |\nabla h| u + d \mathrm{tr} \mathbf{H} h \\ &\leq \psi'(u|x|/d) - \psi''(u|x|/d) \\ &= -1. \end{aligned}$$

We therefore get $\mathbf{E}^x \tau \leq h(x)$. Note that the bound can be written

$$\mathbf{E}^x \tau \leq \frac{R^2}{2d} \frac{e^{\mathrm{Pe}} - \mathrm{Pe} - 1}{\mathrm{Pe}^2/2}$$

where the Péclet number is Pe $= uR/d$. The first term is the diffusive time scale in one dimension (Section 11.2.2), and the second term is a correction, greater than 1, which takes the drift into account. ∎

Note also that the proof establishes that the expected time to exit $\mathbf{E}^x \tau$ vanishes as the initial point approaches the boundary, $|x| \to R$. This should come as no surprise, considering the regularity of the diffusion.

11.3 ABSORBING BOUNDARIES

We now consider the stopped process $\{Y_t = X_{t \wedge \tau} : t \geq 0\}$. The setting is as in subsection 11.2.4: The domain Ω is an open bounded connected subset of $\mathbf{R}^n$, the diffusion is regular, and we stop the process upon exit, i.e., at $\tau = \inf\{t \geq 0 : X_t \notin \Omega\}$. Thus, the boundary is absorbing: Once the process $\{X_t\}$ hits the boundary, the stopped process $\{Y_t\}$ remains there.

We aim to characterize the transition probabilities of the stopped process $\{Y_t\}$. As we will see in the following, they satisfy the usual Kolmogorov equations on the domain, with Dirichlet boundary conditions. Consider first the backward Kolmogorov equation. Following Section 9.4, we pose a terminal reward $h(Y_t)$, where h is a smooth function on $\bar{\Omega}$, and define

$$k(x, s) = \mathbf{E}^{Y_s = x} h(Y_t).$$

Arguing as in Section 9.4, we find that k satisfies the same backward Kolmogorov equation

$$\frac{\partial k}{\partial s} + Lk = 0 \text{ for } x \in \Omega,$$

with the terminal condition $k(x, t) = h(x)$ and the Dirichlet boundary condition

$$k(x, s) = h(x) \text{ for } x \in \partial\Omega.$$

This holds because $Y_s = x \in \partial\Omega$ implies that $s = \tau$ and $Y_t = x$. It can be shown that k is continuous also at the boundary since the diffusion is regular.

We next turn to the forward Kolmogorov equation. $Y_s = X_{s \wedge \tau}$ will be distributed over Ω as well as over its boundary $\partial\Omega$. We first pursue the distribution over the interior, assuming that it admits a density $\phi(x, s)$ so that

$$\mathbf{P}\{Y_s \in A\} = \int_A \phi(x, s) \, dx \text{ for } A \subset \Omega.$$

If the terminal reward function h vanishes on the boundary, then

$$\mathbf{E}h(Y_t) = \mathbf{E}\mathbf{E}\{h(Y_t | \mathcal{F}_s\} = \mathbf{E}k(Y_s, s) = \int_\Omega \phi(x, s)k(x, s) \, dx$$

since $k(x, s)$ vanishes on the boundary, so there are no contributions from the boundary to the expectation. As in Section 9.5, we differentiate w.r.t. s to find

$$\int_\Omega \dot{\phi}k + \phi\dot{k} \, dx = 0.$$

We now use $\dot{k} + Lk = 0$, and apply the divergence theorem twice, retaining the boundary terms:

$$\int_\Omega k \left(\dot{\phi} + \nabla \cdot (u\phi - D\nabla\phi) \right) \, dx + \int_{\partial\Omega} [kD\nabla\phi - ku\phi - \phi D\nabla k] \cdot dn(x) = 0.$$

Here, we have used the advection-diffusion characterization of $\{X_t\}$; n is a normal to the boundary $\partial\Omega$ at x, directed outward. Since this equation holds for all k which vanish on the boundary, we see that ϕ must satisfy the usual forward Kolmogorov equation on the interior, i.e.

$$\dot{\phi} = -\nabla \cdot (u\phi - D\nabla\phi) \text{ on } \Omega,$$

as well as the homogeneous Dirichlet boundary condition

$$\phi = 0 \text{ on } \partial\Omega.$$

The intuition behind this boundary condition is that the absorbing boundary voids its neighborhood, since the diffusion is regular: There are almost no particles near the boundary, because they have already been absorbed.

In summary, absorbing boundaries correspond to Dirichlet boundary conditions on the forward and backward Kolmogorov equations. For the forward equation, the Dirichlet boundary condition is homogeneous, $\phi = 0$.

We have not described how Y_t is distributed on the boundary $\partial\Omega$. From the probability density $\phi(x, t)$ on Ω, we can find the flux of probability onto the boundary at a boundary point $x \in \partial\Omega$:

$$(u\phi - D\nabla\phi) \cdot n.$$

The probability distribution of the stopped process on the boundary can therefore be found from this flux, integrating over time. Explicit expressions for these distributions are rare; in low dimensions, numerical analysis is feasible.

What about stationary solutions to the forward Kolmogorov equation? As time $t \to \infty$, the stopped process $Y_t = X_{t \wedge \tau}$ will be distributed only on the boundary $\partial\Omega$: $\mathbf{E}\tau < \infty$ implies that $\mathbf{P}\{\tau \leq t\} \to 1$ and therefore $\mathbf{P}\{Y_t \in \partial\Omega\} \to 1$. So the stationary density is trivial: $\phi(\cdot, t) \to 0$ as $t \to \infty$. Instead, we may search for a *quasi-stationary density* $\rho(x)$, corresponding to an eigensolution of the forward Kolmogorov equation:

$$\phi(x, t) = \rho(x)e^{-\lambda t}.$$

Here, $\lambda > 0$ is the decay rate, and the eigenvalue problem is

$$\lambda\rho = \nabla \cdot (u\rho - D\nabla\rho) \text{ on } \Omega, \quad \rho = 0 \text{ on } \partial\Omega.$$

The quasi-stationary density ρ is the principal eigenfunction, which is positive on Ω and can be normalized to a probability density function on Ω, i.e.,

$\int_\Omega \rho(x)\ dx = 1$. If the initial condition X_0 is random and distributed according to ρ, then the time to absorption is exponentially distributed with rate parameter λ:

$$\mathbf{P}\{\tau > t\} = \int_\Omega \phi(x, t)\ dx = e^{-\lambda t}.$$

The conditional distribution of Y_t, given that $t < \tau$, is time invariant:

$$\mathbf{P}\{Y_t \in dx | \tau > t\} = \frac{\phi(x, t)\ dx}{\mathbf{P}\{\tau > t\}} = \frac{e^{-\lambda t} \rho(x)\ dx}{e^{-\lambda t}} = \rho(x)\ dx.$$

If the initial condition X_0 is not distributed according to ρ, faster modes will be present in the density $\phi(x, t)$ but will die out as $t \to \infty$. Thus, the tail of the distribution of the time to absorption will approach the exponential distribution, and the conditional density of Y_t, conditional on not being absorbed, will approach the quasi-stationary distribution. This justifies the interest in the quasi-stationary distribution.

Example 11.3.1 *Consider standard Brownian motion $X_t = B_t$ on $\Omega = (0, 1)$. The forward Kolmogorov equation is*

$$\dot{\phi} = \frac{1}{2}\phi'', \quad \phi(0) = \phi(1) = 0.$$

The eigenvalue problem governing the quasi-stationary density ρ is

$$\lambda\rho = -\frac{1}{2}\rho'', \quad \rho(0) = \rho(1) = 0.$$

Eigenfunctions are $\sin n\pi x$ for $n \in \mathbf{N}$, corresponding to the eigenvalues $-n^2\pi^2/2$. The principal mode is given by $n = 1$, i.e., the slowest mode, and the only mode with a positive eigenfunction. Thus, the (unnormalized) quasi-stationary density is $\rho(x) = \sin \pi x$, corresponding to the decay rate $\lambda = \pi^2/2$.

11.4 THE EXPECTED POINT OF EXIT

We now consider how the absorbed process Y_t will be distributed on the boundary as $t \to \infty$. The setting is a diffusion $\{X_t\}$ governed by the Itô SDE $dX_t = f(X_t)\ dt + g(X_t)\ dB_t$ with the initial condition $X_0 = x \in \bar{\Omega}$, where $\Omega \in \mathbf{R}^n$ is an open and bounded domain. We stop the process at τ when it hits the boundary $\partial\Omega$. To describe the distribution on the boundary, we define a reward function $c(x)$ on the boundary, and aim to evaluate

$$\mathbf{E}^x c(X_\tau) \text{ where } \tau = \inf\{t : X_t \notin \Omega\}.$$

This reward function $c(x)$ can be seen as a vehicle for determining the distribution on the boundary of X_τ. However, there are also applications where it is a quantity of direct interest; for example, for a control mission, τ could indicate that the mission has been completed, and $c(X_\tau)$ could be a measure of how well the mission went.

Theorem 11.4.1 *Assume that a function* $h : \bar{\Omega} \mapsto \mathbf{R}$ *is* C^2 *and satisfies*

$$Lh(x) = 0 \text{ for } x \in \Omega, \text{ and } h(x) = c(x) \text{ for } x \in \partial\Omega.$$

Assume further that $\mathbf{E}^x \tau < \infty$ *for all* $x \in \Omega$. *Then*

$$h(x) = \mathbf{E}^x c(X_\tau).$$

The proof is a straightforward application of Dynkin's formula. Note that we explicitly assume that $\mathbf{E}^x \tau < \infty$; we know from Section 11.2.4 that a sufficient condition is that the diffusion is regular, $g(x)g^\top(x) > dI > 0$.

11.4.1 Does a Scalar Diffusion Exit Right or Left?

Consider Brownian motion with drift, as in Section 11.2.1, i.e., $dX_t = u\,dt + \sigma\,dB_t$ where $X_t \in \mathbf{R}$ and u and σ are positive constants. Let the domain be $\Omega = (0, l)$. As usual, let $\tau = \inf\{t : X_t \notin (0, l)\}$ be the time of exit; we know from Section 11.2.4 that $\mathbf{E}^x \tau < \infty$ for all $x \in \Omega$. We aim to determine the probability that the process exits to the right, i.e., find

$$h(x) = \mathbf{P}^x\{X_\tau = l\}.$$

According to the Dynkin formula, this probability is governed by the boundary value problem

$$uh'(x) + Dh''(x) = 0 \text{ for } x \in (0, l), \quad h(0) = 0, \quad h(l) = 1 \qquad (11.4)$$

where the diffusivity is $D = \sigma^2/2$: If h satisfies this boundary value problem, then $h(x) = \mathbf{E}^x h(X_\tau) = \mathbf{P}^x\{X_\tau = l\}$. The solution of this second order linear ordinary differential equation is

$$h(x) = \frac{1 - \exp(-ux/D)}{1 - \exp(-ul/D)}.$$

We can once again introduce the Péclet number $\mathrm{Pe} = ul/D$; then the solution can be written in terms of Pe and non-dimensional position x/l as

$$h(x) = \frac{1 - \exp(-\mathrm{Pe}\ x/l)}{1 - \exp(-\mathrm{Pe})} \qquad (11.5)$$

Note the exponentially decaying term $\exp(-\mathrm{Pe}\ x/l)$. For large Péclet numbers, i.e., when $D/u \ll l$, a verbal characterization of the solution is a diffusive boundary layer around $x = 0$ in which there is a significant probability of exiting to the left. This boundary layer occupies a fraction $1/\mathrm{Pe}$ of the domain, i.e., it has the width $l/\mathrm{Pe} = D/u$. Outside the boundary layer, the process is nearly certain to exit the right. See Figure 11.1 (left panel), where the solution is plotted for various Péclet numbers.

We now aim to generalize this example to any scalar diffusion process $\{X_t\}$. Assume that the process starts in an interval $\bar{\Omega} = [0, l]$ such that g vanishes nowhere in this interval. Defining, as before, the time of exit

$$\tau = \inf\{t : X_t \neq (0, l)\}$$

we aim to find $h(x) = \mathbf{P}^x(X_\tau = l)$. This h is governed by the boundary value problem

$$Lh(x) = h'f + \frac{1}{2}g^2h'' = 0 \text{ on } (0, l), \ h(0) = 0, \ h(l) = 1 \ .$$

We have previously, in Section 7.6.3, studied this equation, and found that the full solution could be written

$$h(x) = c_1 s(x) + c_2$$

where c_1 and c_2 are arbitrary real constants. Here, s is a scale function

$$s(x) = \int_{x_0}^x \phi(y) \, dy$$

where

$$\phi(x) = \exp\left(\int_{x_0}^x \frac{-2f(y)}{g^2(y)} \, dy\right) \ .$$

Here, the lower limit x_0 of the integration is arbitrary. We fix the coefficients c_1 and c_2 through the boundary conditions $h(0) = 0$, $h(l) = 1$, and find

$$h(x) = \frac{s(x) - s(0)}{s(l) - s(0)}.$$

The following exercises give a few examples.

Exercise 11.3: Consider an unbiased random walk on the interval $\bar{\Omega} = [0, l]$ with spatially varying diffusivity, i.e., $f = 0$ while $g > 0$ is not constant. Determine the function $h(x) = \mathbf{P}^x(X_\tau = l)$.

Exercise 11.4: Consider pure diffusion $dX_t = D'(X_t) \, dt + \sqrt{2D(X_t)} \, dB_t$ on the interval $\bar{\Omega} = [0, l]$ with a diffusivity $D(x)$ which is positive and increases with x. Consider $h(l/2)$, the probability of exit to the right, given that the process starts in the center. Is this probability greater or smaller than $1/2$?

11.5 RECURRENCE OF BROWNIAN MOTION

Recall from Section 4.3 that Brownian motion in one dimension is recurrent: With probability 1, it will reach any given point in space at some point of time. We now ask if a similar property holds for n-dimensional Brownian motion. Let $\{X_t\}$ be a diffusion in $\mathbf{R}^n$ and let $A \subset \mathbf{R}^n$ be an arbitrary subset.

Let τ be the time of first entry into A, i.e., $\tau = \inf\{t : X_t \in A\}$; recall our convention that $\tau = \inf \emptyset = \infty$ if the process never enters A. We then say that A is *recurrent* if $\mathbf{P}\{\tau < \infty | X_0 = x\} = 1$ for any $x \in \mathbf{R}^n$; i.e., regardless of the initial condition, we are certain to enter A at some point. Otherwise A is *transient*. The process itself is said to be recurrent, if any set A with non-empty interior is recurrent.

Now let $\{X_t\}$ be Brownian motion in $\mathbf{R}^n$ with $n \geq 2$. We first investigate if the sphere $\{x : |x| \leq r\}$ is recurrent for given $r > 0$. To this end, stop the process when it either hits the "inner" sphere $\{x : |x| \leq r\}$ or leaves an "outer" sphere $\{x : |x| \leq R\}$ for given $R > r$; later we let $R \to \infty$. Thus, let the domain be $\Omega = \{x \in \mathbf{R}^d : r < |x| < R\}$. Define the following exit times:

$$\tau_r = \inf\{t : |X_t| \leq r\},$$
$$\tau_R = \inf\{t : |X_t| \geq R\},$$
$$\tau = \inf\{t : X_t \notin \Omega\} = \min\{\tau_r, \tau_R\}.$$

Define the probability of hitting the inner sphere first:

$$h(x) = \mathbf{P}^x\{|X_\tau| = r\} = \mathbf{P}^x\{\tau_r < \tau_R\}.$$

Noting that $\mathbf{E}^x \tau < \infty$, Dynkin's formula allows us to determine h by solving the governing equation

$$\nabla^2 h = 0$$

with boundary conditions $h(x) = 1$ on the inner sphere $\{x : |x| = r\}$ and $h(x) = 0$ on the outer sphere $\{x : |x| = R\}$. Due to spherical symmetry h can be a function of $|x|$ only; with an abuse of notation, we write $h = h(|x|)$. Writing the Laplacian in spherical coordinates, h must satisfy

$$h''(|x|) + \frac{d-1}{|x|}h'(|x|) = 0 \text{ for } |x| \in (r, R)$$

along with boundary conditions $h(r) = 1$, $h(R) = 0$. Linear independent solutions to this equation are $h \equiv 1$ and

$$h(|x|) = \begin{cases} \log|x| & \text{for } n = 2, \\ |x|^{2-n} & \text{for } n \geq 3. \end{cases}$$

We obtain the solution to the boundary value problem as a linear combination of these solutions, determining the coefficients so as to satisfy boundary conditions. In two dimensions, $n = 2$, we find

$$h(|x|) = \frac{\log(R/r) - \log(|x|/r)}{\log(R/r)}$$

while in higher dimensions, $n \geq 3$, we find

$$h(|x|) = \frac{R^{2-n} - |x|^{2-n}}{R^{2-n} - r^{2-n}}.$$

These expressions hold when we start between the inner and outer sphere, i.e., for an initial condition x such that $r \leq |x| \leq R$. We can now, for fixed $|x|$ and r, let $R \to \infty$. In $n = 2$ dimensions, we get

$$h(|x|) = \frac{\log(R/r) - \log(|x|/r)}{\log(R/r)} = 1 - \frac{\log(|x|/r)}{\log(R/r)} \to 1 \text{ as } R \to \infty.$$

As the outer sphere goes to infinity, we are certain to hit the inner sphere first. Hence, we are certain to hit the inner sphere at some point:

$$\mathbf{P}^x\{\tau_r < \infty\} \geq \mathbf{P}^x\{\tau_r < \tau_R\} \to 1 \text{ as } R \to \infty.$$

We conclude that in two dimensions, the disk $\{x : |x| \leq r\}$ is recurrent. Now let A be any set with non-empty interior, then A contains a disk. By invariance under translations, this disk is recurrent; thus also A recurrent. Since A was arbitrary, we conclude that Brownian motion in two dimensions is recurrent.

In more than two dimensions, $n \geq 3$, we get

$$h(|x|) = \frac{R^{2-n} - |x|^{2-n}}{R^{2-n} - r^{2-n}} \to \left(\frac{r}{|x|}\right)^{n-2} \text{ as } R \to \infty.$$

Therefore, in three dimensions or more, there is a non-zero probability of never hitting the inner sphere, provided that we start outside it:

$$\mathbf{P}^x\{\tau_r < \infty\} = \lim_{R \to \infty} \mathbf{P}^x\{\tau_r < \tau_R\} = (r/|x|)^{n-2}.$$

Here we have used that $\tau_R \to \infty$ in probability as $R \to \infty$ for fixed x. We conclude that the inner sphere is transient and thus Brownian motion is itself transient, in three dimensions or higher. Thus, Brownian motion, and diffusion in general, behaves quite differently in 3 or more dimensions than in 1 or 2 dimensions.

11.6 THE POISSON EQUATION

So far, we have been concerned with two questions: The time of exit τ, and the expectation of a reward $c(X_\tau)$ which depends on the point of exit. Now, we consider also rewards that are accumulated until exit. Such accumulated rewards occur in many applications: For example, in finance, there may be running dividends or profits, while in control systems, the "reward" is typically negative and specifies a penalty for poor performance or for cost of operation.

With the same setting as in Section 11.4, we have the following result:

Theorem 11.6.1 *Assume that a C^2 function $h : \bar{\Omega} \mapsto \mathbf{R}$ is C^2 satisfies the Poisson equation*

$$Lh(x) + r(x) = 0 \text{ for } x \in \Omega$$

with Dirichlet boundary conditions

$$h(x) = c(x) \text{ for } x \in \partial\Omega.$$

Assume further that $\mathbf{E}^x \tau < \infty$ *for all* $x \in \Omega$. *Then*

$$h(x) = \mathbf{E}^x \left[c(X_\tau) + \int_0^\tau r(X_t) \, dt \right].$$

Also this theorem is a straightforward application of Dynkin's lemma, and it generalizes our previous result, where $r(x) \equiv 0$ or $r(x) \equiv 1$. Once again we assume explicitly that $\mathbf{E}^x \tau < \infty$, knowing that a sufficient condition for this is that the diffusion is regular, $g(x)g^\top(x) > dI > 0$.

With this result in mind, a reasonable question to ask if the expected reward always satisfies the Poisson/Dirichlet problem. To answer this question, a more lengthy discussion is needed; see e.g. (Øksendal, 2010). One of the complicating factors is that the expected reward may not be smooth or even continuous, when the diffusion is not regular.

11.7   ANALYSIS OF A SINGULAR BOUNDARY POINT

We continue the analysis of whether a scalar diffusion process exits right or left. Consider the squared Bessel process $\{X_t\}$ given by the Itô SDE

$$dX_t = \mu \, dt + \sigma \sqrt{X_t} \, dB_t. \tag{11.6}$$

Here, the point $x = 0$ is a *singularity* in the sense that the noise intensity vanishes, $g(0) = 0$. To study this singularity, we approach it carefully, first taking the domain to be the interval $\Omega = (a, b)$ with $0 < a < b$. Since $a > 0$, Theorem 8.3.2 guarantees existence and uniqueness of a solution up to exit from the interval (a, b), when the initial condition $X_0 = x$ is in that interval. The scale function is

$$s(x) = \int^x \exp\left(\int^y -\frac{2\mu}{\sigma^2 z} \, dz\right) dy$$

$$= \int^x \exp\left(-\frac{2\mu}{\sigma^2} \log y\right) dy$$

$$= \int^x y^{-2\mu/\sigma^2} \, dy$$

$$= \frac{1}{\nu} x^\nu \text{ when } \nu := 1 - 2\mu/\sigma^2 \neq 0.$$

So the probability of exit at b before at a is

$$h(x) = \mathbf{P}^x\{X_\tau = b\} = \frac{x^\nu - a^\nu}{b^\nu - a^\nu}$$

when starting at $X_0 = x$.

We now ask what happens when $a \to 0$, so that the domain approaches the singularity. Figure 11.3 shows the function $h(x; a) = \mathbf{P}(X_\tau = b | X_0 = x))$ for $b = 1$, different values of a and for two sets of system parameters. In the

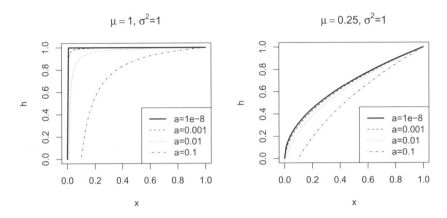

Figure 11.3 The probability of exit to the right of $(a, 1)$ for the squared Bessel process (11.6), as function of the initial condition x, for $b = 1$ and different values of a. Left panel: With $\mu = 1$, $\sigma^2 = 1$, the probability approaches 1 as $a \to 0$. Right panel: With $\mu = 0.25$, $\sigma^2 = 1$, the probability of exit to the right approaches $\sqrt{x}$ as $a \to 0$.

left panel, we have $\mu = 1$, $\sigma^2 = 1$ and therefore $\nu = -1$. We see that $\mathbf{P}(X_\tau = 1 | X_0 = x) \to 1$ as $a \to 0$, for any value of $0 < x < 1$. The interpretation is that when the drift away from the singular point $x = 0$ is stronger than the noise, the singular point will never be reached; when a is very close to the singular point, we are almost certain to exit at 1 rather than at a. From the expression for h, we see that this will be the case when $\nu < 0$, i.e., when $\mu > \frac{1}{2}\sigma^2$.

In contrast, in the right panel we have $\mu = 1/4$, $\sigma^2 = 1$, and thus $\nu = 1 - 2\mu/\sigma^2 = 1/2$. We see that $\mathbf{P}(X_\tau = 1 | X_0 = x) \to \sqrt{x}$ when $a \to 0$. With a slight jump to the conclusion, this implies that with $a = 0$, there is a probability of $\sqrt{x}$ of exiting at the right end 1, and a probability of $1 - \sqrt{x}$ of exiting to the left, at the singular point. The interpretation is now that the drift is relatively weak and random fluctuations due to the noise may cause the state to actually hit the singular point at the origin.

For completeness, we treat also the threshold case $\nu = 0$, which is obtained, for example, with $\sigma = 1$, $\mu = 1/2$. We then get the scale function $s(x) = \log x$ which has a singularity at $x = 0$, so that the origin is not attainable. Specifically, we get $h(x; a) = \log(x/a)/\log(b/a)$ and $h(x; a) \to 1$ as $a \to 0$, for any $0 < x < b$.

To interpret these results, recall from Section 7.5.2 that the squared Bessel process arises as the sum of squares of $n = 4\mu/\sigma^2$ Brownian motions, when this n is integer. In this parametrization, we get $\nu = 1 - n/2$, and we see that the origin is attainable when $n < 2$.

This example was tractable because the drift and noise terms are sufficiently simple that the scale function can be given in closed form. But we see that the conclusion follows from a local analysis of the scale function s near a singular point x_0. Specifically, if the scale function diverges as $x \to x_0$, then the singular point is *repelling* in the sense that as $a \to x_0$, the probability of reaching a vanishes. Conversely, if the scale function has a finite limit as $x \to x_0$, then the singular point x_0 is *attainable* in the sense that there is a positive probability of sample paths which converge to x_0 in finite or infinite time. Moreover, the behavior of the scale function near the singular point depends only on the local behavior of f and g near x_0. Let us illustrate with a couple of examples.

Example 11.7.1 (The Cox-Ingersoll-Ross process)

$$With \ dX_t = \lambda(\xi - X_t) \ dt + \gamma\sqrt{X_t} \ dB_t$$

the origin $x = 0$ is a singular point. Near $x = 0$, the behavior of the scale function is identical to that of equation (11.6) with $\mu = \lambda\xi$, $\sigma = \gamma$. We see that the origin $x = 0$ is attainable if and only if $\lambda\xi < \frac{1}{2}\gamma^2$. Viewing the CIR process as the sum of squares of $n = 4\lambda\xi/\gamma^2$ Ornstein-Uhlenbeck processes (Section 7.5.2), we see that the origin is attainable when $n < 2$; i.e., the same result as for the squared Bessel process.

Example 11.7.2 (Geometric Brownian motion)

$$With \ dX_t = rX_t \ dt + \sigma X_t \ dB_t$$

the scale function agrees with that of (11.6) with $\mu = r$, $\sigma = \sigma$. So we find the same conclusion: The origin is attainable if $r < \frac{1}{2}\sigma^2$. The two models, geometric Brownian motion and the squared Bessel process (11.6), have the same scale function because the two models are time changes of eachother (compare Exercise 7.21).

Example 11.7.3 (Exponential Growth with Demographic Noise)
Here,

$$dX_t = rX_t \ dt + \sigma\sqrt{X_t} \ dB_t$$

models the population dynamics of a bacterial colony, for example. We use the term "demographic noise" to describe this structure where $g(x) \sim \sqrt{x}$, because the variance of the increment is proportional to the state, which is consistent with a view that births and deaths occur randomly and independently at the individual level. We get

$$\phi(x) = \exp\left(\int_0^x \frac{-2ry}{\sigma^2 y} dy\right) = \exp(-2rx/\sigma^2)$$

so that the scale function is

$$s(x) = \exp(-2rx/\sigma^2).$$

Biography: William (Vilibald) Feller (1906–1970)
Born in Croatia, he worked in Germany, briefly in Denmark, then Sweden, before emigrating to the United States in 1939, where he worked at Brown, Cornell and from 1950 at Princeton. His work in probability took offset in the measure-theoretic foundation due to Kolmogorov. Like Kolmogorov, his approach to diffusion processes was based on the semigroup structure of the transition probabilities. His legacy is evident in the notions of Feller continuity, Feller processes, and Feller boundary classification.

Since the scale function has no singularity at 0, the singularity $x = 0$ is attainable: With $r > 0$, $s(x)$ is the probability that the point $x = 0$ is reached at some point. The scale function coincides with that of Brownian motion with drift: These two processes are random time changes of each other.

Example 11.7.4 (Logistic Growth with Demographic Noise) *Expand the previous model by adding a carrying capacity at $x = K$, as in logistic growth:*

$$dX_t = rX_t(1 - X_t/K)\, dt + \sigma\sqrt{X_t}\, dB_t \quad \text{with } r > 0.$$

The carrying capacity K does not affect the scale function near $x = 0$, so the origin is still attainable. However, now the process cannot diverge to $x \to \infty$, so it will revisit the region near $x = 0$ until eventually absorbed at $x = 0$. This process does not admit a stationary distribution on $(0, \infty)$ that can be normalized to integrate to 1, so the stationary distribution is a Dirac delta at $x = 0$. In summary, a population following stochastic logistic growth with demographic noise will eventually go extinct, almost surely.

What we have seen is basic notion of boundary classification due to Feller, i.e. the study of the natural boundaries that arise at singularities. See (Gard, 1988) for precise statements and elaboration, also concerning the time it takes to reach an attainable boundary point.

11.8   DISCOUNTING AND THE FEYNMAN-KAC FORMULA

We now turn to an extension of Kolmogorov's backward equation and Dynkin's formula, which allows to include exponential weights. These weights can be used to discount the value of future gains and losses, which finds direct applications in economy, but can also be used in ecology where animals need to take into account the risk of dying.

As before, we let $\{X_t \in \mathbf{R}^n\}$ be an Itô diffusion which satisfies the stochastic differential equation $dX_t = f(X_t)\, dt + g(X_t)\, dB_t$, and we let L be the backward Kolmogorov operator. We assume that a *discount rate* $\mu : \mathbf{R}^n \mapsto [0, \infty)$ is defined on state space; we consider only non-negative discounting. We also assume that μ is smooth, but in this section we will largely ignore such technical conditions.

We now consider terminal value problems in $h = h(x, t)$ where $x \in \mathbf{R}^n$, $0 \le t \le T$:

$$\dot{h} + Lh - \mu h = 0, \quad h(x, T) = k(x). \tag{11.7}$$

The combined operator $h \mapsto Lh - \mu h$ is sometimes referred to as the *subgenerator*. As for the backward Kolmogorov equation, we allow the notation

$$h(x, t) = \left[e^{(L-\mu)(T-t)} k \right](x)$$

for the solution to such a terminal value problem. The main result is that a solution v to such a terminal value problem has a stochastic interpretation, viz:

$$h(x, t) = \mathbf{E}^{X_t = x} \left\{ e^{-\int_t^T \mu(X_s) \, ds} k(X_T) \right\}. \tag{11.8}$$

Thus, h can be interpreted as a "present day value", i.e., an expected discounted terminal reward, given the initial condition $X_t = x$, where the terminal reward $k(X_T)$ depends on the terminal state, and the reward is discounted along the trajectory with rate $\mu(X_s)$.

To see the connection between the discounted expectation in (11.8) and the terminal value problem (11.7), assume that h satisfies the terminal value problem. Then define the process $\{R_t\}$ which measures the discounting and is given by

$$dR_t = -\mu(X_t) R_t \, dt$$

so that

$$R_t = R_s \exp \left(- \int_s^t \mu(X_u) \, du \right)$$

for $0 \le s \le t \le T$. Next, define the process $\{Y_t\}$ by $Y_t = R_t h(X_t, t)$, for $t \in [0, T]$. Then Itô's lemma gives

$$dY_t = R_t \left[\dot{h} + Lh - \mu h \right] \, dt + R_t \nabla v \, g \, dB_t = R_t \nabla v \, g \, dB_t$$

where we have omitted the arguments (X_t and t) for brevity. Hence, $\{Y_t\}$ is an Itô integral. Now assume that the integrand is well behaved so that $\{Y_t\}$ is a martingale, then

$$\begin{aligned} Y_t &= R_t h(X_t, t) \\ &= \mathbf{E}\{Y_T | \mathcal{F}_t\} \\ &= \mathbf{E}\{R_T h(X_T, T) | \mathcal{F}_t\} \\ &= R_t \mathbf{E}\{e^{-\int_t^T \mu(X_s) \, ds} k(X_T) | \mathcal{F}_t\}. \end{aligned}$$

Thus,

$$h(X_t, t) = \mathbf{E}\{e^{-\int_t^T \mu(X_s) \, ds} k(X_T) | \mathcal{F}_t\}$$

i.e.,

$$h(x,t) = \mathbf{E}^{X_t = x}\{e^{-\int_t^T \mu(X_s)\,ds} k(X_T)\}.$$

While this establishes the connection between the discounted expectation in (11.8) and the terminal value problem (11.7), we have jumped to the conclusion in that we simply assumed that the Itô integral $\{Y_t\}$ is a martingale. Clearly, some technical requirements are needed for this to hold. Øksendal (2010) assumes that the solution h to the terminal value problem is bounded and concludes that it has the interpretation (11.8); conversely, if the terminal value k is smooth and has bounded support, then the expecation (11.8) satisfies the terminal value problem. The bounded support ensures that tail contribuions to the expectation vanish; Karatzas and Shreve (1997) and Rogers and Williams (1994a) user milder requirements. In the following examples, we ignore these issues of regularity and assume that the equivalence between (11.8) and (11.7) holds.

11.8.1 Pricing of Bonds

As an application of the Feynman-Kac formula, we turn to mathematical finance. Here, modeling of interest rates is important; among other reasons, because it allows pricing of bonds. Consider a so-called *zero-coupon bond*, where the issuer pays the holder a fixed sum K of money at the time of maturity T, and no payments are made before that. If the interest rate $\{X_t\}$ were known in advance as a function of time, the fair price of such a bond at time $t < T$ would be

$$Ke^{-\int_t^T X_s\,ds},$$

because the holder could, instead of buying the bond at time t, put the money $K \exp(-\int_t^T X_s\,ds)$ in the bank and let it earn interest - in either case, the holder's portfolio would be worth K at time T. When the interest rate $\{X_s\}$ evolves stochastically between times t and T, the fair price at time t is

$$\mathbf{E}\left\{Ke^{-\int_t^T X_s\,ds}|\mathcal{F}_t\right\}$$

where $\mathcal{F}_t$ is the information available to the market at time t and expectation is with respect to $\mathbf{P}$, the risk neutral measure (Hull, 2014). To evaluate this expectation, a common model for the evolution of interest rates is the Cox-Ingersoll-Ross process

$$dX_t = \lambda(\xi - X_t)\,dt + \gamma\sqrt{X_t}\,dB_t$$

where $\{B_t\}$ is Brownian motion w.r.t. $\mathbf{P}$ and $\{\mathcal{F}_t\}$. Since $\{X_t\}$ is Markov, the fair price depends on the current interest rate only, i.e.,

$$h(X_t,t) = \mathbf{E}\{Ke^{-\int_t^T X_s\,ds}|\mathcal{F}_t\}.$$

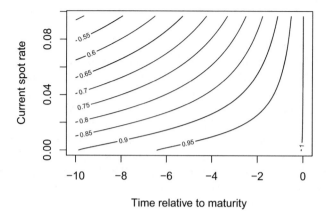

Current spot rate

Time relative to maturity

Figure 11.4   Price $h(x, s - T)$ of a zero-coupon bond as a function of time $s - T$ relative to maturity and current spot rate $X_s = x$. From (11.9) with $K = 1$, $\lambda = 0.1$, $\xi = 0.025$ and $\gamma = 5 \cdot 10^{-5}$.

It then follows from the Feynman-Kac formula that h is governed by the equation

$$\dot{h} + \lambda(\xi - x)h' + \frac{1}{2}\gamma^2 x h'' - xh = 0, \ x \geq 0, \ t < T, \tag{11.9}$$

with terminal condition $h(x, T) = K$. It is possible to solve this partial differential equation analytically (Hull, 2014) but here we solve it numerically using the methods from Section 9.11.5; see Figure 11.4. We see that the price of the bond is greater, the closer we are to the time of maturity, and always smaller than $K = 1$ since rates can never be negative in the Cox-Ingersoll-Ross model. At any given time t, the price is a decreasing function of the current spot rate: If the current spot rate is high, then we expect it to remain high for some time, and therefore the price is discounted much. Conversely, when the current spot rate is nearer 0, we expect less discounting until the time of maturity, and therefore the fair price of the bond is higher.

11.8.2   Darwinian Fitness and Killing

As another application of the Feynman-Kac formula, we turn to behavioral ecology. Here, we model the internal state of an individual animal, and aim to assess its Darwinian fitness; that is, its expected number of future descendants. When computing this expectation, we need to take into account the possibility that the animal dies before producing offspring; this leads to a discounting as in (11.8), as we will see in the following.

We consider an animal the state of which is $\{X_t\}$; our interest is the time interval $t \in [0, T]$, where T is a fixed time of breeding. If the animal survives to time T, we let $k(X_T)$ denote the number of its offspring, but if the animal dies before time T, it produces no offspring. We define the fitness $h(x,t)$ as the expected number of offspring, given that the animal is alive at time t and in state $X_t = x$. To make the connection to the discounted expectation in (11.8), we assume that death occurs with a state-dependent mortality rate $\mu(X_t)$, and define the cumulated mortality

$$M_t = \int_0^t \mu(X_s)\, ds.$$

Then, the discount factor $R_t = \exp(-M_t)$ is the probability that the animal is still alive at time t, conditional on the internal state $\{X_s : 0 \le s \le t\}$. For example, with constant mortality $\mu(x) \equiv m$, the probability that the animal survives to time t is $\exp(-mt)$; i.e., the lifetime is exponentially distributed with rate parameter m. When the mortality is not constant, we take a random variable U, uniformly distributed on $(0,1)$ and independent of the filtration $\{\mathcal{F}_t\}$. We now define the time of death τ implicitly through the equation

$$M_\tau = -\log U. \tag{11.10}$$

One interpretation of this is that the animal is assigned a random number $-\log U$ of life points at time 0. The mortality gradually eats up these life points; the animal dies if and when it runs out of life points. We then have

$$\mathbf{P}\{\tau > t\} = \mathbf{P}\{-\log U > M_t\} = \mathbf{P}\{U < \exp(-M_t)\} = \mathbf{E}\exp(-M_t).$$

We can now give a precise characterization of the fitness $h(x,t)$ of an animal, which is alive at time t and in state $X_t = x$:

$$h(x,t) = \mathbf{E}\{k(X_T)\mathbf{1}(\tau > T)|X_t = x, \tau > t\}.$$

To evaluate this expectation, we first condition also on $\mathcal{F}_T$, i.e. the state trajectory of the animal, so that the only source of randomness is U. We use (11.10) to substitute $\tau > T$ with $M_T < -\log U$, and find:

$$
\begin{aligned}
& \mathbf{E}\{k(X_T)\mathbf{1}(M_T < -\log U)|\mathcal{F}_T, M_t < -\log U\} \\
= \ & k(X_T)\mathbf{P}\{e^{-M_T} > U|\mathcal{F}_T, e^{-M_t} > U\} \\
= \ & k(X_T)e^{M_t - M_T} \\
= \ & k(X_T)e^{-\int_t^T \mu(X_s)\, ds}.
\end{aligned}
$$

Thus, using the Tower property, we obtain

$$h(x,t) = \mathbf{E}^{X_t = x}\left[k(X_T)e^{-\int_t^T \mu(X_s)\, ds} \right],$$

and therefore the Feynman-Kac formula gives us that the fitness h satisfies the terminal value problem (11.7), which we restate:

$$\dot{h} + Lh - \mu h = 0, \quad h(x,T) = k(x).$$

Notice that fitness h spreads in state space according to the usual backward Kolmogorov equation, but in addition is lost with rate μh as time goes backwards: The closer the time t is to the terminal time T, the smaller is the probability that the animal dies in the time interval $[t, T]$, so the greater is the fitness. This equation allows us to assess the fitness of animals, for different states and for different state dynamics. We can therefore identify desirable and undesirable states, investigate the trade-offs that animals face, and evaluate if one behavioral strategy holds an evolutionary advantage over another. We shall see an example of this shortly (Section 11.8.4) and elaborate further in Chapter 13, where we identify optimal behavioral strategies.

At this point, it is illuminating to look at the (formal) adjoint of the subgenerator $L - \mu$, i.e., the forward equation in $\phi = \phi(x, t)$

$$\dot\phi = L^*\phi - \mu\phi.$$

This equation describes how probability is redistributed in space through advection and diffusion, *and* lost with rate $\mu(x)\phi(x, t)$. Thus, the solution $\phi(x, t)$ governs the probability that the animal is alive at time t and near state x.

11.8.3 Cumulated Rewards

So far, we have considered rewards issued at a fixed terminal time T, but in many situations the rewards are cumulated over time (compare Section 11.6). In the ecological example, many animals may produce offspring at any point in time; similarly, bonds may have coupons and pay interest until the time of maturity. Both situations lead us to consider value functions such as

$$h(x, t) = \mathbf{E}\left\{ \int_t^T e^{-\int_t^s \mu(X_u)\, du} r(X_s)\, ds \,\middle|\, X_t = x \right\}. \tag{11.11}$$

In the following we show that this expected cumulated and discounted reward h is governed by the partial differential equation

$$\dot h + Lh - \mu h + r = 0, \quad h(x, T) = 0 \tag{11.12}$$

for $x \in \mathbf{R}^n$, $t \in [0, T]$. In our discussion, we interpret this as the fitness of an animal with a state-dependent fecundity $r(x)$, i.e. the rate with which offspring are produced. Let an animal be alive at time t and in a random state X_t which has probability density $\phi(x, t)$. Then we can compute the expected number (say, J) of offspring produced in the interval $[t, T]$ by conditioning on the position at time t:

$$J := \int_{\mathbf{R}^n} \phi(x, t)h(x, t)\, dx.$$

An alternative "forward" view is to use the distribution $\phi(x, s)$ to find the expected fecundity at time s, and then integrate over $s \in [t, T]$. This yields:

$$J = \int_t^T \int_{\mathbf{R}^n} \phi(x, s) r(x) \, dx \, ds$$

These two expressions for J must agree. Differentiating w.r.t. t, we get

$$-\int_{\mathbf{R}^n} \phi(x, t) r(x) \, dx = \int_{\mathbf{R}^n} [\dot{\phi}(x, t) h(x, t) + \phi(x, t) \dot{h}(x, t)] \, dx.$$

Using $\dot{\phi} = L^* \phi - \mu \phi$ and the duality $\int [h \, L^* \phi - \phi \, Lh] \, dx = 0$, we get

$$\int_{\mathbf{R}^n} \phi \left[\dot{h} + Lh - \mu h + r \right] \, dx = 0.$$

Since $\phi(\cdot, t)$ can be chosen arbitrarily, the expected cumulated and discounted reward h given by (11.11) must satisfy the terminal value problem (11.12).

In many applications, the problem does not specify the terminal T; rather, our interest is in the "infinite horizon" situation where $T - t \to \infty$. In that case, we pursue an equilibrium for this backward equation, viz.

$$Lh - \mu h + r = 0,$$

in the hope that such an equilibrium $h(x)$ would have the characterization

$$h(x) = \mathbf{E}^{X_0 = x} \left[\int_0^\infty e^{-\int_0^t \mu(X_s) \, ds} r(X_t) \, dt \right].$$

Of course, technical requirements are needed for this equivalence to hold, e.g. that μ is bounded away from 0 and r is bounded.

Exercise 11.5: Consider the linear-quadratic version of these equations with system dynamics $dX_t = AX_t \, dt + G \, dB_t$, a running reward $r(x) = x^\top Q x$ and a constant mortality/discount rate, $\mu(x) = m > 0$. Find the stationary equation for $h(x)$, using the Ansatz $h(x) = x^\top P x + p$. Here Q and P are symmetric. When does the stationary equation have a solution which equals the expected discounted reward over an infinite horizon?

11.8.4 Vertical Motion of Zooplankton

We now consider a specific example: Diel vertical movements of zooplankton, i.e., small animals in the ocean. We assume that the animal moves vertically according to pure diffusion $\{X_t\}$ with a diffusivity of 5000 m^2/day, driven by the random movements of the animal as well as of the water. The animal is reflected at the surface $x = 0$ and at the bottom at $x = H = 100$ m. The animal harvests more energy closer to the surface, i.e., we have a harvest rate

$$r(x) = \frac{1}{1 + \exp((x - \bar{y})/w)} \quad [1/\text{day}].$$

The maximum harvest rate of 1 per day is arbitrary. $\bar{y} = 50$ m marks the *nutricline*, i.e., the extent of nutrient-rich surface waters, and $w = 10$ m governs the width of the transition zone. The animals are hunted by fish which rely on vision and therefore light, so we have

$$\mu(x, t) = \frac{\mu_1 e^{-kx}}{1 + \exp(A\cos(2\pi t/T))} + \mu_0.$$

Here, $\mu_1 = 0.5$ per day is the peak predation mortality, which occurs at noon ($t = T/2$) and at the surface ($x = 0$). $k = 0.05$ m^{-1} is the absorption coefficient of light, so that deep waters are dark and safe. $T = 1$ day is the period, and $A = 13$ controls the difference between light levels at day and at night. The base mortality $\mu_0 = 0.01$ per day is independent of light.

Note that the mortality depends explicitly on time, while the previous sections considered time invariant dynamics only. Going through the arguments, we see that time-varying dynamics can be included straightforwardly.

Figure 11.5 shows the harvest rate r and the mortality μ, as well as model results: the fitness h found with the Feynman-Kac formula, and the quasi-stationary distribution of the animals. We have found the periodic solution numerically (Section 9.11.5), adding time as a state variable. To interpret the fitness, recall that the maximum energy harvest $r(x)$ is 1 per day, so a fitness of 8 corresponds to 8 days of harvest at the surface. Notice that the fitness varies with depth in a way that itself depends on time. This gives the animal an incentive to move vertically, aiming to track the fitness maximum. We will explore this in Section 13.10, where we derive fitness optimizing movement strategies. For the quasi-stationary distribution, we see that fluctuations in the density are small, but that the density is lowest at the surface and in particular at dusk, where the day's mortality has taken its toll.

11.9 CONCLUSION

Between the forward and backward Kolmogorov equations, the forward one probably seems most accessible and intuitive at first glance, even if the two equations represent dual views on the same phenomenon - namely, transition probabilities - and therefore two sides of the same coin. In this chapter, we have seen a number of applications that involve the backward operator. The familiarity we have gained with the backward operator will be useful in the next chapters concerning stability analysis and control.

The theme of "exit problems" has been recurrent, i.e., the idea of stopping a process upon exit from a given domain and then asking where and when the process exits. We have seen that these statistics are governed by partial differential equations which involve the backward operator.

We next used this technique to explore singularities in the model, i.e., points where the diffusion vanishes. In many applications, these singularities are at the origin. In the scalar case, we saw that the scale function can be used to determine if the process will ever hit such a singularity; it is possible

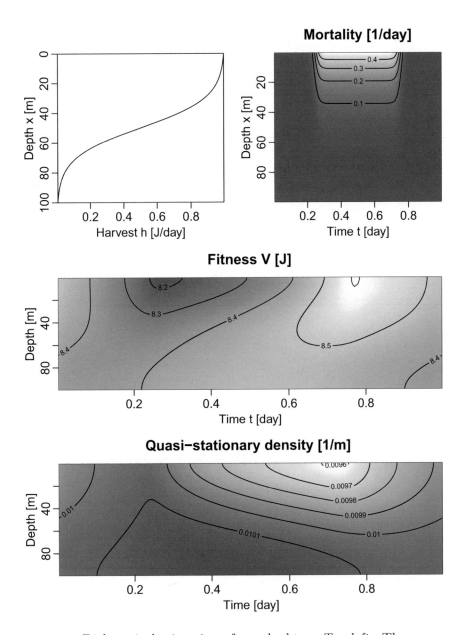

Figure 11.5 Diel vertical migration of zooplankton. *Top left:* The energy harvest $h(x)$ plotted against depth. *Top right:* Mortality $\mu(x,t)$ as a function of time and depth. *Middle:* Resulting fitness $V(x,t)$. *Bottom:* Quasi-stationary distribution $\rho(x,t)$.

to extend this analysis to also determine if the process reaches the singularity in finite or infinite time. It is important to be aware of singularities in one's model, and their properties. Numerical analysis can be sensitive and give misleading results, if one implicitly makes wrong assumptions about the qualitative nature of singularities; for example, by searching for smooth solutions to the Poisson equation. So when models contain singularities, the toolbox for analyzing such singularities is critical.

When we discussed discounting and the Feynman-Kac formula, we took the terminal time to be fixed. We could have combined these two concepts: Stopping a process upon exit, and discounting along the path. This would lead to boundary value problems which involve the sub-generator $L - \mu$. It is possible to form other combinations of the elements we have introduced; for example, stopping processes *either* upon exit *or* at a fixed terminal time, whichever happens first. Some of these extensions are presented in the following exercises, without technicalities.

11.10 NOTES AND REFERENCES

A seminal study that connected partial differential equations and expectations to the future was done by Shizuo Kakutani in 1944 where he studied the Poisson equation with Dirichlet boundary conditions. This motivated Doob to pursue the connection between classical potential theory and stochastic analysis (Doob, 2001).

Dynkin's lemma appeared in (Dynkin, 1965, p. 133).

The study of singularities of their classification was initiated by Feller for scalar processes in 1952 and further in 1957; extensions to the multivariate case was considered by Venttsel (sometimes spelled Wentzell) in 1959.

In this chapter, we have considered classical solutions to the partial differential equations (or inequalities), and their interpretation in terms of expectations to the future. We have not discussed converse statements, i.e., when these expectations to the future satisfy the partial differential equations. See (Doob, 2001) and (Øksendal, 2010). Things are simple when domains are bounded and diffusions are regular (so that the expected time exit is also bounded). Beyond this, technicalities become more challenging beyond this; in particular, the notion of viscosity solutions (Crandall et al., 1992) becomes central.

11.11 EXERCISES

Exercise 11.6: Consider shifted Brownian motion $dX_t = dB_t$, $X_0 = x$, and the first time $\tau = \inf\{t : X_t \geq 1\}$ the process exceeds 1. With $h(x) = x$, show that $\mathbf{E}^x h(X_\tau) = x \wedge 1 = \min(x, 1)$. Next show that $h(x) + \mathbf{E}^x \int_0^\tau Lh(X_t)\, dt = x$. Why does Dynkin's formula does not apply here?

Exercise 11.7: Let $\{X_t : t \geq 0\}$ be Brownian motion with drift, $dX_t = u\,dt + \sigma\,dB_t$ where u and σ are positive constants, and with $X_0 = x > 0$. Let $\tau = \inf\{t : X_t \leq 0\}$. Show that $\mathbf{P}\{\tau < \infty\} = \exp(-ux/D)$ with $D = \sigma^2/2$. Then define $S = \sup\{x - X_t : t \geq 0\}$. Show that S is exponentially distributed with mean D/u. *Note:* Compare with the hitting time distribution from Exercise 9.14.

Exercise 11.8 Numerical Analysis of Scalar Backward Equations: The backward equations are posed as boundary value problems. Here, we consider the scalar case where we can recast them as initial value problems. The motivation for this is that we have more numerical solvers available for initial value problems.

1. Consider the exit point of Brownian motion with drift as in Section 11.4.1. Take $l = 2$, $u = 3$, $D = 4$. Solve the boundary value problem (11.4) numerically as follows: Replace the boundary conditions with $h(0) = 0$, $h'(0) = 1$. Solve this equation numerically on the interval $[0, l]$, rewriting it as two coupled first order equations. Denote the solution $\bar{h}(x)$. Then, rescale the solution, i.e., set $h(x) = \bar{h}(x)/\bar{h}(1)$. Compare with the analytical solution.

2. Repeat for the time to exit: First solve the initial value problem $uh' + Dh'' + 1 = 0$, $h(0) = 0$, $h'(0) = 1$. Then shift the obtained solution with the scale function found in the previous question to obtain a solution to the original boundary value problem.

3. Consider the process leading to the von Mises distribution (Exercise 9.7): $dX_t = -\sin(x)\,dt + \sigma\,dB_t$. For $X_0 = 0$ and $\tau = \inf\{t : |X_t| \geq 2\pi\}$, find $\mathbf{E}\tau$ numerically for $\sigma \in \{2, 1, 1/2, 1/4\}$. *Hint:* Use symmetry to only consider the problem for $x \in [0, 2\pi]$. You may want to first verify your code by applying it to the Ornstein-Uhlenbeck process (Section 11.2.3).

Exercise 11.9 The Expected Exit Point for Unbiased Random Walks: Consider the process $dX_t = g(X_t)\,dB_t$ in a bounded domain $\Omega \subset \mathbf{R}^n$ with $g(x)g^\top(x) > dI > 0$ on $\bar{\Omega}$ and let τ be the time of exit. For a given vector $c \in \mathbf{R}^n$, show that
$$\mathbf{E}^x c X_\tau = cx.$$

Exercise 11.10 The Mean-Square Exit Time: Consider a regular diffusion $\{X_t\}$ on a domain Ω, as in Section 11.2.4, and let $h(x) = \mathbf{E}^x\tau$ be the expected time to exit, which satisfies $Lh + 1 = 0$ on Ω and $h = 0$ on the boundary. Let $k : \Omega \mapsto \mathbf{R}$ satisfy $Lk + 2h = 0$ on Ω and $k = 0$ on $\partial\Omega$. Show that $\mathbf{E}^x\tau^2 = k(x)$. *Note:* This can be extended to moments of arbitrary order; see (Gihman and Skorohod, 1972).

Exercise 11.11 Expected Lifetime in a Growing Population: Consider the stochastic logistic growth model from Example 7.7.1, viz. $dX_t = X_t(1 - X_t)\,dt + \sigma X_t\,dB_t$. Assume that the mortality of an individual animal depends on the abundance of the population, i.e., we have $\mu(x) > 0$.

1. Let $h(x)$ be the expected remaining lifetime of an individual which is alive when the population abundance is $X_0 = x$. Show that this h is governed by the equation

$$x(1-x)h'(x) + \frac{1}{2}\sigma^2 x^2 h''(x) - \mu(x)\,h(x) + 1 = 0.$$

2. Solve the equation numerically for $\mu(x) = \mu_0 \cdot (1+x)$, using parameters $\sigma = 0.5$, $\mu_0 = 0.5$. Truncate the domain to $(0,6)$ and use grid cells of width 0.01. Use reflection at the boundaries. Plot the result.

3. To investigate the effect of μ_0, plot the normalized lifetime $\mu_0 h(x)$ for $\mu_0 \in \{0.01, 0.1, 0.5, 1, 10, 100\}$ and comment. Include for reference the function $1/\mu(x)$.

CHAPTER 12

Stochastic Stability Theory

Stability theory concerns the sensitivity of a model: If we change the model slightly, will it give markedly different predictions? The simplest change to a model is a perturbation of the initial conditions, and stability theory addresses the qualitative effects of such a perturbation.

For ordinary differential equations, stability towards perturbed initial conditions has far-reaching consequences: It implies that exogenous perturbations have bounded effect, as will a slight change in system dynamics. Stability analysis also describes qualitatively the long-term behavior of the system: Whether it comes to rest at an equilibrium, cycles periodically, displays chaotic fluctuations, or diverges to infinity. These connections explain the central position of stability theory in the field of deterministic dynamic systems.

Here, we describe stability analysis for stochastic differential equations. As we will see, some of the deterministic theory generalizes nicely to the stochastic case, even if the technicalities become more demanding. For example, *stochastic Lyapunov exponents* measure how fast nearby trajectories converge or diverge, on average, and *stochastic Lyapunov functions* can be used to guarantee stochastic stability.

For stochastic systems, stability can be understood in several different ways, just as can convergence of random variables. A system may be stable in one sense and unstable in another, and noise may act stabilizing in one sense and one system and destabilizing in others. These contradicting results emphasizes that we must be careful about which notion of stability is relevant in a given situation. Also, equilibrium solutions are less common in stochastic systems than in their noise-free counterparts, so stability analysis of equilibria plays a less pivotal role in the stochastic theory.

Therefore, we also discuss other qualitative properties, such as boundedness, which can be studied using similar techniques, and which give important insight into the qualitative behavior of a stochastic dynamic system.

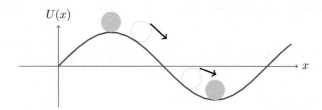

Figure 12.1 Stability for a rolling ball with friction. The hilltop forms an unstable equilibrium, while the bottom forms a stable one. Almost all solutions are stable and converge to the bottom.

12.1 THE STABILITY PROBLEM

Let us first recap the motivation for the stability problem, and the central concepts, in the deterministic context. For a mathematical model of a real-world system, a crucial question is to which degree the predictions from the model agree with the behavior of the real-world system. One of the most basic predictions from dynamic models concern their equilibria: If the initial condition is at an equilibrium, then the state remains there. For a dynamic model, the first thing we ask is typically: What are the equilibria?

Consider a ball moving on a surface subject to gravity and friction (Figure 12.1). A mathematical model of this system has two coupled ordinary differential equations in position X_t and velocity V_t:

$$\dot{X}_t = V_t , \quad \dot{V}_t = -U'(X_t) - \lambda V_t. \tag{12.1}$$

Here, $U(x)$ is the mass-specific potential while λ describes friction. Equilibria in this model are points (x, v) in state space where $v = 0$ and where x is a stationary point for the potential, i.e., $U'(x) = 0$. Most importantly, the ball may stay at rest at the hilltop, or at the bottom of the valley.

However, it is a fragile prediction that the ball may lie at the hilltop: If the initial condition is "close" to the top, then the ball does not remain "close". Even if the ball is at rest there, then a little push from the wind can cause it to roll away. Also, if we place the ball where the map says there is a hilltop, but the map is a little off, then we may lose the ball. In contrast, it is robust prediction that the ball may be at rest at the bottom: If we place the ball close to the bottom, then the ball stays close, and wind or map error will only shift the ball slightly. We say that the top equilibrium is unstable, while the bottom is stable. Note that stability concerns sensitivity to the initial condition, but has implications also for the sensitivity to exogenous perturbations (the wind) and errors in the model (the map).

Stability analysis often focuses on equilibria, but we can also ask non-equilibrium solutions would look radically different, if we changed the initial

Biography: Aleksandr Mikhailovich Lyapunov (1857–1918)
Born in present-day Ukraine; then Russia. Lyapunov studied mathematics and mechanics at the university of St. Petersburg, with Markov and initially under the guidance of Chebyshev. His pioneering contributions to stability theory were part of his doctorate thesis, which he defended in 1892. Among his other contributions was a proof of the Central Limit Theorem based on characteristic functions. Lyapunov ended his life in 1918 after his wife had died from tuberculosis.

condition slightly. In our example, if a solution eventually comes to rest at a local minimum, then that solution is stable.

In applications, the stability of equilibria is often of pivotal importance: Civil engineering constructions are mostly designed to stay at equilibrium, and lack of stability can be disastrous. The collapse in 1940 of the Tacoma Narrows bridge (which is thoroughly documented on the internet) belongs to the common reference frame of engineers. For a chemical production plant, the stability of the operating point is critical. For any biological species living in isolation, an equilibrium is that the species is extinct. If this equilibrium is stable, then a small number of individuals are doomed to extinction, but if the equilibrium is unstable, then these individuals may be able to establish a thriving population. If the species consist of bacteria infecting a human, or virus spreading among human hosts, then the stability problem is relevant to the humans as well.

12.2   THE SENSITIVITY EQUATIONS

We now aim to quantify how sensitive the trajectories are to perturbations in the initial condition. Consider the stochastic differential equation

$$dX_t = f(X_t, t)\ dt + g(X_t, t)\ dB_t. \tag{12.2}$$

where $X_t \in \mathbf{R}^n$ and $B_t \in \mathbf{R}^m$. In this chapter, we assume that the conditions for existence and uniqueness hold (Theorem 8.3.2), unless stated explicitly. We introduce the state transition map

$$\Phi_t(x)$$

which allows us to examine the dependence on initial conditions. For each initial condition x, $\{\Phi_t(x)\}$ is a stochastic process which gives the solution X_t, i.e.,

$$d\Phi_t(x) = f(\Phi_t(x), t)\ dt + g(\Phi_t(x), t)\ dB_t, \quad \Phi_0(x) = x. \tag{12.3}$$

Define the sensitivity $S_t(x) \in \mathbf{R}^{n \times n}$:

$$S_t(x) = \frac{\partial \Phi_t(x)}{\partial x},$$

assuming that $\Phi_t(x)$ is differentiable in x - see (Has'minskĭ, 1980) for a discussion of this point. An infinitesimal perturbation δx of the initial condition will perturb the solution to $\Phi_t(x + \delta x) \approx \Phi_t(x) + S_t(x)\,\delta x$. The sensitivities $\{S_t\}$ satisfies the *sensitivity equations*

$$dS_t(x) = \nabla f(\Phi_t(x), t) \cdot S_t(x)\, dt + \sum_{i=1}^{m} \nabla g_i(\Phi_t(x), t) \cdot S_t(x)\, dB_t^{(i)}, \qquad (12.4)$$

with the initial condition $S_0(x) = I$, the identity, where g_i is the i'th column in the matrix g. To see that these equations hold, differentiate each term in (12.3) and use the chain rule. Alternatively, write up the integral formulation of (12.3), consider the two equations for $\Phi_t(x)$ and $\Phi_t(y)$ and let $y \to x$.

Note that the sensitivity equations are a linear system of stochastic differential equations, in which the nominal solution $X_t = \Phi_t(x)$ determines the coefficients.

Example 12.2.1 (The sensitivities for a narrow-sense linear system)
Consider again the narrow-sense linear system

$$dX_t = AX_t\, dt + G\, dB_t, X_0 = x.$$

The sensitivity equation is
$$dS_t = AS_t\, dt$$
for all $X_0 = x$. With $S_0 = I$, we find $S_t = \exp(At)$. This could also have been found directly from the solution (Exercise 7.4). Note that the sensitivity S_t is deterministic and independent of the initial condition x.

Example 12.2.2 (Sensitivity of a Wide-Sense Linear System) *With*

$$dX_t = AX_t\, dt + \sum_{i=i}^{m} G_i X_t\, dB_t^{(i)}, \qquad X_0 = x,$$

where $B_t \in \mathbf{R}^m$, the sensitivity equations are:

$$dS_t = AS_t\, dt + \sum_{i=i}^{m} G_i S_t\, dB_t^{(i)}, \qquad S_0 = I,$$

i.e., the same equations except that $\{S_t\}$ is a matrix-valued process. S_t is stochastic but independent of x, and $\Phi_t(x) = S_t x$. We do not have explicit solutions for the sensitivities, expect in the scalar case, or when the matrices commute (compare Exercise 7.23). In general, the sensitivity equations must be solved numerically by simulation.

Example 12.2.3 (Sensitivity of an Equilibrium Solution) *If x^* is an equilibrium of a time-invariant stochastic differential equation (12.2), i.e., $f(x^*) = 0$, $g(x^*) = 0$, then the sensitivity equations are*

$$dS_t(x^*) = AS_t(x^*)\, dt + \sum_{i=1}^{n} G_i S_t(x^*)\, dB_t^{(i)}$$

where

$$A = \frac{\partial f}{\partial x}(x^*), \quad G_i = \frac{\partial g_i}{\partial x}(x^*).$$

These are the same equations as in the previous example.

In numerical analysis, it is convenient to solve the sensitivity equations simultaneously with the nominal system. To do this, we would create an extended state vector X_t^e, which contains both $\Phi_t(x)$ and (the columns of) $S_t(x)$, after which we could use our usual numerical solver.

Example 12.2.4 (Sensitivity of Stochastic Logistic Growth)

$$With\ dX_t = rX_t(1 - X_t/K)\, dt + \sigma X_t\, dB_t, \quad X_0 = x, \tag{12.5}$$

the sensitivity equation is

$$dS_t = r(1 - 2X_t/K)S_t\, dt + \sigma S_t\, dB_t. \tag{12.6}$$

The state and sensitivity combine to an extended state (X_t, S_t) which satisfies

$$\begin{pmatrix} dX_t \\ dS_t \end{pmatrix} = \begin{pmatrix} rX_t(1 - X_t/K) \\ r(1 - 2X_t/K)S_t \end{pmatrix} dt + \sigma \begin{pmatrix} X_t \\ S_t \end{pmatrix} dB_t.$$

Figure 12.2 shows a realization of the solution X_t and the sensitivity S_t for $x = 0.01$, $\sigma = 0.25$, $r = 1$, $K = 1$, found using the Euler-Maruyama method. Initially, the sensitivity S_t grows exponentially with the population X_t. Then, once the state X_t approaches the carrying capacity K, the sensitivity decays, and after 15 time units, the state has all but forgotten where it began, and the current position follows the stationary distribution and is determined by the recent history of the driving noise.

It is also possible to obtain the sensitivities automatically, i.e., without solving the sensitivity equations, using the technique of automatic differentiation when simulating the trajectory.

12.3 STOCHASTIC LYAPUNOV EXPONENTS

For multivariate states x, we summarize the matrix S_t with a single number, taking a worst-case view: If we allow the perturbation δx to have any direction in space, what is its largest amplification? Thus, we are interested in

$$\sup_{\delta x \neq 0} \frac{\|S_t(x)\, \delta x\|}{\|\delta x\|}.$$

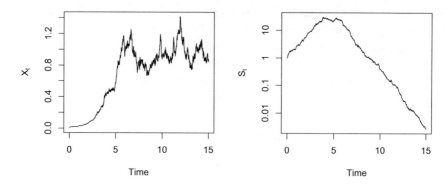

Figure 12.2 Left panel: A realization of logistic growth; compare equation (12.5). Right panel: The sensitivity S_t - note the log scale.

With $\| \cdot \|$ being the Euclidean norm, the answer to this worst-case question is given by the largest singular value $\bar{\sigma}(S_t(x))$, i.e. the spectral norm, which equals the square root of the largest eigenvalue of $S_t^\top S_t$. We define the *stochastic finite-time Lyapunov exponent* as

$$\lambda_t = \frac{1}{t} \log \bar{\sigma}(S_t(x)),$$

i.e., λ_t is the average specific growth rate of $\bar{\sigma}(S_t(x))$ on the time interval $[0,t]$. In general, λ_t is stochastic and depends on the initial condition x.

Consider again Figure 12.2 (right panel). Notice that the sensitivity appears to approach a linear asymptote in the log-domain, i.e., eventually decays exponentially. Then, the finite-time stochastic Lyapunov exponent would converge to the slope of this line, as the time tends to infinity. This leads us to define:

Definition 12.3.1 (Stochastic Lyapunov Exponent) *With the setup just described, we define the stochastic Lyapunov exponent*

$$\bar{\lambda} = \limsup_{t \to \infty} \lambda_t \ (almost \ sure \ limit).$$

Although $\bar{\lambda}$ is certain to be well defined, it is in general a random variable which may also depend on the initial condition x of the nominal solution. In many situations $\bar{\lambda}$ is deterministic in the sense that it attains the same value for almost all realizations, and also independent of the initial condition x. This question belongs to the topic of ergodic theory (Section 9.11.2) and we will not go further into it.

Definition 12.3.2 (Stochastic Stability from Lyapunov Exponents)
We say that the nominal solution $\{X_t\}$ is

1. *stable, if $\bar\lambda < 0$ w.p. 1,*

2. *unstable, if $\bar\lambda > 0$ w.p. 1,*

3. *marginally stable, if $\bar\lambda = 0$ w.p. 1.*

Example 12.3.1 (The narrow-sense linear systems)

$$With\ dX_t = AX_t\ dt + G\ dB_t, \quad X_0 = x,$$

where the sensitivity is $S_t = \exp(At)$ (Example 12.2.1), it follows from some matrix analysis that the stochastic Lyapunov exponent is $\bar\lambda = \max_i \mathbf{Re}\lambda_i$ where $\{\lambda_i\}$ are the eigenvalues of A. See Exercise 12.1. So all solutions are stable if all eigenvalues of A have negative real parts.

Example 12.3.2 (Geometric Brownian Motion)

$$With\ dX_t = rX_t\ dt + \sigma X_t\ dB_t, \quad X_0 = x,$$

the solution is $X_t = S_t x$ where the sensitivity is $S_t = \exp((r - \sigma^2/2)t + \sigma B_t)$. The finite-time stochastic Lyapunov exponent is therefore

$$\lambda_t = r - \frac{1}{2}\sigma^2 + \sigma\frac{1}{t}B_t.$$

Applying the Law of the Iterated Logarithm (Theorem 4.3.4), i.e., using that Brownian motion scales with the square root of time, we obtain

$$\bar\lambda = r - \frac{1}{2}\sigma^2, \quad w.p.\ 1.$$

The Lyapunov exponent $\bar\lambda$ is a decreasing function of the noise intensity σ, and the zero solution is stable when $r < \sigma^2/2$. In particular, when $r > 0$, the individual sample paths diverge to infinity for small σ and converge to 0 for large σ. This may seem counter-intuitive from a technical perspective, where we typically think of noise as a destabilizing factor. Then, let X_t be the size of a bacterial colony: If the noise σ is strong enough, then random fluctuations will eradicate the colony. We will elaborate on this example in Section 12.6; see also Exercise 12.2 for the corresponding Stratonovich equation.

Example 12.3.3 (Stochastic Logistic Growth) *We first examine the trivial solution $X_t \equiv 0$ of (12.5). The sensitivity equation (12.6) becomes $dS_t = rS_t\ dt + \sigma S_t\ dB_t$; i.e., the linearization at the equilibrium. So, from the previous Example 12.3.2, a small population will go extinct iff $\bar\lambda = r - \sigma^2/2 < 0$.*

Now assume that $\bar{\lambda} = r - \sigma^2/2 > 0$ and that $x > 0$. We rewrite the sensitivity equation with Itô's lemma:

$$d\log S_t = r(1 - 2X_t/K)\, dt - \sigma^2/2\, dt + \sigma\, dB_t$$

so that

$$\log S_t = (r - \sigma^2/2)t - \frac{2r}{K}\int_0^t X_t\, dt + \sigma B_t.$$

As $t \to \infty$, X_t will converge in distribution to a Gamma distribution with mean $K(1 - \sigma^2/2r)$ (Example 9.8.2), and $\{X_t\}$ will be ergodic so that

$$\frac{1}{t}\log S_t \to r - \frac{1}{2}\sigma^2 - \frac{2r}{K}\mathbf{E}X_t = -r + \frac{1}{2}\sigma^2, \quad w.p.\ 1.$$

This was negative by the assumption $r - \sigma^2/2 > 0$. Thus, the population is stable, and the ultimate Lyapunov exponent λ equals minus that of the trivial solution. Compare with the symmetry in Figure 12.2 (right panel).

12.3.1 Lyapunov Exponent for a Particle in a Potential

Consider the scalar system

$$dX_t = -U'(X_t)\, dt + \sigma\, dB_t, \quad X_0 = x,$$

where $U : \mathbf{R} \mapsto \mathbf{R}$ is a smooth potential. The sensitivity equation is

$$dS_t = -U''(X_t)S_t\, dt, \quad S_0 = 1,$$

which has the solution

$$S_t = \exp\left(-\int_0^t U''(X_s)\, ds\right)$$

and therefore the Lyapunov exponent is

$$\lambda = -\liminf_{t\to\infty} \frac{1}{t}\int_0^t U''(X_s)\, ds.$$

For non-quadratic U, we typically cannot find the Lyapunov exponent in closed form, but if U is strictly convex, then the Lyapunov exponent is negative, while strictly concave potentials lead to positive Lyapunov exponents. This is consistent with the ball moving in a potential (Section 12.1).

Consider now the case where the potential $U(x)$ is not necessarily convex, but a stationary distribution exists, viz. the canonical distribution

$$\phi(x) = Z^{-1}\exp(-U(x)/D) \text{ with } D = \sigma^2/2$$

where $Z = \int_{-\infty}^{+\infty} \exp(-U(x)/D)\, dx$ is the partition function. Replace the time average in the Lyapunov exponent with an expectation:

$$\lambda = -\liminf_{t\to\infty} \frac{1}{t}\int_0^t U''(X_s)\, ds = -\int_{-\infty}^{+\infty} \phi(x)U''(x)\, dx, \qquad (12.7)$$

almost surely, using ergodicity. The stationary distribution ϕ puts more weight on states with low potential where the curvature $U''(x)$ is positive, so this integral is positive and the Lyapunov exponent is negative. To see this:

$$
\begin{aligned}
\lambda &= -\int_{-\infty}^{+\infty} \phi(x)U''(x)\,dx \\
&= \int_{-\infty}^{+\infty} \phi'(x)U'(x)\,dx \\
&= \int_{-\infty}^{+\infty} -\phi(x)\frac{|U'(x)|^2}{D}\,dx \\
&< 0,
\end{aligned}
$$

almost surely. Here we first use integration by parts and next detailed balance, i.e., $U'\phi + D\phi' = 0$ (Section 9.9). So when a stationary distribution exists, the Lyapunov exponent is negative, and the solution is stable.

Compare this with deterministic systems, where maxima of the potential correspond to unstable equilibria, and almost all non-equilibrium solutions are stable. For stochastic systems, the equilibrium solutions do not exist, and almost all solutions are stable, even if there are local potential maxima. So local maxima of the potential do not give rise to unstable solutions, but to local minima of the stationary distribution.

12.4   EXTREMA OF THE STATIONARY DISTRIBUTION

To explore the connection between equilibria, stability and extrema further, consider the double well model given by the Itô equation

$$
dX_t = (\mu X_t - X_t^3)\,dt + \sigma\,dB_t. \tag{12.8}
$$

Without noise, i.e., $\sigma = 0$, the stability analysis of the system is as follows: For $\mu \leq 0$, the system has a single equilibrium at $x = 0$, which is stable. For $\mu > 0$, the system has three equilibria: An unstable one at $x = 0$ and two stable ones at $x = \pm\sqrt{\mu}$. The qualitative change at $\mu = 0$ is the standard example of a pitchfork bifurcation.

With noise $\sigma > 0$, there are no equilibria, and all solutions have negative Lyapunov exponents, regardless of μ and the initial condition (Section 12.3.1). In stead, the qualitative change that occurs at $\mu = 0$ is that the stationary distribution changes from being unimodal to being bimodal. See Figure 12.3. For $\mu > 0$, the stationary behavior of the system is to fluctuate in one well and occasionally make the transition to the other well.

Consider now equilibria for the drift, and extrema for the stationary distribution, in the scalar equation

$$
dX_t = f(X_t)\,dt + g(x)\,dB_t
$$

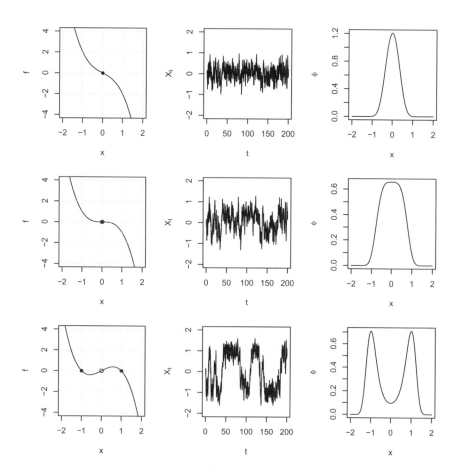

Figure 12.3 A pitchfork bifurcation in the double well model (12.8) with $\sigma = 1/2$. *Left panels:* The drift $f(x)$. Stable equilibria of f are marked with solid circles and unstable equilibria are marked with open circles. *Middle panels:* Simulations starting at $X_0 = 0$; the same realization of Brownian motion is used in the three panels. *Right panels:* The stationary probability density. *Upper panels:* $\mu = -1$. *Middle panels:* $\mu = 0$. *Lower panels:* $\mu = 1$.

where $g(x) > 0$ for all x. If the stationary density $\phi(x)$ exists, then it satisfies detailed balance: $u\phi - D\phi' = 0$, with diffusivity $D = g^2/2$ and advection $u = f - D'$ (Sections 9.7 and 9.9). Thus, stationary points x^* of the stationary distribution (i.e., with $\phi'(x^*) = 0$) are those where the advection vanishes, $u(x^*) = 0$. Moreover, the stationary forward Kolmogorov equation is

$$(u\phi - D\phi')' = u'\phi - D\phi'' + (u - D')\phi' = 0$$

so that at a stationary point x^*, we have

$$u'(x^*)\phi(x^*) = D(x^*)\phi''(x^*).$$

Thus, a stationary point x^* is a strict local maximum for the stationary density ϕ, i.e., $\phi''(x^*) < 0$, if and only if it is a stable equilibrium for the flow (i.e., $u'(x^*) < 0$). Conversely, unstable equilibria for the flow have $u'(x^*) > 0$ and correspond to local minima for the stationary density, i.e., $\phi''(x^*) > 0$. Keep in mind that these results depend on the choice of coordinate system; neither equilibria for the drift, or modes of the stationary distribution, are preserved under coordinate transformations.

In the multivariate case, if the stationary distribution satisfies detailed balance $u\phi - D\nabla\phi = 0$, then stationary points $\nabla\phi(x^*) = 0$ are still those where the advective flow vanishes, $u(x^*) = 0$. Without detailed balance, there is no such result: Stationary densities often have extrema at points where the flow does not vanish. Also the connection between stability properties of the flow field u, and extrema of the stationary density ϕ, is weaker, but a simple observation is as follows: The stationary density satisfies $\nabla \cdot (u\phi - D\nabla\phi) = 0$, and evaluated at a stationary point $\nabla\phi(x^*) = 0$, this simplifies to

$$\phi\nabla \cdot u - \text{tr}(D\mathbf{H}\phi) = 0,$$

where $\mathbf{H}\phi$ is the Hessian of ϕ. So the sign of $\nabla \cdot u$ equals the sign of $\text{tr}(D\mathbf{H}\phi)$. For example, if x^* is a local maximum point of ϕ, then $\nabla \cdot u < 0$ must hold at x^*, i.e., the flow is convergent at x^*. However, the flow needs not vanish at x^*, and even if it does, x^* needs not be a stable node. The following section will illustrate this with an example.

12.5 A WORKED EXAMPLE: A STOCHASTIC PREDATOR-PREY MODEL

We now demonstrate a stability analysis, using as an example a predator-prey system consisting of two coupled stochastic differential equations

$$dN_t = rN_t(1 - N_t/K)\,dt - \frac{cN_tP_t}{N_t + \tilde{N}}\,dt + \sigma_N N_t\,dB_t^{(1)}, \quad (12.9)$$

$$dP_t = \epsilon\frac{cN_tP_t}{N_t + \tilde{N}}\,dt - \mu P_t\,dt + \sigma_P P_t\,dB_t^{(2)}. \quad (12.10)$$

Here, N_t is the prey abundance and P_t is the predator abundance. All parameters r, K, $\bar{N}$, c, ϵ, μ, σ_N, σ_P are positive.

The objective of the stability analysis is to examine the three qualitatively different classes of solutions: The *trivial* solutions where $N_t = P_t = 0$, the *prey-only* solutions where $N_t > 0$ but $P_t = 0$, and the *co-existence* solutions where $N_t > 0$, $P_t > 0$. For which values of the parameters will the different classes of solutions be prominent?

Let us first discuss the model structure. Without predators ($P_t = 0$), the prey display stochastic logistic growth (Example 12.2.4). The noise-free model is the *Rosenzweig-MacArthur* model from ecology, which is an elaboration of the Lotka-Volterra predator-prey model. The term $cN_t/(N_t + \bar{N})$ determines the consumption of prey per predator; this is known as a Holling type II functional response in ecology. At low prey densities, the consumption is linear in the prey with slope $c/\bar{N}$, but at high prey densities, it saturates at c, the maximum consumption per predator. An efficiency ϵ determines how prey biomass is converted to predator biomass, and finally the predators suffer a constant mortality μ. The noise is multiplicative and diagonal, so it models rapid unpredictable environmental fluctuations that affect the two species independently. This choice is mostly for simplicity.

With the state $X_t = (N_t, P_t)$, we can write the system in standard form

$$dX_t = f(X_t)\,dt + g_1(X_t)\,dB_t^{(1)} + g_2(X_t)\,dB_t^{(2)}$$

with

$$f(x) = \begin{pmatrix} rn(1 - n/K) - \frac{cnp}{n+\bar{N}} \\ \epsilon\frac{cnp}{n+\bar{N}} - \mu p \end{pmatrix}, \quad g_1(x) = \begin{pmatrix} \sigma_N n \\ 0 \end{pmatrix}, \quad g_2(x) = \begin{pmatrix} 0 \\ \sigma_P p \end{pmatrix}.$$

The sensitivity equations are given by the Jacobians

$$\nabla f(n, p) = \begin{bmatrix} r(1 - 2n/K) - p\gamma'(n) & -\gamma(n) \\ \epsilon\gamma'(n)p & \epsilon\gamma(n) - \mu \end{bmatrix}$$

with the shorthand $\gamma(n) = cn/(n + \bar{N})$, and

$$\nabla g_1(n, p) = \begin{bmatrix} \sigma_N & 0 \\ 0 & 0 \end{bmatrix}, \quad \nabla g_2(n, p) = \begin{bmatrix} 0 & 0 \\ 0 & \sigma_P \end{bmatrix}.$$

Sensitivity of the Trivial Solution

For the *trivial* solution $N_t \equiv 0$, $P_t \equiv 0$, the sensitivity equations are

$$dS_t = \begin{bmatrix} r & 0 \\ 0 & -\mu \end{bmatrix} S_t\,dt + \begin{bmatrix} \sigma_N & 0 \\ 0 & 0 \end{bmatrix} S_t\,dB_t^{(1)} + \begin{bmatrix} 0 & 0 \\ 0 & \sigma_P \end{bmatrix} S_t\,dB_t^{(2)}.$$

With the initial condition $S_0 = I$, the off-diagonal elements remain 0, and the two diagonal elements evolve independently as geometric Brownian motion:

$$S_t = \begin{bmatrix} \exp((r - \frac{1}{2}\sigma_N^2)t + \sigma_N B_t^{(1)}) & 0 \\ 0 & \exp(-(\mu + \frac{1}{2}\sigma_P^2)t + \sigma_P B_t^{(2)}) \end{bmatrix}.$$

Therefore, the origin is stable if and only if $r - \sigma_N^2/2 < 0$ (Example 12.3.2). Thus, the stability of the origin depends only on the prey: If the noise σ_N is sufficiently large, it will drive the prey population to extinction, and the origin is stable. If the noise is weak so that $r - \sigma_N^2/2 > 0$, the prey population will grow, and the origin is unstable. At the origin, the predator population will die out in absence of prey, so the predators cannot destabilize the origin.

Sensitivity of Prey-Only Solutions

Now assuming $r - \sigma_N^2/2 > 0$, we turn to a *prey-only* solution with initial condition $N_0 = n > 0$, $P_0 = 0$. Then, the sensitivity equations are

$$dS_t = \begin{bmatrix} r(1 - 2N_t/K) & -\gamma(N_t) \\ 0 & \epsilon\gamma(N_t) - \mu \end{bmatrix} S_t \, dt$$
$$+ \begin{bmatrix} \sigma_N & 0 \\ 0 & 0 \end{bmatrix} S_t \, dB_t^{(1)} + \begin{bmatrix} 0 & 0 \\ 0 & \sigma_P \end{bmatrix} S_t \, dB_t^{(2)}.$$

The sensitivity S_t remains upper tridiagonal: If the $(2,1)$ element in S_t is 0, then also the $(2,1)$ element in dS_t is 0. The explanation is that a perturbation in the initial prey population cannot create predators out of the blue. The $(1,1)$ element of the sensitivity matrix evolves on its own:

$$dS_t^{(1,1)} = r(1 - 2N_t/K)S_t^{(1,1)} \, dt + \sigma_N S_t^{(1,1)} \, dB_t^{(1)}.$$

This is the sensitivity equation for logistic growth (compare Figure 12.2 and Examples 12.2.4 and 12.3.3). Thus, the prey population in itself has a stochastic Lyapunov exponent $-r + \sigma_N^2/2 < 0$ and is stable by assumption.

We proceed to the $(1,2)$-element of S_t, which satisfies the equation

$$dS_t^{(1,2)} = r(1 - 2N_t/K)S_t^{(1,2)} \, dt - \gamma(N_t)S_t^{(2,2)} \, dt + \sigma_N S_t^{(1,2)} \, dB_t^{(1)}.$$

This is the same linear equation as for the $S_t^{(1,1)}$ element, except for the source term $-\gamma(N_t)S_t^{(2,2)} \, dt$. We know that this equation is stable, so if the source vanishes, $S_t^{(2,2)} \to 0$, then also $S_t^{(1,2)} \to 0$ w.p. 1. This should not be surprising: If a small predator population dies out, then it will not have a large and lasting effect on the prey population.

Thus any instability in the system must come from the $(2,2)$-element of S_t. This element satisfies the equation

$$dS_t^{(2,2)} = (\epsilon\gamma(N_t) - \mu)S_t^{(2,2)} \, dt + \sigma_P S^{(2,2)} \, dB_t^{(2)}$$

and using Itô's lemma, we find

$$d\log S_t^{(2,2)} = (\epsilon\gamma(N_t) - \mu - \frac{1}{2}\sigma_P^2) \, dt + \sigma_P \, dB_t^{(2)}.$$

Using ergodicity of $\{N_t\}$, the solution is unstable if and only if

$$\lambda_P := \epsilon \mathbf{E}\gamma(N_t) - \mu - \frac{1}{2}\sigma_P^2 > 0, \qquad (12.11)$$

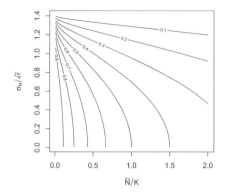

Figure 12.4 Contour plot of the non-dimensional function $\eta(\frac{\bar{N}}{K}, \frac{\sigma}{\sqrt{r}})$ which determines the expected feeding rate of a predator.

where N_t is sampled from the stationary Gamma distribution (Example 9.8.2). Thus, the *expected* growth rate of the predators determine their stability. The expression for $\mathbf{E}\gamma(N_t)$ is long and hardly useful, but can be written

$$\mathbf{E}\gamma(N_t) = c\eta(\bar{N}/K, \sigma_N/\sqrt{r})$$

where η is a non-dimensional function of the two non-dimensional parameters $\bar{N}/K$ and $\sigma_N/\sqrt{r}$. The function η describes the *feeding level* of the predators, i.e., their uptake relative to their maximal uptake c (Figure 12.4). This lets us determine the Lyapunov exponent λ_P and identify bifurcations when the dimensionless parameters are varied. For example, assume that $\epsilon = 0.25, c = 1$, and $\mu + \sigma_P^2/2 = 0.1$. From (12.11), we see that predators can invade the system provided their feeding level is high enough, $\eta > 0.4$, and the bifurcation curve $\eta = 0.4$ can be identified in Figure 12.4. In absence of noise on the prey, $\sigma_N = 0$, the predators must be close enough to satiation at the carrying capacity: $\bar{N}/K < 1.5$. With positive prey noise $\sigma_N > 0$, the average prey population decreases below K, so that the predators must be closer to satiation at K, i.e. lower $\bar{N}/K$. At $\sigma_N = \sqrt{2r}$, the prey population collapses so that predators are unable to invade regardless of their feeding efficiency.

Coexistence Solutions and Their Lyapunov Exponents

It remains to examine *coexistence* solutions where both species are present, i.e., $N_t > 0$ and $P_t > 0$. Figure 12.5 shows one sample path obtained with parameters $r = 1, K = 1, c = 1, \bar{N} = 0.6, \epsilon = 0.5, \mu = 0.1$, and $\sigma_N = \sigma_P = 0.02$, for which the prey-only solution is unstable. The noise-free system has an equilibrium at

$$N^* = \frac{\bar{N}}{\frac{\epsilon c}{\mu} - 1}, \quad P^* = \epsilon \bar{N} \frac{r}{\mu} \frac{\frac{c\epsilon}{\mu} - 1 - \frac{\bar{N}}{K}}{(\frac{c\epsilon}{\mu} - 1)^2}$$

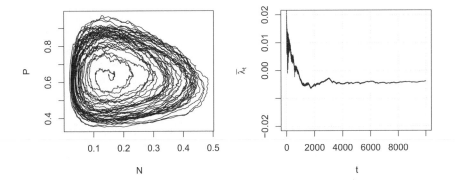

Figure 12.5 *Left panel:* Simulation of the stochastic Rosenzweig-MacArthur model starting at the unstable equilibrium of the noise-free system. *Right panel:* The finite-time Lyapunov exponent $\bar{\lambda}_t$ vs. time.

which is positive and unstable with these specific parameters; a small perturbation away from the equilibrium would grow with rate 0.01 while oscillating with period 24, until it approached a limit cycle. Here, including noise, we initialize the system right at the unstable equilibrium. Figure 12.5 shows the first part of the trajectory, up to time $t = 1,000$, in the phase plane, as well as the finite-time stochastic Lyapunov exponent $\bar{\lambda}_t$ up to time 10,000. The limit cycle is clearly visible, even if blurred by noise. The finite-time Lyapunov exponent initially fluctuates substantially, but seems to be well approximated by the deterministic prediction 0.01. However, when the system settles down in the stochastic steady state, the sensitivity decreases, and the stochastic Lyapunov exponent, i.e., the limit as $t \to \infty$, appears to be near -0.005. The convergence is very slow and requires at least 100 cycles, i.e., 2,400 time units.

In conclusion, the stochastic Lyapunov exponent is quite tractable for the trivial solution and with a little effort also for the prey-only solution, and for these situations the Lyapunov exponent gives useful information about when the system can be invaded. For the coexistence solution, starting where the drift f vanishes in a region where the advective flow is divergent, the finite-time Lyapunov exponent initially grows, as the solution is pushed away from the stationary point by noise and spirals out. Then, the sensitivity decreases, as the trajectory follows the limit cycle and eventually forgets its starting point. If we aim to diagnose the initial instability, we need the finite-time stochastic Lyapunov exponent, while the (infinite-time) stochastic Lyapunov exponent indicates the mixing time along the limit cycle.

Stationary Solutions

Finally, we examine the stationary behavior of the coexistence solution for the situation where the prey-only solution is unstable. Figure 12.6 shows stationary probability density as well as a sample path, for different values of the efficiency ϵ. The other parameters are $r = 1$, $K = 1$, $c = 1$, $\mu = 0.15$, $\bar{N} = 1/4$, $\sigma_N^2 = \sigma_P^2 = 0.002$. For these parameters, the predators are able to invade the system if $\epsilon \geq 0.19$, roughly. With $\epsilon = 0.2$, the two species coexist and display negatively correlated fluctuations. With $\epsilon = 0.22$, we begin to see the first appearance of predator-prey cycles. With $\epsilon = 0.25$, these cycles are clear, but they are still irregular. With $\epsilon = 0.3$, the limit cycle is obvious even if perturbed by noise, and the stationary density displays a marked local minimum inside the limit cycle. The bifurcation structure from the deterministic system is visible although obscured by noise.

12.6 GEOMETRIC BROWNIAN MOTION REVISITED

Stochastic stability analysis has an extra layer of complexity compared to deterministic analysis, which is not revealed by the stochastic Lyapunov exponents. Recall Example 12.3.2 concerning geometric Brownian motion $X_t = x \exp((r - \frac{1}{2}\sigma^2)t + \sigma B_t)$ and its governing equation $dX_t = rX_t \, dt + \sigma X_t \, dB_t$, $X_0 = x$: There, we found that a stochastic Lyapunov exponent $\bar{\lambda} = r - \sigma^2/2$. We now turn to the moments of $\{X_t\}$ which are (Exercise 7.1):

$$\mathbf{E}X_t = xe^{rt}, \quad \mathbf{E}X_t^2 = x^2 e^{(2r+\sigma^2)t}, \text{ so:}$$

1. $\mathbf{E}X_t \to 0$ iff $r < 0$; we say that the zero solution is *stable in mean.*

2. $\mathbf{E}X_t^2 \to 0$ iff $r + \sigma^2/2 < 0$; we say that the zero solution is *stable in mean square.*

The threshold between stability and instability is therefore $r - \sigma^2/2$, if we focus on sample paths; $r = 0$, if we focus on the mean, or $r + \sigma^2/2 = 0$, if we focus on the mean square. If $r < 0$ but $r + \sigma^2/2 > 0$, then the mean is stable but the mean square is unstable. Similarly, if $r > 0$ but $r - \sigma^2/2 < 0$, then sample paths converges to 0 (w.p. 1) but the mean diverges. In the stochastic case, even for a linear time-invariant system, we have (at least) 3 notions of stability, differing in the mode of convergence in the statement $X_t \to 0$.

For fixed r, we see that the mean is insensitive to the noise level σ, but that the variance grows with σ, and indeed diverges to $+\infty$ as $t \to \infty$ when $\sigma^2 > -2r$. To summarize, for geometric Brownian motion, Itô noise stabilizes sample paths, does not affect the mean, and destabilizes the mean square.

If your intuition objects to sample paths that converge to 0 while moments diverge to infinity, consider (again) Exercise 4.15. Also, Figure 12.7 displays 5 sample paths of geometric Brownian motion with parameters $r = 1$, $\sigma = 2$, $X_0 = x = 1$, for which $r - \sigma^2/2 = -1$, so trajectories converge to 0. However, the sample paths display wild fluctuations before converging. The figure suggests that EX_t grows on the interval $t \in [0, 1]$ which agrees with $r > 0$. To see that it continues to grow on the interval $t \in [0, 10]$, we would need a lot more than 5 sample paths, because the continued growth of the

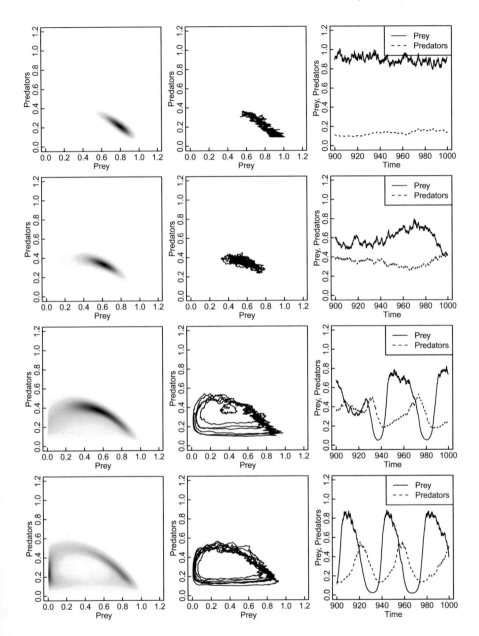

Figure 12.6 Behavior of the stochastic Rosenzweigh-MacArthur model for different values of the efficiency ϵ. *Left panels:* The stationary joint probability density of prey and predators. *Center panels:* A simulated trajectory in the phase plane. *Right panels:* The simulated trajectory as time series. *First row:* $\epsilon = 0.20$. *Second row:* $\epsilon = 0.22$. *Third row:* $\epsilon = 0.25$. *Fourth row:* $\epsilon = 0.3$.

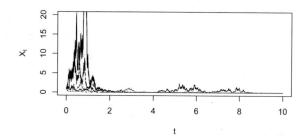

Figure 12.7 5 sample paths of Geometric Brownian Motion starting at $X_0 = 1$. The parameters are $r = 1$ and $\sigma = 2$, for which the system is stochastically stable but unstable in mean and in mean square.

mean relies on fewer and fewer realizations. To characterize the magnitude of the transient fluctuations, we can use that the maximum $\max\{Y_t : t \geq 0\}$ is exponentially distributed with mean $1/(1 - 2r/\sigma^2) = 2$ (Exercise 11.7).

This example demonstrates the different notions of stability in the context of stochastic differential equations. If one wishes to do a stability analysis for a given application, one needs to consider carefully which notion of stability is relevant. In the given situation, would behavior such as in Figure 12.7 be considered stable or unstable?

12.7   STOCHASTIC LYAPUNOV FUNCTIONS

A different approach to stochastic stability employs auxiliary functions on state space, called Lyapunov functions. For many deterministic mechanical systems, think of the Lyapunov function as the mechanical energy: If there is no external energy supply, then the mechanical energy cannot increase in time - energy cannot be created, but friction can convert mechanical energy to thermal energy. Therefore trajectories must be confined to a region in state space, where the mechanical energy is no greater than the initial energy. This simple observation turns out to be immensely powerful.

For stochastic systems, there is the complication that the noise typically supplies energy so that the energy in the system is no longer a non-increasing function of time. As a result, we can only bound the trajectories in probability. A basic and central result, which assumes that the energy decreases *on average* and yields a bound on the probability of escape, is the following:

Theorem 12.7.1 *Let x^* be an equilibrium point of the stochastic differential equation (12.2), i.e., $f(x^*) = 0$, $g(x^*) = 0$. Assume that there exists a function $V(x)$ defined on a domain D containing x^*, such that:*

1. *V is C^2 on $D\backslash\{x^*\}$.*

2. *There exist continuous strictly increasing functions a and b with $a(0) = b(0) = 0$, such that $a(|x - x^*|) \leq V(x) \leq b(|x - x^*|)$.*

3. *$LV(x) \leq 0$ for $x \in D\backslash\{x^*\}$.*

Here L is the backward Kolmogorov operator, as usual. Then

$$\lim_{x \to x^*} \mathbf{P}^x \{ \sup_{t \geq 0} |X_t - x^*| \geq \epsilon \} = 0$$

for any $\epsilon > 0$.

We call $V(\cdot)$ a stochastic Lyapunov function. Let us first explain the conclusion: Let $\epsilon > 0$ be given, then the set $\{ \xi : |\xi - x^*| \leq \epsilon \}$ is a "permitted region" in state space containing the equilibrium point. For each initial condition x, there is a probability $\mathbf{P}^x \{ \sup_{t \geq 0} |X_t - x^*| > \epsilon \}$ that the process ever leaves the permitted region. For $x = x^*$ this probability is obviously 0. The conclusion says that this probability is a continuous function of x at x^*: A small perturbation from equilibrium implies a small probability of escape.

The theorem gives a sufficient condition in terms of existence of a Lyapunov function. In many cases, we aim to establish stability, and therefore aim to find a Lyapunov function. The condition is necessary under regularity conditions; if x^* is stable then $V(x) := \mathbf{P}^x \{ \sup_{t \geq 0} |X_t - x^*| \geq \epsilon \}$ is a candidate Lyapunov function. There are also converse theorems (Has'minskiǐ, 1980).

Now let us examine the sufficient condition in the theorem. Condition 2 implies that we can bound $|x - x^*|$ in terms of $V(x)$ and conversely. In absence of noise, the third condition $LV \leq 0$ implies that the "energy" $V(X_t)$ is a non-increasing function of time. In the stochastic case, random fluctuations may inject energy into the system, but the energy will decrease *on average*. To capture this property, we introduce a new class of processes, which are similar to martingales but have decreasing expectation:

Definition 12.7.1 (Supermartingale) *We say that a process $\{ M_t : t \geq 0 \}$ is a* supermartingale *with respect to an underlying probability measure $\mathbf{P}$ and filtration $\{ \mathcal{F}_t : t \geq 0 \}$, if the following three conditions hold:*

1. *The process $\{ M_t : t \geq 0 \}$ is adapted to the filtration $\{ \mathcal{F}_t : t \geq 0 \}$.*

2. *$\mathbf{E} |M_t| < \infty$ for all $t \geq 0$.*

3. *$\mathbf{E} \{ M_t | \mathcal{F}_s \} \leq M_s$, almost surely, whenever $0 \leq s \leq t$.*

Compared to martingales (Definition 4.5.1), the only difference is the inequality in the third property. Our primary interest is supermartingales that generalize the idea of energy decreasing with time, so we focus on non-negative supermartingales which have continuous sample paths. Such supermartingales satisfy probabilistic bounds:

Theorem 12.7.2 (Supermartingale inequality) *Let $\{ M_t : t \geq 0 \}$ be a non-negative supermartingale with continuous sample paths. Then*

$$\mathbf{P} \{ \sup_{0 \leq s \leq t} M_s \geq c \} \leq \frac{M_0}{c}$$

holds for all $t \geq 0$ and all $c > 0$.

Proof: The argument is analogous to the proof of the martingale inequality (Theorem 4.5.2). With the stopping time $\tau = \inf\{t > 0 : M_t \geq c\}$, we have

$$\mathbf{E}M_{t\wedge\tau} \geq c \cdot \mathbf{P}(M_{t\wedge\tau} = c) = c \cdot \mathbf{P}(\sup_{0\leq s\leq t} M_s \geq c).$$

As in Lemma 4.5.1, we may show that the stopped process $M_{t\wedge\tau}$ is also a supermartingale, so $\mathbf{E}M_{t\wedge\tau} \leq M_0$. Combining, we get

$$c \cdot \mathbf{P}(\sup_{0\leq s\leq t} M_s \geq c) \leq M_0$$

which was to be shown. ■

We now have the ingredients necessary to prove Theorem 12.7.1.

Proof: Without loss of generality, take $x^* = 0$. Define the stopping time $\tau = \inf\{t : X_t \in D\}$ and $Y_t = V(X_t)$, then $\{Y_{t\wedge\tau}\}$ is a supermartingale since $LV \leq 0$ on $D\backslash\{0\}$. Here, we are neglecting the possible singularity at 0; this can be dealt with by stopping the process if it gets too close to 0. With the supermartingale inequality (Theorem 12.7.2), we conclude that

$$\mathbf{P}^x\{\sup_{0\leq t\tau} Y_t \geq c\} \leq \frac{\mathbf{E}^x Y_t}{c} \leq \frac{Y_0}{c} \text{ for any } c > 0.$$

Let ϵ be small enough that the set $\{\xi : |\xi| \leq \epsilon\}$ is contained in D. Next, $|x| \geq \epsilon$ implies that $V(x) \geq a(|x|) \geq a(\epsilon)$, and so

$$\mathbf{P}^x\{\sup_{t\geq 0} |X_t| \geq \epsilon\} \leq \mathbf{P}^x\{\sup_{t\geq 0} Y_t \geq a(\epsilon)\} \leq \frac{Y_0}{a(\epsilon)} \leq \frac{b(|x|)}{a(|\epsilon|)}.$$

Thus, $x \to 0$ implies that $\mathbf{P}^x\{\sup_{t\geq 0} |X_t| \geq \epsilon\} \to 0$. ■

Example 12.7.1 (Geometric Brownian Motion (again)) *Consider again geometric Brownian motion* $\{X_t : t \geq 0\}$ *given by the Itô equation*

$$dX_t = r \ X_t \ dt + \sigma X_t \ dB_t.$$

To analyse stability of the equilibrium $X_t \equiv 0$, *we try a Lyapunov function*

$$V(x) = |x|^p \text{ where } p > 0.$$

This candidate satisfies all requirements from Theorem 12.7.1 except possibly $LV \leq 0$. *We find* $LV(x) = p(r+\sigma^2(p-1)/2)|x|^p$. *So V is a Lyapunov function if* $r + \sigma^2(p-1)/2 \leq 0$, *or equivalently* $p \leq 1 - 2r/\sigma^2$. *Such a* $p > 0$ *exists iff* $1 - 2r/\sigma^2 > 0$, *which is therefore a sufficient condition for stochastic stability in the sense of Theorem 12.7.1. Note that this is equivalent to the Lyapunov exponent being negative (Example 12.3.2 and Section 12.6), and that we may need* $p < 1$ *to show stability; i.e., V is not differentiable at* $x = 0$.

12.8 STABILITY IN MEAN SQUARE

Stability in mean square aims to bound the variance of perturbations. Next to negative Lyapunov exponents, it is probably the most used notion of stability. We briefly mentioned this mode of stability when discussing geometric Brownian motion (Section 12.7.1); the general definition is:

Definition 12.8.1 *Let x^* be an equilibrium point of the stochastic differential equation (12.2), i.e., $f(x^*) = 0$, $g(x^*) = 0$. We say that x^* is exponentially stable in mean square, if there exist constants C, $\lambda > 0$, such that*

$$\mathbf{E}^x |X_t - x^*|^2 \leq C |x - x^*|^2 \exp(-\lambda t)$$

for all $t \geq 0$ and all x.

The following theorem gives a sufficient condition in terms of a Lyapunov function:

Theorem 12.8.1 *If there exists a C^2 Lyapunov function V such that*

$$\forall x : k_1 |x - x^*|^2 \leq V(x) \leq k_2 |x - x^*|^2 \text{ and } LV(x) \leq -k_3 |x - x^*|^2$$

where $k_1, k_2, k_3 > 0$, then x^* is exponentially stable in mean square.

Proof: Without loss of generality, assume $x^* = 0$. By the properties of V,

$$k_1 \mathbf{E}^x |X_t|^2 \leq \mathbf{E}^x V(X_t) \leq k_2 |x|^2 - k_3 \int_0^t \mathbf{E}|X_s|^2 \, ds$$

and by existence and uniqueness, these are all finite. By Dynkin's formula, $\mathbf{E}^x V(X_t) = V(x) + \mathbf{E}^x \int_0^t LV(X_s) \, ds$. Here, we are omitting the step of localization; i.e., first we stop the process upon exist of an R-sphere, and then we let $R \to \infty$. Define $a_t = \mathbf{E}^x |X_t|^2$, then this reads

$$k_1 a_t \leq k_2 a_0 - k_3 \int_0^t a_s \, ds.$$

By Gronwall's inequality, we find that $a_t \leq \frac{k_2}{k_1} a_0 \exp\left(-\frac{k_3 t}{k_1}\right)$. Inserting a_t and a_0, this is the exponentially decaying bound on $\mathbf{E}^x |X_t|^2$. ■

Example 12.8.1 (Geometric Brownian Motion) *Consider again the equation $dX_t = r X_t \, dt + \sigma X_t \, dB_t$. As candidate Lyapunov function, take*

$$V(x) = \frac{1}{2} x^2 \text{ so that } LV(x) = (r + \frac{1}{2}\sigma^2) x^2.$$

We see that this Lyapunov function proves exponential stability in mean square, if and only if $r + \sigma^2/2 < 0$. This is, of course, consistent with our previous analysis of geometric Brownian motion (Section 12.6).

Example 12.8.2 (The multivariate wide-sense linear system) *With*

$$dX_t = AX_t \, dt + \sum_{i=1}^{m} G_i X_t dB_t^{(i)}$$

where $X_t \in \mathbf{R}^n$ and $(B_t^{(1)}, \ldots, B_t^{(m)})$ is m-dimensional standard Brownian motion, take as candidate Lyapunov function

$$V(x) = x'Px$$

where $P > 0$ is a symmetric positive definite n-by-n matrix. We find

$$LV(x) = x' \left(PA + A'P + \sum_{i=1}^{m} G_i'PG_i \right) x.$$

There are (at least) two ways to search for such a V. First, the set $\{P \in \mathbf{R}^{n \times n} : P = P', P > 0\}$ is a convex cone, as is the set $\{P \in \mathbf{R}^{n \times n} : P = P', PA + A'P + \sum_{i=1}^{m} G_i'PG_i < 0\}$. So convex search may identify a point in the intersection of these two cones, or establish that they are disjoint. A benefit of this approach is that additional (convex) constraints on P can be included. Convex search and optimization are important computational tools in systems and control theory (Boyd et al., 1994).

Alternatively, if the system is exponentially stable in mean square, then a quadratic Lyapunov function $V(x) = x'Px$ exists such that $LV(x) + |x|^2 = 0$. This Lyapunov function is in fact

$$V(x) = \mathbf{E}^x \int_0^\infty |X_t|^2 \, dt$$

which must be finite if the system is exponentially stable in mean square, and a quadratic function of x due to linearity of system dynamics. Therefore, we may solve the linear algebraic matrix equation

$$A'P + PA' + \sum_{i=1}^{m} G_i'PG_i + I = 0$$

in the unknown symmetric P. A solution $P > 0$ establishes a Lyapunov function and that the system is exponentially stable. If there is no such solution, the system is not exponentially stable in mean square. This includes the situation where there is a unique solution P with at least one negative eigenvalue, as well as the situation where there is no solution P. In the scalar case of geometric Brownian motion $(A = r, G = \sigma)$, these two possibilities correspond to $r + \sigma^2/2 > 0$ and $r + \sigma^2/2 = 0$, respectively.

Notice that exponential stability in mean square is a *global* property: The mean square $\mathbf{E}^x |X_t|^2$ depends on system dynamics in the entire state space. So exponential stability in mean square can not be verified from the linearization of the system dynamics around the equilibrium, and a Lyapunov function V on a bounded domain only shows local stability in a certain sense.

Example 12.8.3 (A System with a Stabilizing Non-Linearity)

$$With \ dX_t = -(X_t + X_t^3) \ dt + X_t \ dB_t$$

the linearization around $x = 0$ is $dX_t = -X_t \ dt + X_t \ dB_t$ which is exponentially stable in mean square. To see that the non-linearity $-X_t^3$ is stabilizing, take as Lyapunov function $V(x) = x^2$:

$$LV(x) = -2x(x + x^3) + x^2 = -x^2 - 2x^4 \leq -x^2$$

which shows that the zero solution of the nonlinear equation is exponentially stable in mean square.

12.9 STOCHASTIC BOUNDEDNESS

Equilibrium solutions are less common in stochastic differential equations than in their deterministic counterparts. For non-equilibrium solutions, stability to perturbations in initial conditions is more difficult to analyse - linear systems in the wide or narrow sense make the notable exceptions to this rule (Examples 12.2.1 and 12.2.2). Stability is also less crucial, because the driving noise perturbs the trajectory, too. Therefore, this section focuses on the more modest ambition of bounding the trajectories in different ways. Due to noise, bounds are typically only probabilistic, and it is typically the qualitative result that is most interesting, i.e., that a bound exists, rather than what the specific bound is.

The results use a Lyapunov-type function $V(x)$ which generalizes energy, and we aim to show that the process $\{X_t\}$ spends most of the time in states where $V(x)$ is low, and that very high-energy states constitute rare events. This turns out to be the case if the expected energy decreases initially, when the process starts in a high-energy state. To illustrate this, think of the Ornstein-Uhlenbeck process $dX_t = -\lambda X_t \ dt + \sigma \ dB_t$, with $\lambda > 0$ and a quadratic Lyapunov function $V(x) = x^2$. Then

$$LV(x) = -2\lambda x^2 + \sigma^2.$$

Note that $LV(x) \leq 0$ iff $V(x) \geq \sigma^2/(2\lambda)$: In low-energy states, close to the equilibrium $x = 0$ of the drift $-\lambda x$, the noise pumps energy into the system, so the system will never come to rest there. But far from equilibrium, dissipation dominates, which will bring the system towards the equilibrium. The result is persistent fluctuations around the equilibrium. Any bound will be broken, eventually, but the process will still be concentrated in low-energy states.

To generalize this example, consider the stochastic differential equation

$$dX_t = f(X_t) \ dt + g(X_t) \ dB_t \ \text{with} \ X_t \in \mathbf{R}^n \quad (12.12)$$

with backward Kolmogorov operator L. In the following, we assume that Lyapunov functions $V : \mathbf{R}^n \to \mathbf{R}$ are C^2 and *proper*, i.e., the preimage $V^{-1}([a, b])$ of any bounded interval is bounded. This means that we can bound the energy $V(x)$ in terms of the state $|x|$ and vise versa. A first result is:

Lemma 12.9.1 *Let* $\{X_t\}$ *satisfy (12.12) and let the Lyapunov function* V *be proper,* C^2 *and such that*

$$LV(x) \leq 0 \text{ for all } x \text{ such that } V(x) \geq K.$$

Let $M > K$ *and define* $\tau = \inf\{t : V(X_t) \notin [K, M]\}$. *Then, for* $V(x) \geq K$

$$\mathbf{P}^x[\sup\{V(X_t) : 0 \leq t \leq \tau\} = M] \leq \frac{V(x) - K}{M - K}.$$

The lemma bounds the energy $V(X_t)$ in the stochastic sense that as the upper threshold M is increased, it becomes more probable that we reach the lower threshold K first. Keep the Ornstein-Uhlenbeck example in mind: After time τ, if $V(X_\tau) = K$, energy may be injected back into the system.
Proof: Define $Y_t = V(X_t) - K$. Then, for $V(x) \geq K$, $Y_{t \wedge \tau}$ is a non-negative continuous supermartingale. It follows from the supermartingale inequality (Theorem 12.7.2) that

$$\mathbf{P}[\sup\{Y_t : 0 \leq t \leq \tau\} \geq M - K] \leq \frac{Y_0}{M - K}$$

which can be re-written as in the theorem. ■

Remark 12.9.1 *We may not be interested in the Lyapunov function* $V(X_t)$ *per se, but use it to bound the state* $|X_t|$. *This is possible, since* V *is proper: Define* $a(r) = \inf\{V(x) : |x| \geq r\}$ *and* $b(r) = \sup\{V(x) : |x| \leq r\}$. *Then* a *and* b *are continuous non-decreasing functions with* $a(|x|) \leq V(x) \leq b(|x|)$. *Moreover,* $a(|x|) \to \infty$ *when* $|x| \to \infty$, *because* V *is proper. Now, assume that there is an inner sphere with radius* $\xi > 0$ *such that* $LV(x) \leq 0$ *holds whenever* $|x| \geq \xi$. *Take an outer sphere with radius* $\chi > \xi$ *and stop when we reach either of the spheres, i.e., at* $\tau = \inf\{t : |X_t| \notin [\xi, \chi]\}$. *Then*

$$\mathbf{P}^x[\sup\{|X_t| : 0 \leq t \leq \tau\} \geq \chi] \leq \frac{b(|x|) - a(\xi)}{a(\chi) - a(\xi)}.$$

The bound may be conservative, but note that this probability vanishes as $\chi \to \infty$. *We say that sample paths are stochastically bounded (Gard, 1988).*

With a stronger assumption, we can bound the expected time it takes for the system to move from an initial high-energy state to the low-energy region:

Theorem 12.9.2 *With the setting as in Lemma 12.9.1, assume that there is a* $\lambda > 0$ *such that* $LV(x) + \lambda \leq 0$ *holds whenever* $V(x) \geq K$. *Let* $\tau = \inf\{t : V(X_t) \leq K\}$. *Then, for* $V(x) \geq K$:

$$\mathbf{E}^x \tau \leq \lambda^{-1}(V(x) - K).$$

Proof: This follows from Dynkin's formula, but to ensure that the assumptions for Dynkin's formula are met, we first localize: Let $M > 0$ and define $\tau_M = \inf\{t : V(X_t) \geq M\}$. Then, using the bounds and Dynkin's formula,

$$K \leq \mathbf{E}^x V(X_{t \wedge \tau \wedge \tau_M})$$

$$= V(x) + \mathbf{E}^x \int_0^{t \wedge \tau \wedge \tau_M} LV(X_s)\, ds$$

$$\leq V(x) - \lambda \mathbf{E}^x(t \wedge \tau \wedge \tau_M).$$

Now let $M \to \infty$, then $\mathbf{P}^x\{\tau_M < t\} \to 0$, so $\mathbf{E}^x(t \wedge \tau \wedge \tau_M) \to \mathbf{E}^x(t \wedge \tau)$, and

$$K \leq V(x) - \lambda \mathbf{E}^x(t \wedge \tau).$$

Now let $t \to \infty$ to see that $K \leq V(x) - \lambda \mathbf{E}^x \tau$. ∎

More interesting than the specific bound $\mathbf{E}^x \tau$ is often simply that $\mathbf{E}^x \tau < \infty$. In the terminology of Markov processes, the low-energy states $\{x : V(x) \leq K\}$ are *positively* (or *non-null*) recurrent. This is a key step in ensuring that a stationary distribution exists and therefore in ergodic theory for stochastic differential equations (Has'minskiĭ, 1980; Rey-Bellet, 2006).

We now sketch a couple of other applications of this machinery. First, a fundamental way to bound solutions is to ensure that explosions do not occur in finite time. Recall that Theorem 8.3.1 assumed a linear growth bound on the dynamics, and that this ruled out explosions in finite time. A way to obtain similar results using Lyapunov function is to first assume that f, g are locally Lipschitz continuous, which guarantees uniqueness and existence up to a possible explosion. Next, if there exists Lyapunov function V which is C^2, proper, positive and satisfies

$$LV(x) \leq \lambda V(x)$$

where $\lambda \geq 0$, then $\mathbf{E}^x V(X_t)$ can grow at most exponentially in time, which rules out explosions. See (Has'minskiĭ, 1980; Gard, 1988) for precise statements. The criterion from Theorem 8.3.1 can be seen in this light; it guarantees that a Lyapunov function in the form $V(x) = |x^2| + 1$ applies, with $\lambda = 3C$. This technique can be refined to concern escape from bounded sets also: Exercise 12.7 shows that the stochastic logistic growth process $\{X_t\}$ (Example 12.2.4) remains positive if $X_0 > 0$, i.e. neither hits 0 or ∞ in finite time.

As a final application of Lyapunov-type arguments, assume that V is proper and that $LV(x) \leq -\lambda V(x)$ holds for some $\lambda > 0$, whenever $V(x) > K$. That is, energy dissipates exponentially while in the high-energy region. If the diffusion is regular, this implies that the probability density converges *exponentially* to the unique stationary density (Rey-Bellet, 2006), i.e., the process mixes exponentially and is therefore ergodic. Moreover, since $\mathbf{E}LV(X) = 0$ when X is sampled from the stationary distribution, we have $\mathbf{E}V(X) \leq \gamma/\lambda$ where γ is such that $\gamma \geq LV(x) + \lambda V(x)$ everywhere. Even if we only hint these results, they illustrate the versatility of the technique of Lyapunov functions.

12.10   CONCLUSION

If you have a background in deterministic dynamical systems, you will be well aware of the central importance of stability in that field. It is then obvious to examine which notions of stability are relevant in the stochastic case, and which techniques are applicable. Here, we have shown how classical deterministic Lyapunov theory can be extended to stochastic differential equations.

For deterministic systems, it is a standard exercise to identify equilibria, linearize the dynamics around them, and compute eigenvalues to conclude on stability. The stochastic situation is less straightforward, even if stochastic Lyapunov exponents can be defined analogously: Equilibrium solutions are rare, and Lyapunov exponents for non-equilibrirum solutions generally depend on the entire global dynamics. The framework is perhaps most operational and relevant when analyzing the stability of equilibria and, more generally, invariant low-dimensional manifolds. The predator-prey system (Section 12.5) is a typical example: The stability analysis of the equilibrium $n = p = 0$, and also of the prey-only solution $n > 0$, $p = 0$, is illuminating. However, linearizing the dynamics around equilibria gives rise to wide-sense linear sensitivity equations, which - in the general multivariate case - cannot be analyzed in terms of linear algebra.

The machinery of Lyapunov functions also generalizes to the stochastic case. A bottleneck is actually finding Lyapunov functions. Quadratic Lyapunov functions for mean-square stability of wide-sense linear systems are tractable. Beyond this, the structural properties of the system may provide a hint, such as energy considerations or that the system consists of interacting subcomponents which can be analyzed one at a time.

In the stochastic case, there are several definitions of stability, and one must carefully specify which notion is relevant for the study at hand. The different definitions may very well lead to different conclusions. For example, the noise in geometric Brownian motion is stabilizing in the sense that trajectories converge to zero, if the noise is strong enough, but destabilizing in the sense that the mean square diverges, if the noise is strong enough. The results for other models are different (Exercise 12.4), and also for the corresponding Stratonovich equation (Exercise 12.2).

For real-world stochastic systems, it is hard to envision experiments where we perturb the initial conditions but subject the system to the same realization of noise. That means that the real-world relevance of stability to initial conditions is greatest when the nominal trajectory is insensitive to noise, e.g., an equilibrium. In general, the sensitivity to the noise is of equal importance, e.g., the magnitude of fluctuations around an equilibrium of the drift. Recall the double well (Section 12.4), where we showed that the Lyapunov exponent is always negative; this Lyapunov exponent describes the mixing time within wells, at least when diffusivity is low. One can imagine applications where this time scale is less important than the stationary distribution and the rate of transition between wells. So the stability analysis rarely stands alone but is accompanied by other analysis.

A strength of the machinery of stochastic Lyapunov functions is that they address not just stability, but also boundedness, and the existence and properties of a stationary distribution, without actually finding this stationary distribution. This is feasible because the Lyapunov functions are only required to satisfy inequalities. We have seen basic results to this end.

In summary, stability analysis of stochastic differential equations is a rich field, where several notions and approaches coexist. We must carefully consider which questions to ask about a given system, giving thought both to what is relevant in the situation at hand, and which analyses are tractable.

12.11 NOTES AND REFERENCES

Stability theory for stochastic differential equation was developed in the 1960's; classical references are (Kushner, 1967) and (Has'minskiĭ, 1980); see also (Gard, 1988).

12.12 EXERCISES

Exercise 12.1: Let A be a matrix and define $S_t = \exp(At)$. Show that the largest singular value $\bar{\sigma}(S_t)$ of S_t can be approximated with $\exp \lambda t$ as $t \to \infty$, where $\lambda = \max_i \mathbf{Re}\lambda_i$ and $\{\lambda_i\}$ are the eigenvalues of A. More precisely, show that $\frac{1}{t} \log \bar{\sigma}(S_t) \to \lambda$.

Exercise 12.2: For the Stratonovich equation $dX_t = rX_t \, dt + \sigma X_t \circ dB_t$ with initial condition $X_0 = x$, for which parameters (r, σ^2) does it hold that

1. the mean $\mathbf{E}X_t$ stays bounded as $t \to \infty$?

2. the mean square $\mathbf{E}X_t^2$ stays bounded as $t \to \infty$?

3. the individual sample path X_t stays bounded as $t \to \infty$, w.p. 1?

Exercise 12.3: Let $\{X_t\}$ be geometric Brownian motion given by the Itô equation $dX_t = rX_t \, dt + \sigma X_t \, dB_t$ with the initial condition $X_0 = 1$. Assume that the parameters are such that $X_t \to 0$ as $t \to \infty$, w.p. 1. Find the distribution of $S := \sup\{X_t : t \geq 0\}$. *Hint:* Recall the result of 11.7 regarding the extremum of Brownian motion with drift. Next, confirm the result with stochastic simulation for suitably chosen parameters.

Exercise 12.4: Consider the two-dimensional wide-sense linear equation driven by scalar Brownian motion (Mao, 2008)

$$dX_t = -X_t \, dt + GX_t \, dB_t \quad \text{with } G = \begin{bmatrix} 0 & -2 \\ 2 & 0 \end{bmatrix}.$$

Show that the stochastic Lyapunov exponent is $\lambda = 1$, using the solution (Exercise 7.23). *Note:* We see that Itô noise is not always stabilizing.

Exercise 12.5 Lyapunov Exponents for the Double Well System: For the parameters in Figure 12.3, compute the Lyapunov exponent in two ways: First, by integration over state space using (12.7), and next, by stochastic simulation.

Exercise 12.6 Lyapunov Exponents of the van der Pol Oscillator: For the van der Pol system of Section 6.2.4, write up the sensitivity equations. Solve the system and the sensitivity equations numerically on the time interval $[0,100]$ with initial condition $X_0 = V_0 = 0$. Plot the finite-time Lyapuov exponent on this time interval.

Exercise 12.7 Stochastic Logistic Growth: With equation (12.5)

$$dX_t = X_t(1 - X_t)\, dt + \sigma X_t\, dB_t \text{ with } X_0 = x > 0,$$

consider the candidate Lyapunov function $V(x) = x^2 - \log x$ defined for $x > 0$.

1. Show that V is proper, that $V(x) > 0$ and that $LV(x)$ is bounded above, for $x > 0$.

2. Find $\delta > 0$ such that $LV(x) \leq V(x) + \delta$ holds for all $x > 0$.

3. Show that $\{Y_t = [V(X_t) + \delta]\exp(-t)\}$ is a supermartingale.

4. Show that $\mathbf{E}^x V(X_t) < \infty$ for all t. In particular, the solution does not diverge to 0 or to ∞ in finite time.

5. Assume that $\sigma^2 < 2$. Show that $LV(x)$ is negative for $x > 0$ sufficiently close to 0, or sufficiently large. Conclude that there exists an interval $[a, b]$ with $0 < a < b$ which is positively recurrent.

6. With $\sigma^2 < 2$, show that the candidate Lyapunov function $V(x) = -\log x$ can be used to assess the time it takes the time for the population to recover from a near-extinction ($x \ll 1$). *Hint:* Show that there exist $\gamma, a > 0$ such that $LV(x) + \gamma \leq 0$ for $0 < x < a$. Then use $V(x)$ to bound the expected time τ it takes to travel from $X_0 = x$ to $X_\tau = a$.

7. With $\sigma^2 < 2$, consider the candidate Lyapunov function $V(x) = x^p$. Show that, with $p = 2$, $LV(x)/V(x)$ is negative for x sufficiently large. Then show, with $p = (1 - 2/s^2)/2$, $LV(x)/V(x)$ is negative for $x > 0$ sufficiently close to 0. Then, with this p, use $V(x) = x^2 + x^p$ to show that the distribution of X_t converges exponentially to the stationary distribution.

Exercise 12.8 The Stochastic Rosenzweig-MacArthur Model: Consider (12.9) and (12.10). We take for granted that the positive quadrant $(0, \infty)^2$ is invariant. Let $V(n, p) = \epsilon n + p$. Show that $LV(n, p) = rn(1 - n/K) - \mu p$ and sketch the region in state space where $LV \leq 0$. Conclude that the system is stochastically sample path bounded and that for some $v > 0$, the region given by $V(x) \leq v$ is positively recurrent.

CHAPTER 13

Dynamic Optimization

One application for stochastic differential equations is stochastic control, also known as stochastic dynamic optimization.

There are numerous examples of these problems, across all domains of application: Control engineers build feedback control systems to reduce harmful oscillations in the blades of wind turbines. Process engineers design chemical plants to operate with maximum output without jeopardizing the plant. In quantitative finance, investors continuously manage portfolios to obtain a high expected income and a low risk. Every living organism, from bacteria to humans, has been selected through evolution to maximize the ability to generate offspring, which explains the traits and behaviors we observe. In summary, the theory of dynamic optimization helps us to take decisions, design new systems, as well as to analyze the behavior of existing systems and decision makers.

The approach in this chapter is based on the so-called *value function*, which depends on the state and on time. This value function quantifies the expected future performance of the controlled process, under the assumption that the control is optimal. This value function generalizes the expectations to the future that we studied earlier in Chapter 11, and satisfies the so-called Hamilton-Jacobi-Bellman equation, a "backward" partial differential equation which generalizes the backward Kolmogorov equation. This approach to optimization is called *dynamic programming*.

In stochastic dynamic optimization, some technicalities become critical. These concern if the value function is smooth, if an optimal decision exists or only near-optimal ones, and whether the decisions depend smoothly on state and time. These technicalities may overshadow the conceptual simplicity of the approach. To avoid this, we first consider a discrete Markov Decision Problem where these technicalities do not appear. This illuminates the structure of the solution. Thereafter, we apply this structure to our original problem of dynamic optimization for stochastic differential equations.

DOI: 10.1201/9781003277569-13

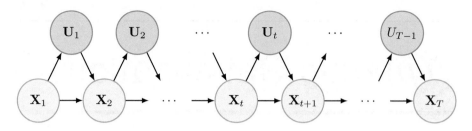

Figure 13.1 Probabilistic graphical model illustrating the interdependence between the random variables in the Markov Decision Problem.

13.1 MARKOV DECISION PROBLEMS

> It is perfectly true, as the philosophers say, that life must be understood backwards. But they forget the other proposition, that it must be lived forwards.
>
> *- Søren Kierkegaard, Journals IV A 164 (1843).*

The simplest case of stochastic dynamic optimization is when time is discrete, and the state space is discrete and finite. The resulting problem of dynamic optimization is commonly referred to as a Markov decision problem (Puterman, 2014), and lets us focus on the core in the problem and its solution, rather than the technicalities of diffusions. Also, dynamic optimization problems for diffusions can be discretized to Markov decision problems (Section 13.12.2), so an understanding of these discrete problems are helpful for numerical algorithms.

Consider a stochastic process $\{X_t : t \in \mathbf{T}\}$ with state space $\mathbf{X} = \{1, \ldots, N\}$ and in discrete time, $\mathbf{T} = \{1, 2, \ldots, T\}$. At each time t we choose a decision variable $U_t \in \mathbf{U}$. We require that the set $\mathbf{U}$ of permissible controls is compact; e.g., a finite set or an interval of reals. We must choose U_t based on the current state, i.e., $U_t = \mu(X_t, t)$ for some *control strategy* $\mu : \mathbf{X} \times \mathbf{T} \mapsto \mathbf{U}$.

The initial state X_1 has a prescribed distribution. Conditional on the current state and control, the next state is random with probabilities

$$\mathbf{P}\{X_{t+1} = j \mid X_t = i, U_t = u\} = P_{ij}^u$$

and independent of past states and controls X_s and U_s for $s = 1, \ldots, t-1$. We assume that these transition probabilities P_{ij}^u are continuous functions of $u \in \mathbf{U}$, for each i, j (when $\mathbf{U}$ is continuous). The interaction between the random variables is illustrated in Figure 13.1.

We next turn to the control objective: A reward is issued at each time $t \in \mathbf{T}$, which depends on the state X_t and the control U_t and on time t itself. That is, we assume a reward function

$$h : \mathbf{X} \times \mathbf{U} \times \mathbf{T} \mapsto \mathbf{R}$$

Biography: Richard Ernest Bellman (1920–1984)

An American applied mathematician who was central to the rise of "modern control theory", i.e., state space methods in systems and control. His most celebrated contribution is dynamic programming and the principle of optimality, which both concern dividing complex decision-making problems into more, but simpler, sub-problems. A Ph.D. from Princeton, he spend the majority of his career at RAND corporation. Photograph ©ShutterStock.

which is continuous in $u \in \mathbf{U}$ everywhere, such that the total reward is

$$\sum_{t \in \mathbf{T}} h(X_t, U_t, t).$$

The objective is to maximize the expected reward, i.e., we aim to identify

$$\max_{\mu} \mathbf{E}^{\mu} \sum_{t \in \mathbf{T}} h(X_t, U_t, t) \tag{13.1}$$

and the maximizing argument μ. Here the superscript in $\mathbf{E}^{\mu}$ indicates that the distribution of X_t and U_t depends on μ. The maximum is over all functions $\mu : \mathbf{X} \times \mathbf{T} \mapsto \mathbf{U}$. This is a compact set, and the expected reward is a continuous function of μ, so the maximum is guaranteed to exist.

The difficulty in the Markov decision problem (13.1) is that the space of controls $\mu : \mathbf{X} \times \mathbf{T} \mapsto \mathbf{U}$ can be very large. The celebrated *dynamic programming* solution, due to Richard Bellman, tackles this difficulty with an iteration that breaks the large problem (13.1) into smaller sub-problems: We first consider the problem with $s = T$, and then let the initial time s march backward, ending at $s = 1$. At each step s in this iteration, the optimization problem is tractable, given the solution to the previous problem $s + 1$.

To see this, let the *value* $V(x, s)$ be the highest expected reward, when starting in state $x \in \mathbf{X}$ at time $s \in \mathbf{T}$; that is:

$$V(x, s) = \max_{\mu} \mathbf{E}^{\mu, X_s = x} \sum_{t=s}^{T} h(X_t, U_t, t). \tag{13.2}$$

Theorem 13.1.1 (Dynamic programming) *The value function V is the unique solution to the recursion*

$$V(x, s) = \max_{u \in \mathbf{U}} \mathbf{E}^{X_s = x, U_s = u} \left[h(x, u, s) + V(X_{s+1}, s+1) \right]. \tag{13.3}$$

with the terminal condition

$$V(x, T) = \max_{u \in \mathbf{U}} h(x, u, T) , \qquad (13.4)$$

The optimal value $V(x, 1)$ is obtained with any strategy μ such that

$$\mu(x, s) \in \arg\max_{u \in \mathbf{U}} \mathbf{E}^{X_s = x, U_s = u} \left[h(x, u, s) + V(X_{s+1}, s+1) \right].$$

In terms of the transition probabilities, the dynamic program (13.3) is:

$$V(x, s) = \max_{u \in \mathbf{U}} \left[h(x, u, s) + \sum_{y \in \mathbf{X}} P^u_{xy} V(y, s+1) \right]. \qquad (13.5)$$

The dynamic programming equation (13.3) contains two critical statements: First, when an optimal decision maker is at state $X_s = x$, their decision does not depend on how they got there. That is, we maintain a Markov structure: Given the present, the optimal decision which affects the future is independent of the past. Second, if they know the value function at time $s + 1$, then they do not need to look further into the future: The optimal decision at time s is a trade-off between immediate gains (maximizing $h(x, u, s)$) and moving into favorable regions in state space (maximizing $\mathbf{E}^u V(X_{s+1}, s + 1)$). This trade-off determines the value function at time s. It is easy to solve the Dynamic Programming equation (13.3) numerically as a terminal value problem, if we at each state x and time s can maximize over u. The solution provides us with the value V of the Markov Decision Problem (13.1) and the optimal control strategy μ.

Proof: Let $W : \mathbf{X} \times \mathbf{T} \to \mathbf{R}$ be the solution to the recursion given in the theorem. We claim that $W = V$ where V is given by (13.2). Clearly this holds for $s = T$ due to the terminal value (13.4). So assume that $W = V$ on $\mathbf{X} \times \{s + 1\}$; we aim to show that $W = V$ on $\mathbf{X} \times \{s\}$. Write

$$\mathbf{E}^{\mu, X_s = x} \sum_{t=s}^{T} h(X_t, U_t, t) = h(x, u, s) + \mathbf{E}^{\mu, X_s = x} \sum_{t=s+1}^{T} h(X_t, U_t, t)$$

where $u = \mu(x, s)$. Condition on X_{s+1} and use the Law of Total Expectation:

$$\mathbf{E}^{\mu, X_s = x} \sum_{t=s+1}^{T} h(X_t, U_t, t) = \mathbf{E}^{U_s = u, X_s = x} \mathbf{E}^{\mu^T_{s+1}} \left[\sum_{t=s+1}^{T} h(X_t, U_t, t) \Bigg| X_{s+1} \right]$$

The notation $\mathbf{E}^{\mu^T_{s+1}}$ indicates that the conditional expectation only depends on the restriction of μ to $\mathbf{X} \times \{s + 1, \ldots, T\}$ and is independent of the initial condition $X_s = x$, due to the Markov property and conditioning on X_{s+1}. So:

$$V(x,s) = \max_{u,\mu_{s+1}^T} \left(h(x,u,s) + \mathbf{E}^{U_s=u,X_s=x} \mathbf{E}^{\mu_{s+1}^T} \left[\sum_{t=s+1}^{T} h(X_t, U_t, t) \middle| X_{s+1} \right] \right)$$

$$= \max_u \left(h(x,u,s) + \mathbf{E}^{U_s=u,X_s=x} \max_{\mu_{s+1}^T} \mathbf{E}^{\mu_{s+1}^T} \left[\sum_{t=s+1}^{T} h(X_t, U_t, t) \middle| X_{s+1} \right] \right)$$

$$= \max_u \left(h(x,u,s) + \mathbf{E}^{U_s=u,X_s=x} V(X_{s+1}, s+1) \right)$$

$$= \max_u \left(h(x,u,s) + \mathbf{E}^{U_s=u,X_s=x} W(X_{s+1}, s+1) \right)$$

$$= W(x,s)$$

as claimed. So $V = W$ on $\mathbf{X} \times \{s\}$ and, by iteration, on $\mathbf{X} \times \mathbf{T}$.   ■

13.2   CONTROLLED DIFFUSIONS AND PERFORMANCE OBJECTIVES

We now return to diffusion processes, aiming to pose a dynamic optimization problem similar to the Markov decision problem, and to determine the solution using dynamic programing. We consider a *controlled* diffusion $\{X_t : 0 \le t \le T\}$ taking values in $\mathbf{X} = \mathbf{R}^n$ and with a finite terminal time $T > 0$, given by

$$dX_t = f(X_t, U_t)\, dt + g(X_t, U_t)\, dB_t$$

with initial condition $X_0 = x$. Here $\{B_t : 0 \le t \le T\}$ is multivariate standard Brownian motion, and $\{U_t : 0 \le t \le T\}$ is the *control signal*, which we consider a decision variable: At each time $t \in [0, T]$ we must choose U_t from some specified set $\mathbf{U}$ of permissible controls. We restrict attention to *state feedback* controls

$$U_t = \mu(X_t, t)$$

where the function $\mu : \mathbf{X} \times [0, T] \mapsto \mathbf{U}$ is such that the *closed-loop* system

$$dX_t = f(X_t, \mu(X_t, t))\, dt + g(X_t, \mu(X_t, t))\, dB_t \tag{13.6}$$

satisfies the conditions in Chapter 8 for existence and uniqueness of a solution $\{X_t\}$. This solution will then be a Markov process, so another name for state feedback controls $U_t = \mu(X_t, t)$ is *Markov controls*.

The controlled system terminates when the state exits a set $G \subset \mathbf{X}$, or when the time reaches T, i.e., we have a stopping time

$$\tau = \min\{T, \inf\{t \in [0, T] : X_t \notin G\}\}$$

At this point a reward is issued

$$k(X_\tau, \tau) + \int_0^\tau h(X_t, U_t, t)\, dt$$

The first term, $k(X_\tau, \tau)$, the *terminal reward*, while the integrand is the *running reward*, which is accumulated along the trajectory until termination.

For a given Markov control strategy $\{U_t = \mu(X_t, t) : 0 \le t \le T\}$, and a given initial condition $X_s = x$, we can assess the performance objective given by

$$J(x, \mu, s) = \mathbf{E}^{\mu, X_s = x} \left[k(X_\tau, \tau) + \int_s^\tau h(X_t, U_t, t) \, dt \right]$$

The control problem we consider is to maximize $J(x, \mu, s)$ w.r.t. the control signal $\{U_t\}$, or equivalently w.r.t. the control strategy μ. Our original interest is the initial time zero, $s = 0$, but we include the initial time s as a parameter to prepare for a dynamic programming solution.

Instrumental in the solution is the generator of the controlled diffusion. For a fixed control $u \in \mathbf{U}$, define the generator L^u as

$$(L^u V)(x) = \frac{\partial V}{\partial x}(x) \cdot f(x, u) + \frac{1}{2} \mathrm{tr} \left[g^T(x, u) \frac{\partial^2 V}{\partial x^2}(x) g(x, u) \right]$$

while for a control strategy $\mu : \mathbf{X} \mapsto \mathbf{U}$, define the "closed-loop" generator L^μ:

$$(L^\mu V)(x) = \frac{\partial V}{\partial x}(x) \cdot f(x, \mu(x)) + \frac{1}{2} \mathrm{tr} \left[g^T(x, \mu(x)) \frac{\partial^2 V}{\partial x^2}(x) g(x, \mu(x)) \right].$$

Finally, let h^μ be the resulting running reward; that is:

$$h^\mu(x, t) = h(x, \mu(x), t).$$

13.3 VERIFICATION AND THE HAMILTON-JACOBI-BELLMAN EQUATION

We now state the *verification theorem* which lets us conclude that we have solved the optimization problem, if we have found a solution to the dynamic programming equation.

Theorem 13.3.1 *Let the domain G be open and bounded, let $V : G \times [0, T] \mapsto \mathbf{R}$ be $C^{2,1}$ and satisfy the Hamilton-Jacobi-Bellman equation*

$$\frac{\partial V}{\partial t} + \sup_{u \in \mathbf{U}} [L^u V + h] = 0 \tag{13.7}$$

on $G \times [0, T]$, along with boundary and terminal conditions

$$V = k \text{ on } \partial(G \times [0, T])$$

Let $\mu^* : G \times [0, T] \mapsto \mathbf{U}$ be such that

$$\sup_{u \in \mathbf{U}} [L^u V + h] = L^{\mu^*} V + h$$

on $G \times [0,T]$, and assume that with this μ^*, the closed-loop system (13.6) satisfies the conditions in Theorem 8.3.2. Then, for all $x \in \mathbf{G}$ and all $s \in [0,T]$

$$V(x,s) = \sup_{\mu} J(x,\mu,s) = J(x,\mu^*,s)$$

so that μ^* is the optimal strategy.

This theorem provides a strategy for solving stochastic dynamic optimization problems: First, try to solve the Hamilton-Jacobi-Bellman equation, and in doing so, identify the optimal control strategy μ^*. If this succeeds, then the theorem states that we have solved the optimization problem. Because the solution involves a terminal value problem governing the value function, just as was the case for the Markov Decision Problem of Section 13.1, we refer to the solution provided by Theorem 13.3.1 as a dynamic programming solution.

We have required that G is bounded for simplicity, so that the proof that we are about to give can use Dynkin's formula. This condition can be relaxed; see (Øksendal, 2010; Touzi, 2013) and the example in the next section.

Proof: Let $\mu : \mathbf{X} \times [0,T] \mapsto \mathbf{U}$ be a given control strategy such that the closed loop dynamics (13.6) satisfy the conditions for existence and uniqueness in Theorem 8.3.2. Let $V : \mathbf{X} \times [0,T] \mapsto \mathbf{R}$ be $C^{2,1}$ and satisfy

$$\frac{\partial V}{\partial t} + L^{\mu}V + h(x,\mu(x,t),t) = 0 \text{ for } (x,t) \in G \times [0,T] \qquad (13.8)$$

along with the boundary condition $V = k$ on $\partial(G \times [0,T])$. Then Dynkin's lemma, Theorem 11.1.1, states that

$$\mathbf{E}^{\mu,X_t=x}V(X_\tau,\tau) = V(x,t) + \mathbf{E}^{\mu,X_t=x}\int_0^\tau \frac{\partial V}{\partial t} + L^{\mu}V \ dt$$

and therefore, using the PDE (13.8) and the associated boundary condition, we get

$$V(x,t) = \mathbf{E}^{\mu,X_t=x}\int_t^\tau h(X_s,U_s,s) \ ds + k(X_\tau,\tau).$$

In particular, let μ be the control μ^* in the verification Theorem 13.3.1 and let V be the value function. Then V is the expected pay-off with the control μ^*.

Now, let μ_1 be any other control strategy such that the closed loop system satisfies the usual assumptions. Then the HJB equation ensures that

$$\frac{\partial V}{\partial s} + L^{\mu_1}V + h(x,\mu_1(x,s),s) \le 0$$

By Dynkin's lemma, it follows that

$$\mathbf{E}^{\mu_1,X_s=x}\int_s^\tau h(X_t,U_t,t) \ dt + k(X_\tau,\tau) \le V(x,s)$$

It follows that this control strategy μ_1 results in an expected pay-off which is no greater than what is obtained with the strategy μ^*. We conclude that the strategy μ^* is optimal. ■

13.4 PORTFOLIO SELECTION

Dynamic optimization is widespread in mathematical finance. A simplified portfolio selection problem (Merton, 1969) considers an investor whose wealth at time t is W_t. She invests a fraction P_t of this in a stock, the price X_t of which is geometric Brownian motion given by

$$dX_t = \alpha X_t \, dt + \sigma X_t \, dB_t,$$

where $\{B_t\}$ is standard Brownian motion. The rest of her wealth, $(1 - P_t)W_t$, she invests in a risk-less asset which earns a steady interest with rate r. Her objective is to maximize her *utility* at a fixed time $T > 0$, specifically

$$k(W_T) \text{ where } k(w) = \frac{w^{1-\gamma}}{1-\gamma}.$$

Here, $\gamma > 0$, $\gamma \neq 1$, is a measure of risk aversion. Recall that this specific utility is called *iso-elastic* or *constant relative risk aversion*.

To frame this problem in our standard formulation, we could let the state at time t be (X_t, W_t) while the control is P_t. In an infinitesimal time interval dt, the return from the risky asset is $P_t W_t(\alpha \, dt + \sigma \, dB_t)$, while the return from the riskless asset is $(1 - P_t)W_t r \, dt$, so W_t evolves as

$$dW_t = W_t(r + P_t(\alpha - r)) \, dt + \sigma P_t W_t \, dB_t.$$

Note that this equation does not involve X_t, so we can exclude X_t from the state vector; then W_t is the state. We therefore seek a value function $V(w,t)$ which satisifies the Hamilton-Jacobi-Bellman equation, i.e.,

$$\dot{V} + \sup_p \left[V'w(r + p(\alpha - r)) + \frac{1}{2}\sigma^2 p^2 w^2 V'' \right] = 0, \quad V(w, T) = k(w).$$

A qualified guess is that V is itself a power function of w at each time t, i.e.

$$V(w,t) = b(t)\frac{w^{1-\gamma}}{1-\gamma} \text{ where } b(t) > 0.$$

With this, the terminal condition is $b(T) = 1$ and, after division with $w^{1-\gamma}$,

$$\frac{1}{1-\gamma}\dot{b} + b\sup_p \left[(r + p(\alpha - r)) - \frac{1}{2}\sigma^2 p^2 \gamma \right] = 0.$$

The bracket is concave in p and does not involve time or the state, so the optimal policy has $P_t = p$ constant and is found as a stationary point:

$$\alpha - r - \sigma^2 p\gamma \text{ or } P_t \equiv p = \frac{\alpha - r}{\gamma\sigma^2}.$$

The investor should constantly buy or sell stock to maintain the constant fraction p. This fraction p depends on parameters: It increases with the extra return $\alpha - r$ of the risky asset and decreases with volatility σ^2 and with the risk aversion γ. If $\alpha > r$ (a reasonable assumption), then $p > 0$. If the risky asset is very attractive or the risk aversion γ is small, we may find find $p > 1$, so that our investor borrows money to invest in the stock. If we require that $P_t \in [0, 1]$, then the optimal strategy is $P_t \equiv 1$ when $(\alpha - r) > \gamma\sigma^2$.

We can now solve for $b(t)$:

$$b(t) = e^{\lambda(t-T)} \text{ where } \lambda = (1-\gamma)\left(r + \frac{1}{2}\frac{(\alpha-r)^2}{\gamma\sigma^2}\right).$$

This establishes a solution to the Hamilton-Jacobi-Bellman equation. To address that the domain in this case is unbounded, assume that the investor retires if her wealth ever reaches $\bar{w}$ and then receives a reward $V(\bar{w}, \tau)$ where $\tau = \inf\{t : W_t \geq \bar{w}\}$. Then, the verification Theorem 13.3.1 assures us that the strategy $\{P_t\}$ maximizes the expected reward. Now let $\bar{w} \to \infty$; then early retirement happens with probability 0, so we have verified that the optimal value and strategy also applies on this unbounded domain.

13.5 MULTIVARIATE LINEAR-QUADRATIC CONTROL

An important special case of control problems is the linear-quadratic regulator (LQR) problem, where system dynamics are linear

$$dX_t = AX_t \, dt + FU_t \, dt + G \, dB_t \tag{13.9}$$

while the control objective is to minimize the quadratic functional

$$J(x, U) = \mathbf{E}^x \int_0^T \frac{1}{2}X_t^\top QX_t + \frac{1}{2}U_t^\top RU_t \, dt + \frac{1}{2}X_T^\top PX_T. \tag{13.10}$$

Here, $X_t \in \mathbf{R}^n$, $U_t \in \mathbf{R}^m$, while $\{B_t : t \geq 0\}$ is l-dimensional standard Brownian motion. The terminal time $T > 0$ is fixed. The matrix dimensions are $A \in \mathbf{R}^{n\times n}$, $F \in \mathbf{R}^{n\times m}$, $G \in \mathbf{R}^{n\times l}$, $Q \in \mathbf{R}^{n\times n}$, $P \in \mathbf{R}^{n\times n}$, $R \in \mathbf{R}^{m\times m}$. We assume $Q \geq 0$, $R > 0$, $P \geq 0$.

We guess that the value function is quadratic in the state:

$$V(x, t) = \frac{1}{2}x^\top W_t x + w_t$$

where $W_t \in \mathbf{R}^{n\times n}$, $W_t \geq 0$, while w_t is scalar. The HJB equation then reads

$$\frac{1}{2}x^\top \dot{W}_t x + \dot{w}_t + \inf_u \left[x^\top W_t(Ax + Fu) + \frac{1}{2}\text{tr}G^\top W_t G + \frac{1}{2}x^\top Qx + \frac{1}{2}u^\top Ru\right] = 0$$

The bracket is convex in u, so the minimizing u is found as a stationary point:

$$x^\top W_t F + u^\top R = 0$$

Factbox: [Maximization vs. minimization] Optimization problems can concern maximization, i.e., $\max_x f(x)$ or $\sup_x f(x)$, or minimization, i.e., $\min_x f(x)$ or $\inf_x f(x)$. We can shift between the two; for example:

$$\min_{x\in\mathbf{X}} f(x) = -\max_{x\in\mathbf{X}}[-f(x)].$$

Convexity is important in optimization: A convex minimization problem is to minimize a convex function f over a convex set $\mathbf{X}$. This is equivalent to maximizing a concave function $-f$ over a convex set.

---

i.e., the optimal control signal is

$$u^* = -R^{-1}F^\top W_t x$$

where we have used that $R = R^\top$ and $W_t = W_t^\top$. This optimal control u^* depends linearly on the state x, with a *gain* $-R^{-1}F^\top W_t$ which depends on the value function, i.e., on W_t. Inserting in the HJB equation and collecting terms, we get

$$\frac{1}{2}x^\top \left[\dot{W}_t + 2W_t(A - FR^{-1}F^\top W_t) + Q + W_t F^\top R^{-1}FW_t\right]x+$$
$$\dot{w}_t + \frac{1}{2}\mathrm{tr}G^\top W_t G = 0$$

while the terminal condition is

$$\frac{1}{2}x^\top Px = \frac{1}{2}x^\top W_T x + w_T.$$

Now notice that $x^\top 2W_t Ax = x^\top(W_t A + A^\top W_t)x$, where the matrix in the bracket is symmetric. Then, recall that if two quadratic forms $x^\top S_1 x$ and $x^\top S_2 x$ agree for all x, and S_1 and S_2 are symmetric, then $S_1 = S_2$. We see that the HJB equation is satisfied for all x,t iff the following two hold:

1. The matrix function $\{W_t : 0 \le t \le T\}$ satisfies

$$\dot{W}_t + W_t A + A^\top W_t - W_t FR^{-1}F^\top W_t + Q = 0 \qquad (13.11)$$

along with the terminal condition $W_T = P$. This matrix differential equation is termed the Riccati equation.

2. The off-set $\{w_t : 0 \le t \le T\}$ satisfies the scalar ordinary differential equation

$$\dot{w}_t + \frac{1}{2}\mathrm{tr}G^\top W_t G = 0$$

along with the terminal condition $w_T = 0$.

We summarize the result:

Theorem 13.5.1 (LQR Control) *The LQR problem of minimizing the quadratic cost (13.10) w.r.t. the control strategy $\{U_t : 0 \leq t \leq T\}$, subject to system dynamics (13.9), is solved by the linear static state feedback control*

$$U_t = \mu(X_t, t) = -R^{-1}F^\top W_t X_t$$

where $\{W_t : 0 \leq t \leq T\}$ is governed by the Riccati equation (13.11), with terminal condition $W_T = P$. The associated cost is

$$\Phi(x) = x^\top W_0 x + w_0$$

where $\{w_t : 0 \leq t \leq T\}$ is found by

$$w_t = \int_t^T \frac{1}{2}\mathrm{tr}G^\top W_s G \; ds.$$

Notice that the Riccati equation (13.11) does not involve the noise intensity G. So the optimal control strategy is independent of the noise intensity, but the noise determines the optimal cost through w_t.

The advantage of the linear quadratic framework is that the problem reduces to the Riccati matrix differential equation (13.11). So instead of a partial differential equation on $\mathbf{R}^n \times [0, T]$, we face $n(n+1)/2$ scalar ordinary differential equations; here we have used the symmetry of W_t and solve only for, say, the upper triangular part of W_t. This reduction from a PDE to a set of ODE's allows one to include a large number of states in the problem: Solving the Hamilton-Jacobi-Bellman partial differential equation (13.7) for a general nonlinear problem becomes numerically challenging even in three dimensions, but an LQR problem can have hundreds of states.

13.6 STEADY-STATE CONTROL PROBLEMS

In many practical applications, the control horizon T is exceedingly large compared to the time constants of the controlled system. In that case, we may pursue the limit $T \to \infty$. This situation is called "infinite horizon control problems". If also the system dynamics f, g and the running pay-off h are independent of time t, they often give rise to steady-state control strategies in which the optimal control strategy μ does not depend on time.

This situation can come in two flavors: *Transient* control problems and *stationary* control problems. In a transient control problem, the control mission ends when the state exits the region G of the state space, and in the infinite horizon case we assume that this takes place before time runs out at $t = T$. For example, for an autopilot that lands an aircraft, the mission ends when the aircraft is at standstill on the runway, and not at a specified time. In this situation, we seek a value function $V : \mathbf{X} \mapsto \mathbf{R}$ which satisfies

$$\sup_{u \in \mathbf{U}} [L^u V + h] = 0 \text{ for } x \in G, \quad V = k \text{ on } \partial G.$$

Example 13.6.1 (Swim Left or Right?) *A swimmer in a river is taken randomly left or right by currents, so that his position is a controlled diffusion*

$$dX_t = U_t \, dt + dB_t$$

on the domain $X_t \in G = (-H, H)$. Here, U_t is his swimming velocity. He aims to get ashore as quickly as possible but with limited effort, i.e., to minimize

$$\int_0^\tau \frac{1}{2} + \frac{1}{2}U_t^2 \, dt$$

where $\tau = \inf\{t : X_t \notin G\}$ and $w > 0$. This reflects a trade-off between time (the constant $1/2$) and effort (the term $U_t^2/2$). The Hamilton-Jacobi-Bellman equation is

$$\inf_u \left[uV' + \frac{1}{2}V'' + \frac{1}{2} + \frac{1}{2}u^2 \right] = 0$$

with boundary conditions $V(-H) = V(H) = 0$. The optimal control strategy is $u = \mu^*(x) = -V'(x)$, and the HJB equation becomes

$$\frac{1}{2}V'' - \frac{1}{2}(V')^2 + \frac{1}{2} = 0 \text{ for } x \in (-H, H).$$

The derivative $W := V'$ satisfies $W' = W^2 - 1$, so $V'(x) = -\tanh x$. The optimal control strategy is $u = \mu^*(x) = \tanh x$ and the value function is

$$V(x) = \log \frac{\cosh H}{\cosh x}.$$

See Figure 13.2. In the middle of the river, the swimmer stays calm ($u = 0$), waiting to see if random currents take him left or right. He then swims towards the nearest bank, with a determination that grows as he approaches the bank, confident that random currents will not take him back to the center again. Far from the center, the noise is irrelevant; the optimal swimming speed approaches 1 for $x = H \to \infty$, which would be the result if there was no noise.

In the stationary control problem, on the other hand, the closed loop system (13.6) with the time-invariant control strategy $U_t = \mu(X_t)$ admits a stationary solution $\{X_t : t \geq 0\}$, and the assumption is that the process mixes sufficiently fast compared to the terminal time T, so that the optimization problem concerns this stationary process. In this case, the boundary ∂G should never be reached; if the domain G is bounded, we may ensure that the boundary is repelling (Section 11.7), or replace the *absorbing* boundary ∂G with a reflecting one (Section 9.11.3). The terminal pay-off k becomes irrelevant. Then, we search for solutions to the HJB equation (13.7) of the form

$$V(x, t) = V_0(x) - \gamma \cdot t$$

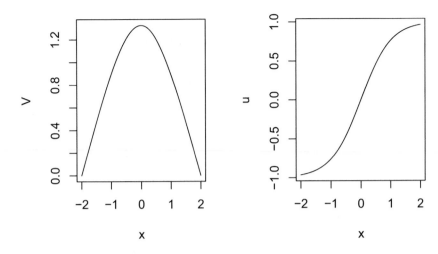

Figure 13.2 Value function (left) and optimal strategy (right) for the "swim left or right" Example 13.6.1.

where $\gamma \in \mathbf{R}$ is the expected running payoff of the stationary process, while the off-set $V_0(x)$ indicates whether a state x is more or less favorable than average. Inserting this particular form into the HJB equation (13.7), we see that this off-set $V_0 : \mathbf{X} \mapsto \mathbf{R}$ must satisfy

$$\sup_{u \in \mathbf{U}} [L^u V_0 + h] = \gamma. \tag{13.12}$$

We seek a matching pair (V_0, γ). In general, there can be many such pairs, and we seek the maximal γ. This equation specifies V_0 at best up to an additive constant; so we may require that V_0 has maximum 0, or that $V_0(X)$ has expectation 0 under the stationary distribution of $\{X_t\}$.

13.6.1   Stationary LQR Control

We consider the stationary version of the linear quadratic regulator problem (compare Section 13.5), starting in the scalar case. Let the controlled scalar diffusion $\{X_t : t \geq 0\}$ be given by the linear SDE

$$dX_t = (aX_t + fU_t)\, dt + g\, dB_t$$

with $f, g \neq 0$, where the performance objective is to minimize

$$\int_0^T \frac{1}{2}qX_t^2 + \frac{1}{2}U_t^2\, dt$$

with $q > 0$. We pursue the limit $T \to \infty$. We guess a solution $V_0(x) = \frac{1}{2}Sx^2$ to the stationary HJB equation (13.12), which then reads

$$\min_u \left[xS(ax + fu) + \frac{1}{2}g^2S + \frac{1}{2}qx^2 + \frac{1}{2}u^2 \right] = \gamma.$$

Minimizing w.r.t. u, inserting this optimal u, and collecting terms which are independent of and quadratic in x, we find

$$u = -Sfx, \quad \gamma = \frac{1}{2}g^2S, \quad Sa - \frac{1}{2}S^2f^2 + \frac{1}{2}q = 0. \tag{13.13}$$

The last equation is quadratic in S. It is termed the *algebraic Riccati equation*. It admits two solutions:

$$S_1 = \frac{a - \sqrt{a^2 + f^2q}}{f^2}, \quad S_2 = \frac{a + \sqrt{a^2 + f^2q}}{f^2}$$

For each of these two solutions, we can compute the corresponding stationary running cost γ. Note that S_1 is negative, regardless of parameters, which should correspond to a negative stationary expected running cost γ. This clearly cannot be the case, since the running cost is non-negative by definition. The reason for this is that the closed-loop system corresponding to S_1 is

$$dX_t = (a - S_1f^2)X_t \, dt + g \, dB_t = \sqrt{a^2 + f^2q}X_t \, dt + g \, dB_t.$$

This system is unstable and therefore does not admit a stationary solution $\{X_t\}$. So the solution S_1 does not help us with the stationary control problem (even if it has relevance for other problems, e.g., finite-time problems of driving the system away from its equilibrium). In turn, inserting the solution S_2, we find a positive expected stationary running cost γ, and closed-loop dynamics

$$dX_t = (a - S_2f^2)X_t \, dt + g \, dB_t = -\sqrt{a^2 + f^2q}X_t \, dt + g \, dB_t$$

which are stable. To summarize, the stationary HJB equation has more than one solution, and the relevant one is the maximal one, which is also the unique one which leads to stable closed-loop dynamics, and thus a stationary controlled process.

The following theorem generalizes this example.

Theorem 13.6.1 *Consider the LQR problem (13.9), (13.10) and let $T \to \infty$. Assume that the pair (A, Q) is detectable and that the pair (A, F) is stabilizable, i.e., for any eigenvalue λ of A with $\mathbf{Re}\lambda \geq 0$, it holds that the corresponding right eigenvector v satisfies $Qv \neq 0$ and the left eigenvector p satisfies $pF \neq 0$. Then, the optimal state feedback is*

$$U_t = \mu(X_t) = -R^{-1}F^\top SX_t$$

where S is the unique positive semidefinite solution to the Algebraic Riccati equation

$$SA + A^{\top}S - SFR^{-1}F^{\top}S + Q = 0 \qquad (13.14)$$

with the property that $A - FR^{-1}F^{\top}S$ is stable. This solution is also maximal in the sense that any other symmetric solution S_1 to the algebraic Riccati equation has $S_1 \leq S$. The associated time-averaged running cost is

$$\frac{1}{2}\mathrm{tr}G^{\top}SG.$$

See e.g. (Doyle et al., 1989). The importance of stabilizability is that any unstable mode (λ, v, p) is affected by the control, so it *can* be stabilized. The importance of detectability is that such an unstable mode *must* be stabilized: If not, the state and the cost will diverge to infinity. Combined, these two requirements assure that the optimal controller results in a stable system. The theory of linear-quadratic control is very complete, both theoretically and numerically, and has been applied to a large suite of real-world problems.

13.7 DESIGNING AN AUTOPILOT

Consider designing an autopilot for a ship, i.e., a control system that maintains the desired course by controlling the rudder. The model is

$$d\omega_t = (k\phi_t - c\omega_t + \tau_t)dt, \quad d\tilde{\theta}_t = \omega_t\, dt, \quad dI_t = \tilde{\theta}_t\, dt.$$

Here, ω_t is the speed of rotation of the ship at time t. The control signal is ϕ_t, the angle of the rudder, while τ_t is an exogenous torque on the ship from wind and waves. The constants k and c determine the dynamics of rotation: With no exogenous torque ($\tau_t \equiv 0$), a constant rudder angle ϕ_t will result in a constant speed of rotation $\omega_t = k\phi_t/c$. If the ship is turning and the rudder is set midships (neutral, $\phi_t = 0$), then the ship will gradually stop rotating over a time scale of $1/c$.

The course of the ship is θ_t, which we aim to keep constant at θ^{REF}, and the error is $\tilde{\theta}_t = \theta_t - \theta^{\mathrm{REF}}$. I_t is the integrated error; we will discuss the logic behind this term later, after having designed the controller.

The state vector is $X_t = (\omega_t, \tilde{\theta}_t, I_t)$. We model the exogenous force τ_t as white noise with intensity g, and define the control signal $U_t = \phi_t$. The system dynamics can then be written in the standard form

$$dX_t = AX_t\, dt + FU_t\, dt + G\, dB_t$$

with

$$A = \begin{bmatrix} -c & 0 & 0 \\ 1 & 0 & 0 \\ 0 & 1 & 0 \end{bmatrix}, \quad F = \begin{bmatrix} k \\ 0 \\ 0 \end{bmatrix}, \quad G = \begin{bmatrix} g \\ 0 \\ 0 \end{bmatrix}.$$

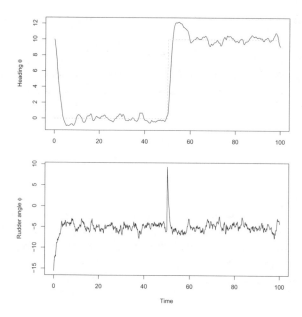

Figure 13.3 Simulation of an autopilot for a ship. The noise signal in the simulation is biased, i.e., the Brownian motion has a drift of 5. *Top panel:* The heading θ_t of the ship. Included is also the desired heading θ^{REF} (dashed), which changes during the simulation. *Bottom panel:* The rudder angle ϕ_t, i.e., the control signal.

We seek a stationary controller which minimizes the steady-state expectation

$$\mathbf{E}\frac{1}{2}X_t^\top Q X_t + \frac{1}{2}R|U_t|^2.$$

We take $R = 1$ and Q to be a diagonal matrix with entries q_{11}, q_{22} and q_{33}. Figure 13.3 shows a simulation where the system has been non-dimensionalized and all parameters taken to be 1, except q_{33} which is 0.1. In the beginning, the desired course is $\theta^{\text{REF}} = 0$ and the ship starts at $\theta_0 = 10$, but the controller achieves the correct course in about 10 time units. Mid-way during the simulation, the desired course changes, and the autopilot again completes the turn in, again, roughly 10 time units.

For this case, the optimal feedback strategy is

$$U_t = \mu(X_t) \approx -1.6\omega_t - 2.5\tilde{\theta}_t - 1I_t.$$

This structure, where the control signal is a linear combination of the error signal $\tilde{\theta}_t$ itself as well as its derivative ω_t and its integral I_t, is termed a PID controller in control engineering, for *proportional-integral-derivative*. The proportional part says that if you are too far to the right (e.g., at time 0 in the simulation), turn the rudder left. The derivative, i.e., the turning rate, acts as a damping: As the ship begins to turn left, the rudder action is decreased. This prevents an excessive overshoot.

The integral term I_t makes sure that the expected error vanishes in steady state, $\mathbf{E}\tilde{\theta}_t = 0$, even if the exogenous torque τ_t is biased, i.e., even if the Brownian motion has a drift. In the simulation, the torque has a positive bias of $+5$, so the rudder fluctuates around $\mathbf{E}U_t = -5$ to compensate for this torque. It is custom in control engineering to include such integrated error terms to compensate for biased noise and achieve robustness towards model errors. An alternative would be to include the bias of the torque in the model and compensate for it, but the integral is popular due to its simplicity. A disadvantage of the integral term is that it leads to some overshoot (Figure 13.3, around times 5 and 55): Since the controller is stabilizing, the integral I_t would converge to 0 as $t \to \infty$ in absence of noise. This implies that an initial positive error must be balanced by a negative error later, i.e., an overshoot.

A relevant question is how to choose the weights in Q and R. Ideally, they could be determined from higher level objectives, e.g., to make the ship reach its destination as quickly as possible. With deviations from the desired course, the ship follows a less direct and therefore longer route, but excessive rudder action slows the ship down. However, in practice it is not always feasible to deduce the weights Q and R. Then, they can be seen as tuning parameters that we adjust until we are satisfied with the resulting control system. An advantage of tuning the controller through the weights Q and R, rather than, say, the feedback policy μ, is that the controlled system is guaranteed to be stable for any choice of weights.

13.8   DESIGNING A FISHERIES MANAGEMENT SYSTEM

We now turn to an economic problem: How to design an optimal fishing policy. Consider a fish stock where the biomass displays stochastic logistic growth:

$$dX_t = [X_t(1 - X_t) - U_t] \ dt + \sigma X_t \ dB_t$$

where $\{X_t : t \geq 0\}$ is the biomass, $\{U_t : t \geq 0\}$ is the catch rate, which is our decision variable, $\{B_t : t \geq 0\}$ is standard Brownian motion, and σ is the level of noise in the population dynamics. This is a non-dimensionalized model; compare Example 7.7.1. We assume that $\sigma^2 < 2$ so that in absence of fishing, the zero solution is unstable; compare Example 12.3.3.

First, the management objective is to maximize the long-term expected catch U_t. The steady-state Hamilton-Jacobi-Bellman equation governing V_0 is

$$\sup_{u \geq 0} \left[(x(1 - x) - u)V_0' + \frac{1}{2}\sigma^2 x^2 V_0'' + u \right] = \gamma.$$

We maximize over only non-negative u, because we cannot unfish. Then, an optimal strategy must satisfy

$$u = \begin{cases} 0 & \text{when } V_0'(x) > 1 \\ \infty & \text{when } V_0'(x) < 1. \end{cases}$$

When $V_0'(x) = 1$, any u is optimal. In words, we should either not fish at all, or fish as hard as possible, depending on $V_0'(x)$. Such a strategy, where the control variable U_t switches discontinuously between the extremes of the permissible values, is termed a *bang-bang* control strategy. They arise when the instantaneous maximization problem in the HJB equation is linear in u, so that optimal controls u always exist on the boundary of **U** - in this case, 0 and ∞. Bang-bang control are often problematic: Mathematically, there is no Markov strategy which satisfies this and at the same time satisfies the assumptions for existence and uniqueness of solutions (Theorem 8.3.2). More practically, they are difficult to implement, and they may drive the system to an extreme where the mathematical model is no longer a valid representation of reality, for example because they excite non-linearities or fast unmodelled dynamics. In the context of fisheries management, most managers will typically deem them unacceptable.

We conclude that our problem formulation was too simplistic, so we try again. We argue that exceedingly large catches u will flood the market and reduce the price, and take this into account by assuming that the cumulated profit is

$$\int_0^T \sqrt{U_t}\, dt.$$

It is important that $\sqrt{u}$ is an increasing concave function of $u \geq 0$, but the specific choice of a square root is rather arbitrary. With this performance criterion, the HJB equation becomes

$$\sup_{u \geq 0} \left[(x(1-x) - u)V_0' + \frac{1}{2}\sigma^2 x^2 V_0'' + \sqrt{u} \right] = \gamma. \tag{13.15}$$

Now, the optimal control is, whenever $V_0'(x) > 0$,

$$u = \mu^*(x) = \frac{1}{4(V_0'(x))^2} \tag{13.16}$$

It is easy to verify that a solution to this equation is

$$V_0(x) = \frac{1}{2}\log x, \quad u = \mu^*(x) = x^2, \quad \gamma = \frac{1}{2}\left(1 - \frac{1}{2}\sigma^2\right).$$

With this strategy, the biomass satisfies the logistic growth equation

$$dX_t = X_t(1 - 2X_t)\, dt + \sigma X_t\, dB_t$$

and will therefore approach a Gamma-distributed steady state (Example 9.8.2). In absence of noise, the biomass will approach the equilibrium state $x = 1/2$. This is the point in state space where the surplus production $x(1-x)$ is maximized, so this point allows the *maximum sustainable yield* (Schaefer, 1957). The noise will perturb the system away from this state, but the optimal control attempts to bring the system back to this state.

A simulation of the system is shown in Figure 13.4. The simulation starts at the unexploited equilibrium, $X_0 = 1$. In addition to the optimal policy $U_t = X_t^2$, the figure includes two simpler policies: First, the constant catch policy $U_t = 1/4$. Without any noise in the system, this policy would be able to maintain Maximum Sustainable Yield and a population at $X_t = 1/2$. However, this equilibrium solution would be unstable, and the noise in the population dynamics (or any other perturbation) drives the system away from the equilibrium and causes the population to crash and reach $X_t = 0$, at which point the population is extinct and the fishery closes.

The figure also includes the *constant-effort* policy $U_t = X_t/2$, where we harvest a fixed fraction of the biomass in each small-time interval. Without noise, this would also lead to Maximum Sustainable Yield, $X_t = 1/2$, $U_t = 1/4$, which would now be a stable equilibrium. With (weak) noise, fluctuations around this equilibrium arise. Compared to the optimal policy $U_t = X_t^2$, the constant effort policy leads to lower biomass most of the time, and consequently also to lower catches and profits. The optimal policy relaxes the fishery effort more in bad years, allowing the biomass to rebuild, and conversely exploits good years to the fullest.

Exercise 13.1: The value function $V_0(x)$ tends to $-\infty$ as $x \searrow 0$. How should this be understood? *Hint:* Consider first the time-varying optimization problem and the corresponding value function $V(x,t)$ for x near 0.

For this particular case, the optimal control and the value function V_0 turn out to be independent of the noise level. But the noise moves the system away from the optimal operating point and reduces the expected payoff: $\gamma = \frac{1}{2}(1-\frac{1}{2}\sigma^2)$. The limit $\gamma = 0$ is reached when $\sigma^2 = 2$, i.e., where the unexploited system becomes unstable (Example 12.3.3). With $\sigma^2 > 2$, the population will crash even without fishing, so that no profit can be obtained in steady state.

13.9 A FISHERIES MANAGEMENT PROBLEM IN 2D

We extend the fisheries example to two dimensions. Aiming to demonstrate that numerical solutions are feasible, we give a concise presentation of the problem. The system is a predator-prey system with harvest:

$$dN_t = N_t\left[r(1 - \frac{N_t}{K}) - \frac{cP_t}{N_t + \bar{N}} - F_t^N\right] dt + \sigma_N N_t \, dB_t^{(1)} \quad (13.17)$$

$$dP_t = P_t\left[\frac{\epsilon c N_t}{N_t + \bar{N}} - \mu_P - F_t^P\right] dt + \sigma_P P_t \, dB_t^{(2)}. \quad (13.18)$$

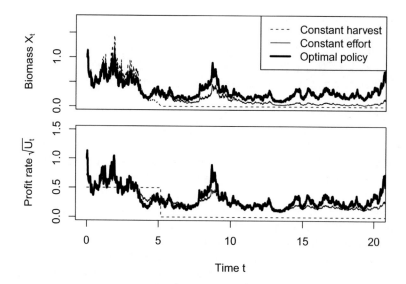

Figure 13.4 Simulation of the fisheries management system with $\sigma^2 = 1/2$ and three management regimes: Constant catch $U_t = 1/4$ until collapse, constant effort $U_t = X_t/2$, and the optimal policy $U_t = X_t^2$. The three simulations are done with the same realization of Brownian motion.

Here, N_t is the prey abundance (think herring or sardines), and P_t is the predator abundance (cod or tuna). This is the Rosenzweig-MacArthur model we posed in Section 12.5, except for the catch terms $(C_t^N = F_t^N N_t, C_t^P = F_t^P P_t)$ which are written in terms of *fishing mortalities* (F_t^N, F_t^P).

We seek fishing mortalities (F_t^N, F_t^P) which maximize the total revenue, assuming a payoff structure as in the previous section, for each fishery:

$$h(N, P, F^N, F^P) = \rho_N \sqrt{NF^N} + \rho_P \sqrt{PF^P}.$$

As before, the square root models price dynamics, or our utility from the catches NF^N, PF^P. The weights ρ_N and ρ_P determine if we prefer to catch prey or predators.

Theoretically, we aim for a solution to the Hamilton-Jacobi-Bellman equation (13.12) with $x = (N, P)$, $u = (F^N, F^P)$ on the entire region $\{(N, P) : N \geq 0, P \geq 0\}$. For numerical purposes, we confine ourselves to the bounded rectangular set $\{(N, P) : N_{\min} \leq N \leq N_{\max}, P_{\min} \leq P \leq P_{\max}\}$. We enforce $F^N = 0$ whenever $N = N_{\min}$; i.e., there will be no fishing on the prey when its abundance drops to $N_{\min}$, and similarly for the predator. Next, we enforce reflection (no-flux) at all boundaries. This approximation simplifies the problem, and is (partially) justified by the axes being unstable (compare Section 12.5) in absence of fishing. Therefore, if we close down the fishery when approaching the axes, the system will never reach the axes.

We use the parameters $r = 1$, $K = 1$, $c = 1$, $\mu = 0.15$, $\bar{N} = 1.5$, $\epsilon = 0.5$, $\frac{1}{2}\sigma_N^2 = 0.01$, $\frac{1}{2}\sigma_P^2 = 0.01$ and discretize the domain, log-transforming the dynamics and using a regular grid in the $(\log N, \log P)$-plane with 150-by-151 cells, with a lower left corner of $(\exp(-4), \exp(-4))$ and upper right corner of (e, e). The Hamilton-Jacobi-Bellman equation is discretized to a Markov Decision Problem which is solved using Policy Iteration (Section 13.12.2). We plot only a subset of the computational domain (Figure 13.5) where the probability is concentrated under the resulting strategy, but a larger computational domain is needed to avoid that numerical artifacts from the boundaries deteriorate the results. The computations take a couple of seconds on a standard laptop.

Results are seen in Figure 13.5 for parameters $\rho_N = 0.05$, $\rho_P = 1$, where the optimal policy ensures that both species co-exist (left panel). At the mode of the distribution, the prey abundance is $N \approx 0.9$ (top left panel) and the prey fishing mortality is $F_N \approx 0.006$ (bottom left panel). This is a much higher abundance, and a much lower effort, than we found in the single-species model the previous section, where the population was held around 0.5 with a mortality of $1/2$. Also, the optimal fishing mortality on the prey depends on the two abundances in a non-trivial way, in particular around the mode of the stationary distribution (bottom left panel). The profit from each of the fisheries fluctuates (top right panel) and in particular the predator fishery shows large fluctuations, which corresponds to the large range of the stationary distribution (top left panel). To some degree, the two fisheries operate in counterphase so that good periods for the predator fishery are bad periods for the prey fishery, whereas predator-prey cycles are not visible (compare Section 12.5).

This model is quite rich and we could elaborate extensively on the results, for example, by examining the effect of varying parameters. However, our objective of this section was simply to demonstrate that highly non-linear two-dimensional control problems can be solved numerically with relatively modest effort, and that optimizing an entire system leads to very different dynamics, compared to optimizing components in isolation.

13.10 OPTIMAL DIEL VERTICAL MIGRATIONS

We reconsider the foraging, mortality, and fitness of zooplankton (Section 11.8.2). The state is the vertical position of the animal, $x \in [0, H]$, but since the mortality is time-varying, we can consider time t an extra state variable. Thus, the problem becomes two-dimensional.

In Section 11.8.2, the animals moved by pure diffusion, and we computed the fitness using the Feynman-Kac formula. We found spatiotemporal patterns in the fitness which gave the animals an incentive to move vertically, tracking the fitness maximum. Therefore, we now consider an individual with

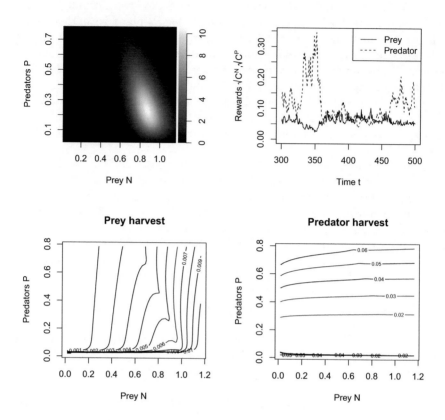

Figure 13.5 Results from the problem of optimal harvest from a two-species system consisting of a prey and a predator. *Top left:* Stationary probability density under the optimal strategy. *Top right:* Simulated time-series of the payoff ($\sqrt{C_t^N}, \sqrt{C_t^P}$) under the optimal strategy. *Bottom left:* Optimal fishing mortality F^N. *Bottom right:* Optimal fishing mortality F^P.

a movement strategy $u(x,t)$, leading to the stochastic differential equation

$$dX_t = u(X_t, t)\ dt + \sigma\ dB_t$$

for the position, still with reflection at the boundaries $x \in \{0, H\}$. Movement comes with a cost, so the animal should aim to maximize

$$\mathbf{E}^{X_t=x, S_t=1} \int_t^\tau \left[h(X_s) - \frac{1}{2}\nu u^2(X_s, s) \right]\ ds$$

where, as in Section 11.8.2, h is the energy harvest rate, and τ is the random time of death, governed by the mortality $\mu(X_s, s)$. ν is a parameter which controls the energetic cost of motion. The value function $V(x, t)$ for this optimization problem is given by the Hamilton-Jacobi-Bellman equation

$$\dot{V} + \sup_u \left[uV' + \frac{1}{2}\sigma^2 V'' + h - \frac{1}{2}\nu u^2 - \mu V \right] = 0. \qquad (13.19)$$

We impose homogeneous Neumann conditions at the boundaries $x \in \{0, H\}$, and pursue the time-periodic solution, $V(x, 0) = V(x, T)$. For a given value function, the optimal velocity is $U_s = u(X_s, s) = V'(X_s, s)/\nu$, so the animals move in direction of increased fitness; this is called *fitness taxis*. We find the solution numerically using the methods in Section 13.12.2; with a 100×120 grid, policy iteration takes half a second on a standard laptop. Results are seen in Figure 13.6. Parameters are as in Section 11.8.2, except the diffusivity is reduced to $D = \sigma^2/2 = 500$ m^2/day, since the animals now move directed rather than purely randomly. The cost of motion is $\nu = 2 \cdot 10^{-5}$ J day/m^2. Note that the optimal movement strategy (middle panel) is to evade the surface around dawn ($t = 0.25$) and stay away from it during the day. During the day, the animals around the nutricline ($x \approx 50$ m) do not move much vertically; this location, it appears, represents the optimal trade-off between energetic gains and risk of dying. At dusk ($t = 0.75$), the animals take advantage of the safety of darkness to move towards the surface, where they can exploit the nutrient-rich environment. These strategies derive from the patterns in the fitness landscape (Figure 13.6, top panel). The resulting quasi-stationary density $\phi(x, t)$ of animals is seen in the bottom panel. Note that the animals concentrate around a fairly narrow band, which is at depth during the day and closer to the surface at night. The distribution appears symmetric in time, although neither the fitness landscape nor the swimming strategy does, since the animals anticipate dawn and dusk rather than respond reactively.

13.11 CONCLUSION

Optimal control problems appear in a range of applications, where the objective is to design a dynamic system which performs optimally. These covers traditional control engineering applications, i.e., technical systems, as well as financial and management problems. They also appear in situations where we do not aim to design a system, but rather to understand an existing decision maker, for example, to predict future decisions.

We have focused on the technique of Dynamic Programming for solving such problems. In the case of stochastic differential equations, dynamic programming amounts to solving the Hamilton-Jacobi-Bellman equation. This equation generalizes the backward Kolmogorov equation and, more specifically, Dynkin's lemma that we studied in Chapter 11: When analyzing a given control system, i.e., when $\mathbf{U}$ is a singleton, we can determine its performance by solving this backward equation. When we include the choice of control,

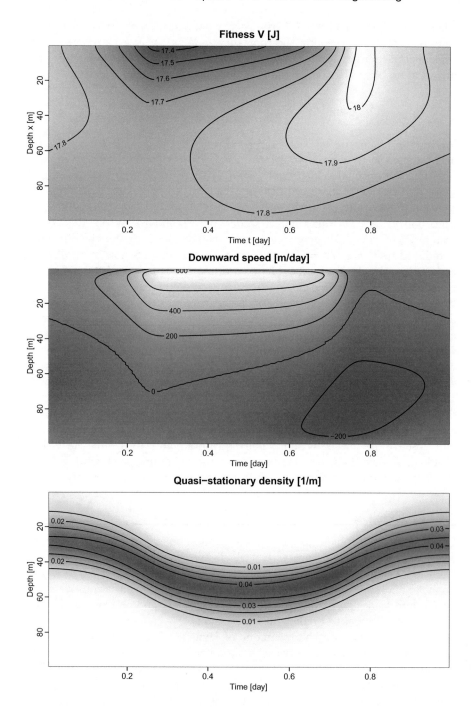

Figure 13.6 Optimal vertical migrations of zooplankton. *Top panel:* Fitness $V(x,t)$ as a function of time and depth, given by (13.19). *Middle panel:* The optimal upwards swimming velocity $u(x,t)$. *Bottom panel:* The quasi-stationary distribution $\phi(x,t)$.

this adds the "sup" in the Hamilton-Jacobi-Bellman equation (13.7), and this equation reduces the problem of dynamic optimization to a family of static optimization problems, where we trade-off instantaneous gains $h(x, u, t)$ against future gains $(L^u V)(x, t)$.

The presentation in this chapter is meant as a first introduction, and a number of important issues have been omitted. An important one is the *characterization theorem*, which states that the value function satisfies the HJB equation, provided it is smooth. See (Øksendal, 2010). In many problems, however, the solution is not smooth. The framework of *viscosity solutions* to partial differential equation addresses this issue; see the following notes.

In some of the examples in this section, we found analytical solutions to the HJB equation. Examples with analytical solutions play a prominent role in the literature, but it should be clear that they are exceptions rather than the rule. From a modeling perspective, it would be an extreme restriction if we have to confine ourselves to models with analytical solutions, and simple numerical methods are an important element of the toolbox. Sections 13.9 and 13.10 demonstrated applications of such numerical analysis for problems in one and two dimenions; the methods are detailed in the following notes. For problems in higher dimenions, techniques from reinforcement learning/neuro-dynamic programming are applicable.

13.12 NOTES AND REFERENCES

Bellman (1957) coined the term *Dynamic Programming* and demonstrated the versatility of the approach through a sequence of papers, mostly in the context of deterministic and stochastic problems with discrete state space. Dynamic optimization for deterministic differential equations was approached with calculus of variations at that time, which lead to Pontryagin formulating his Maximum Principle in the late 1950's. The connection between calculus of variations and partial differential equations of Hamilton-Jacobi type was developed by Carathéodory in 1935, and pointed out by Kalman (1963b). Dynamic programming was soon applied to stochastic differential equations (Kushner, 1967); an established text is (Fleming and Rishel, 1975).

A difficulty is that the value function may not be smooth, when the diffusivity is singular, in which case it cannot satisfy the Hamilton-Jacobi-Bellman equation in the classical sense. Under regularity conditions, it is then a *viscosity* solution (Fleming and Soner, 1993; Pham, 2009; Touzi, 2013). This weaker concept of a solution can be understood as a limiting procedure, where we first add a little white noise to all states, solve the Hamilton-Jacobi-Bellman equation in the classical sense, and then let the intensity of this regularizing noise tend to zero (Lions, 1982; Crandall et al., 1992). Example 13.6.1, the swimmer in the river, is useful, noting that increasing the domain size has the same effect as decreasing the noise intensity and rescaling: Without noise, the optimal control strategy is $u = \operatorname{sign} x$, i.e., discontinuous, and the value function is $V(x) = H - |x|$, which is only piecewise differentiable. This V is a viscosity solution of the HJB equation.

13.12.1 Control as PDE-Constrained Optimization

The optimal control problem can also be seen as optimizing subject to the constrains imposed by system dynamics. We now make this specific in the particular situation of steady-state control, skipping technicalities: Let $U_t = \mu(X_t)$ be a control strategy which leads to the closed-loop generator L^μ and the running reward h^μ. We consider maximizing the expected running reward:

$$\sup_{\mu,\pi} \langle \pi, h^\mu \rangle$$

subject to the constraint that π is the stationary probability distribution resulting from the control strategy μ, i.e., subject to the constraints

$$(L^\mu)^*\pi = 0, \quad \langle \pi, 1 \rangle = 1.$$

We include these constraints using the technique of Lagrange multipliers, and thus reach the Lagrange relaxed optimization problem

$$\sup_{\mu,\pi,V,\gamma} \langle \pi, h^\mu \rangle + \langle (L^\mu)^*\pi, V \rangle + \gamma - \langle \pi, \gamma \rangle.$$

Here, $V : \mathbf{X} \mapsto \mathbf{R}$ and $\gamma \in \mathbf{R}$ are Lagrange multipliers associated with the constraints $(L^\mu)^*\pi = 0$ and $\langle \pi, 1 \rangle = 1$. We employ duality to rewrite $\langle (L^\mu)^*\pi, V \rangle = \langle \pi, L^\mu V \rangle$ and collect terms:

$$\sup_{\mu,\pi,V,\gamma} \langle \pi, L^\mu V + h^\mu - \gamma \rangle + \gamma.$$

This relaxed optimization problem is linear in π, so for a finite supremum to exist, the triple (V, μ, γ) must satisfy

$$L^\mu V + h^\mu - \gamma = 0.$$

Moreover, this term must be maximized w.r.t. μ for each x. Thus, this PDE-constrained optimization problem is equivalent to the Hamilton-Jacobi-Bellman equation, and the value function can also be interpretated as a Lagrange multiplier associated with the stationarity constraint. This formulation opens up for a number of extensions to the problem as well as for a number of different numerical methods.

13.12.2 Numerical Analysis of the HJB Equation

We now describe a technique for numerical analysis of the Hamilton-Jacobi-Bellman equation based on (Kushner and Dupuis, 2001). To simplify and be specific, we consider the stationary, infinite-horizon situation (13.12):

$$\forall x : \sup_{u \in \mathbf{U}} [L^u V(x) + h^u(x)] = \gamma.$$

Assuming that a solution (V, γ) exists, we present a *policy iteration* which identifies it. Starting with a policy $\mu : \mathbf{X} \mapsto \mathbf{U}$, we evaluate its performance V and γ. Based on V, we improve the policy. Specifically:

1. Start with an arbitrary policy μ_1.

2. For each $i = 1, 2, \ldots$:

   (a) Identify the performance (V_i, γ_i) corresponding to μ_i by solving

$$L^{\mu_i} V_i + h^{\mu_i} = \gamma_i. \tag{13.20}$$

   (b) Determine the next policy μ_{i+1} by

$$\mu_{i+1}(x) = \arg \max_u \left[(L^u V_i)(x) + h^u(x) \right]. \tag{13.21}$$

We explicitly assume that at each step in the policy iteration, the closed-loop system with the control μ_i has a unique stationary distribution, say π_i, under which h^{μ_i} has finite expectation.

Lemma 13.12.1 *The sequence $\{\gamma_i\}$ is non-decreasing and bounded by γ.*

Proof: Let (V_i, γ_i) be given such that $L^{\mu_i} V_i + h^{\mu_i} = \gamma_i$. When we identify the next policy μ_{i+1}, it must hold that

$$L^{\mu_{i+1}} V_i + h^{\mu_{i+1}} \geq \gamma_i.$$

We find the next value function V_{i+1} and performance γ_{i+1} from

$$L^{\mu_{i+1}} V_{i+1} + h^{\mu_{i+1}} = \gamma_{i+1}.$$

Let π_{i+1} be the stationary probability distribution for $L^{\mu_{i+1}}$, so that $\langle \pi_{i+1}, L^{\mu_{i+1}} V_{i+1} \rangle = \langle \pi_{i+1}, L^{\mu_{i+1}} V_i \rangle = 0$. Then take inner product of the two preceding equations with π_{i+1} to obtain

$$\gamma_{i+1} = \langle \pi_{i+1}, h^{\mu_{i+1}} \rangle \geq \gamma_i.$$

Thus, the sequence $\{\gamma_i\}$ is non-decreasing. It is clear that $\gamma_i \leq \gamma$, since γ_i is the performance of the policy μ_i while γ is the optimal performance. ∎

In practice we are rarely able to solve these partial differential equations, so we now discretize these equations in a way that ensures that monotonicity is preserved. We consider additive controls, i.e.,

$$dX_t = \left[f_0(X_t) + \sum_{j=1}^m f_j(X_t) U_t^j \right] dt + g(X_t)\, dB_t$$

where the set of admissable controls is

$$U_t = (U_t^1, U_t^2, \ldots, U_t^m) \in \mathbf{U} = [0, \infty)^m.$$

Note that the noise intensity $g(x)$ is independent of the control; this simple structure is common in applications. The backward operator L^u is then

$$L^u = L_0 + \sum_{j=1}^{m} u_j L_j$$

where

$$L_0 V = \nabla V \cdot f_0 + \frac{1}{2}\mathrm{tr}[gg^\top \mathbf{H}V], \quad L_j V = \nabla V \cdot f_j. \tag{13.22}$$

Here, $\mathbf{H}V$ is the Hessian of V, as usual. Since the control are non-negative, the operators $L_j V$ can be discretized using the upwind method (Section 9.11.5). If the original problem allows both positive and negative controls, for example,

$$dX_t = [f_0(X_t) + U_t]\ dt + g(X_t)\ dB_t, \text{ where } U_t \in \mathbf{R},$$

then we rewrite this using two controls as

$$dX_t = \left[f_0(X_t) + U_t^1 - U_t^2\right]\ dt + g(X_t)\ dB_t,$$

where both U_t^1 and U_t^2 are non-negative; i.e., $f_1(X_t) = 1$ and $f_2(X_t) = -1$.

We now discretize each of the generators $L_0, L_1, \ldots, L_m$ as in Section 9.11.5. Let n be the number of grid cells, i.e., the discretized generators are $G_0, G_1, \ldots, G_m$, which are all n-by-n matrices, each being a generator for a continuous-time Markov chain on the state space $\{1, \ldots, n\}$.

For a control strategy $\mu : \mathbf{X} \mapsto [0,\infty)^m$, we let $U \in [0,\infty)^{n\times m}$ denote its discretization, obtained by evaluating the each of the m control variables in the n grid cells. This yields the generator of the discretized controlled system:

$$G^U = G_0 + \sum_{j=1}^{m} \mathrm{diag}(U_{\cdot j})G_j. \tag{13.23}$$

Here, $U_{\cdot j}$ is the j'th column in U, i.e., the j'th control variable evaluated at each grid point. Note that G^U is a generator, since every G_j is, and each element in U is nonnegative. Moreover, G^U is irreducible if G_0 is (this condition is not necessary; the controls can also ensure irreducibility).

We next discretize the pay-off h, abusing notation by also letting h denote the discretized version. This discretized h is a function $[0,\infty)^{n\times m} \mapsto \mathbf{R}^n$; for a given control strategy U it gives the pay-off at each grid cell. We can now write the stationary Hamilton-Jacobi-Bellman equation as

$$\sup_{U\in[0,\infty)^{n\times m}} \left[\left(G_0 + \sum_{j=1}^{m} \mathrm{diag}(U_{\cdot j})G_j\right)V + h(U)\right] = \gamma e \tag{13.24}$$

where e is an n-vector of all ones, i.e., a constant function on state space.

This discrete Hamilton-Jacobi-Bellman equation is a Markov Decision Problem (Puterman, 2014) which we solve with policy iteration. At an iteration i, we have a given policy U, which gives a closed-loop generator G^U from (13.23) and a pay-off $h = h(U)$, and we solve the equation

$$G^U V + h = \gamma e$$

for (V, γ) to determine the performance of this policy U. This equation is underdetermined, since generators G^U are singular. We therefore, somewhat arbitrarily, require that the elements in V sum to 0, and reach the system

$$\begin{bmatrix} G^U & e \\ e^\top & 0 \end{bmatrix} \begin{pmatrix} V \\ \gamma \end{pmatrix} = \begin{pmatrix} -h \\ 0 \end{pmatrix}. \tag{13.25}$$

Lemma 13.12.2 *Assume that G^U is the generator of an irreducible Markov chain. Then the system (13.25) of linear algebraic equations is regular. Moreover, let π be the stationary probability distribution of this Markov chain, then $\gamma = \pi h$, and in particular, if $h \geq 0$, then $\gamma \geq 0$.*

Proof: Assume that there is a row vector (ϕ, ψ) such that

$$(\phi \ \psi) \begin{bmatrix} G & e \\ e^\top & 0 \end{bmatrix} = (0 \ 0) \tag{13.26}$$

or

$$\phi G^U + \psi e^\top = 0, \quad \phi e = 0.$$

Multiply the first equation from the right with e, and use $G^U e = 0$ to reach

$$\psi e^\top e = 0$$

or simply $\psi = 0$. Thus $\phi G^U = 0$, so ϕ is the stationary density, rescaled. But since $\phi e = 0$, we find $\phi = 0$. Since $(\phi, \psi) = (0, 0)$ is the only solution to the homogeneous equation (13.26), we conclude that the system is regular.

Letting π be the stationary probability distribution, we have $\pi e = 1$, $\pi G^U = 0$. We pre-multiply the system (13.25) with $(\pi \ 0)$, to get

$$\pi G^U V - \pi e \gamma = -\pi h$$

or simply $\gamma = \pi h$, as claimed. ■

We can now evaluate the performance γ of a given discretized policy U, and find the associated value function V, and the "comparison" property holds that γ is increasing in h. We next specify the update of the policy U.

The policy update: Let $V^{(i)}$ and $\gamma^{(i)}$ be given and define the next policy $U^{(i+1)}$ as follows: Its k'th row is found as

$$U^{(i+1)}_{k\cdot} = \mathrm{Arg} \max_{u \in [0,\infty)^m} \left[\left(e_k G_0 + \sum_{j=1}^{m} u_j e_k G_j \right) V^{(i)} + h(x_k, u) \right]$$

where e_k is a (row) vector with a 1 in position k and 0 elsewhere. Notice that this optimization problem is done at each state x_k separately, and structurally, it is the same problem as the continuous-space version (13.21). In many application, $h(x, u)$ is concave in u for each x, so that we may find a solution to this optimization problem through the stationarity condition

$$\frac{\partial}{\partial u_j} h(x_k, u) + e_k G_j V^{(i)} = 0$$

although the requirement $u_j \geq 0$ must also be taken into account.

It is now easy to see, arguing as in the continuous case, that the sequence $\{\gamma^{(i)}\}$ is non-decreasing under policy iteration and hence convergent. We include the specific statement for completeness.

Theorem 13.12.3 *Assume that under policy iteration, the closed-loop system is an irreducible Markov chain at each iteration. Then the performance $\{\gamma^{(i)}\}$ is non-decreasing. If the algorithm has converged, then the policy cannot be outperformed by any other policy.*

Recall that a *sufficient* condition for irreducibility is that the open-loop system G_0 is irreducible.

Proof: Let $U^{(i)} \in [0, \infty)^{n \times m}$ be a policy, let $G^{(i)}$ be the corresponding closed-loop generator, let $h^{(i)}$ be the corresponding payoff, and finally let $(V^{(i)}, \gamma^{(i)})$ be the performance of that strategy. The equation (13.26) can then be written as

$$G^{(i)} V^{(i)} + h^{(i)} = \gamma^{(i)} e.$$

We find the next policy $U^{(i+1)}$ using the policy update; let $G^{(i+1)}$ and $h^{(i+1)}$ be the corresponding generator and payoff, so that

$$G^{(i+1)} V^{(i)} + h^{(i+1)} \geq \gamma^{(i)} e.$$

Let $V^{(i+1)}, \gamma^{(i+1)}$ be the performance of strategy $U^{(i+1)}$, given by

$$G^{(i+1)} V^{(i+1)} + h^{(i+1)} = \gamma^{(i+1)} e$$

Let $\pi^{(i+1)}$ be the stationary distribution corresponding to strategy $U^{(i+1)}$, so that $\pi^{(i+1)} G^{(i+1)} = 0$. We then get

$$\begin{aligned}
\gamma^{(i+1)} &= \pi^{(i+1)} (G^{(i+1)} V^{(i+1)} + h^{(i+1)}) \\
&= \pi^{(i+1)} (G^{(i+1)} V^{(i)} + h^{(i+1)}) \\
&\geq \pi^{(i+1)} \gamma^{(i)} e \\
&= \gamma^{(i)}.
\end{aligned}$$

It follows that the sequence $\{\gamma^{(i)}\}$ is non-decreasing. We can repeat this reasoning to show that if the algorithm has converged, then the policy cannot be outperformed by any other policy. ■

In practice, this algorithm is often straightforward to implement and performs well. It is implemented in the SDEtools package; see the functions PolicyIterationRegular and PolicyIterationSingular. Similar techniques can be devised for other computational problems using the same principles; see (Kushner and Dupuis, 2001).

13.13 EXERCISES

Exercise 13.2 Linear-Quadratic Tracking: Consider controlled Brownian motion $dX_t = U_t\, dt + \sigma\, dB_t$, where the aim is to track a reference $\{Y_t = sW_t\}$, where $\{B_t\}$ and $\{W_t\}$ are independent standard Brownian motions. $\sigma > 0$ and $s > 0$ are parameters. The objective is to find a Markov control $U_t = \mu(X_t, Y_t)$ such as to minimize $\frac{1}{2}\mathbf{E}(q(X_t - Y_t)^2 + U_t^2)$ in steady state.

1. Defining the tracking error $Z_t = X_t - Y_t$, write $\{Z_t\}$ as a controlled Itô process, and rewrite the problem in terms of $\{Z_t\}$.

2. Find the optimal control law $U_t = u(Z_t)$ and the corresponding cost.

3. Now consider the original problem in (X_t, Y_t). Explain why Theorem 13.6.1 does not apply.

4. Undeterred, show that

$$S = \sqrt{q}\begin{bmatrix} 1 & -1 \\ -1 & 1 \end{bmatrix}$$

satisfies the algebraic Riccati equation (13.14) and gives rise to the control law from question 2. Find the two eigenvalues of the controlled system.

Exercise 13.3 Fisheries management: This exercises reproduces Section 13.8 by solving the Hamilton-Jacobi-Bellman equation numerically, following Section 13.12.2.

1. Write up the generator L_0 for the uncontrolled system and the generator L_1 for the control, as in (13.22).

2. Truncating the state space to the interval $[0,3]$ and discretizing it in 600 equidistant bins, identify the discretized generators G_0 and G_1 as in (13.23).

3. Identify the pay-off (reward) function $h(u)$ as in (13.24). *Note:* In the discretized version, this function takes as argument a vector of controls, one for each grid cell, and returns the reward obtained in each grid cell. To avoid problems at the lower boundary, this should always return 0 for the first grid cell (heuristically, you cannot profit from fishing if there are no fish).

4. Identify the optimal control. *Note:* In the discretized version, this takes a vector $G_1 V$ and returns a vector containing the optimal control in each grid cell. It should be "robustified", so that it always returns a non-negative real number which is not absurd large, and 0 in the first grid cell.

5. Solve the stationary control problem, e.g. using `PolicuIteration Singular` from `SDEtools`. Plot the value function, the optimal control, and the resulting stationary p.d.f. of $\{X_t\}$. Compare with the analytical results from Section 13.12.2 and comment.

CHAPTER 14

Perspectives

The more you know,
the more you realize you don't know.

Attributed to both Aristotle (384–322 BC)
and to Albert Einstein (1879–1955).

This book was written for a first course on stochastic differential equations for scientists and engineers. Even if it contains more material than can be covered in detail in a 5 ECTS course, its ambition is only to serve as an entry point into the vast realm of stochastic differential equations. We now end this book with some suggestions for next steps, reflecting personal bias and limitations.

We have treated several topics superficially. These include ergodicity, Girsanov's theorem, estimation of parameters, and optimal control; Malliavin calculus we only mentioned in passing. In several places we have sacrificed mathematical rigor for pace. We have already given references to more in-depth treatments in the literature; an obvious next step is to pursue these.

We have only presented the simplest numerical methods. This applies to simulation of sample paths, to Kolmogorov's equations and other partial differential equations, as well as to estimation and optimization. The reasoning is that it is important for a generalist to have access to a broad numerical toolbox, for the purpose of pilot studies and experimenting with different models. In contrast, to master a particular type of numerical analysis (e.g., simulation of sample paths, or filters for non-linear systems) requires specialized studies that are more appropriate in a advanced dedicated course than in a first introductory one.

We have only considered continuous diffusions, but in many applications, it is relevant to include discrete jumps. These jumps can be viewed as internal to the state space model; we saw a first example of this when discussing killing in Section 11.8.2. When the state can jump to different positions in state space, the Kolmogorov equations become integro-differential equations, where the integral terms reflect how probability is redistributed non-locally through jumps (Gardiner, 1985; Björk, 2021). If the jump rate does not depend on the state, then jumps can be viewed as imposed on the state from the outside world. For example, the state dynamics may be driven by a Poisson process, or a Gamma process, in addition to the Brownian motion, so that the sample paths involve Itô integrals with respect to these processes.

A different class of driving input is fractional Brownian motion (Metzler and Klafter, 2000). These are self-similar processes, with statistics that satisfy power laws, but where the variance does not grow linearly with time, but rather with a non-integer exponent. These processes give rise to anomalous diffusion and the resulting Kolmogorov equations are fractional partial differential equations, which are most accessible in the linear case where frequency domain methods apply. They can describe subdiffusion in porous media, where particles may "get stuck" in pores for long times, or superdiffusion in turbulence, where variance grows super-linearly due to eddy structure. In general, they can describe phenomena with long-range dependence.

Stochastic differential equations can be extended to spatiotemporal phenomena, leading to stochastic partial differential equations. A useful image is that of a guitar string in a sand storm (Walsh, 1986): Without noise, the motion of the string is governed by a wave equation; the sand particles which hit the string are represented by space-time white noise. These equations are most easily understood by discretizing space, which leads to large systems of coupled ordinary stochastic differential equations (compare Exercise 5.10). Passing to the fine-discretization limit is not trivial (Pardoux, 2021), in particular in more than one spatial dimension. Stochastic partial differential equations are also relevant for spatial modeling and statistics (Krainski et al., 2019).

One background for spatio-temporal phenomena is interacting particle systems, even if this term is mostly used for discrete-space systems (Liggett, 1985; Lanchier, 2017). The so-called *superprocesses* (Etheridge, 2000) involve particles that are born and die, and move by diffusion during their lifespan. Such a system can be seen as a Markov process where the state is a spatial point pattern, and can display interesting spatial structure and patterns. When particles also affect the motion of other particles, mean field theory (Muntean et al., 2016) can be applied. When the particles are agents that each pursue their own interests, mean field games emerge (Lasry and Lions, 2007; Bensoussan et al., 2013), which find applications in economy as well as ecology.

Finally, each domain of applications has its particular motivations, classical models, and refined methods. It is worthwhile to study these carefully, which will allow you to recognize where and how the state of the art can be advanced with stochastic differential equations.

In conclusion: My hope for this book is that it serves as a useful introduction to stochastic differential equations, and that it motivates you, the reader, to dive deeper into this theory, and to explore the broader connections to neighboring fields. Wherever you go, there will be no shortage of fascinating phenomena to study.

Bibliography

Batchelor G (1967) *An Introduction to Fluid Dynamics* Cambridge Mathematical Library.

Bellman R (1957) *Dynamic Programming* Princeton University Press, Princeton, N.J.

Bensoussan A, Frehse J, Yam P (2013) *Mean Field Games and Mean Field Type Control Theory* Springer.

Benzi R, Sutera A, Vulpiani A (1981) The mechanism of stochastic resonance. *Journal of Physics A: mathematical and general* 14:L453.

Billingsley P (1995) *Probability and Measure* Wiley-Interscience, third edition, New York.

Björk T (2021) *Point Processes and Jump Diffusions: An Introduction with Finance Applications* Cambridge University Press, Cambridge.

Black F, Scholes M (1973) The pricing of options and corporate liabilities. *Journal of Political Economy* 81:637–654.

Boyd S, El Ghaoui L, Feron E, Balakrishnan V (1994) *Linear Matrix Inequalities in System and Control Theory* Studies in Applied Mathematics. Siam, Philadelphia.

Burrage K, Burrage P, Tian T (2004) Numerical methods for strong solutions of stochastic differential equations: an overview. *Proceedings of the Royal Society of London. Series A: Mathematical, Physical and Engineering Sciences* 460:373–402.

Cox J, Ingersoll Jr J, Ross S (1985) A theory of the term structure of interest rates. *Econometrica: Journal of the Econometric Society* pp. 385–407.

Crandall MG, Ishii H, Lions PL (1992) User's guide to viscosity solutions of second order partial differential equations. *Bull. Amer. Math. Soc.* 27:1–67.

Doob J (1953) *Stochastic Processes* John Wiley & Sons, New York.

Doob J (2001) *Classical potential theory and its probabilistic counterpart* Springer, New York.

Doyle J, Glover K, Khargonekar P, Francis B (1989) State-space solutions to standard $\mathcal{H}_2$ and $\mathcal{H}_\infty$ control problems. *IEEE Transactions on Automatic Control* 34:831–847.

Dynkin EB (1965) *Markov processes*, Vol. 1 and 2 Springer, New York.

Einstein A (1905) On the motion of small particles suspended in liquids at rest required by the molecular-kinetic theory of heat. *Annalen der Physik* 17:549–560 Translated into English by A.D. Cowper and reprinted by Dover (1956).

Etheridge A (2000) *An introduction to superprocesses* American Mathematical Society, Providence, Rhode Island.

Feller W (1951) Two singular diffusion problems. *Annals of Mathematics* 54:173–182.

Fleming W, Rishel R (1975) *Deterministic and Stochastic Optimal Control* Springer, New York.

Fleming W, Soner H (1993) *Controlled Markov Processes and Viscosity Solutions* Springer, New York.

Gard T (1988) *Introduction to Stochastic Differential Equations*, Vol. 114 of *Monographs and textbooks in pure and applied mathematics* Marcel Dekker, New York.

Gardiner C (1985) *Handbook of Stochastic Models* Springer, New York, second edition.

Gihman I, Skorohod A (1972) *Stochastic Differential Equations* Springer, New York.

Giles M (2008) Multilevel Monte Carlo path simulation. *Operations Research* 56:607–617.

Grimmett G, Stirzaker D (1992) *Probability and Random Processes* Oxford University Press, Oxford, second edition.

Harvey AC (1989) *Forecasting, structural time series models and the Kalman filter* Cambridge University Press, Cambridge.

Has'minskĭ R (1980) *Stochastic Stability of Differential Equations* Sijthoff & Noordhoff, Alphen aan den Rijn.

Higham D (2001) An algorithmic introduction to numerical simulation of stochastic differential equations. *SIAM Review* 43:525–546.

Hull JC (2014) *Options, futures and other derivatives* Pearson, Boston, ninth edition.

Iacus S (2008) *Simulation and inference for stochastic differential equations: with R examples* Springer Verlag, New York.

Itô K (1944) Stochastic integral. *Proceedings of the Japan Academy, Series A, Mathematical Sciences* 20:519–524.

Itô K (1950) Stochastic differential equations in a differentiable manifold. *Nagoya Mathematical Journal* 3:35–47.

Itô K (1951a) On a formula concerning stochastic differentials. *Nagoya Mathematical Journal* 3:55–65.

Itô K (1951b) On stochastic differential equations. *Mem. Amer. Math. Soc* 4:1–51.

Jazwinski A (1970) *Stochastic processes and filtering theory* Academic Press, New York.

Kalman R (1960) A new approach to linear filtering and prediction problems. *Journal of Basic Engineering* pp. 35–45.

Kalman R (1963a) Mathematical description of linear dynamical systems. *JSIAM Control* 1:152–192.

Kalman R (1963b) The theory of optimal control and the calculus of variations. *Mathematical optimization techniques* 309:329.

Kalman R, Bucy R (1961) New results in linear filtering and prediction theory. *Journal of Basic Engineering* pp. 95–108.

Karatzas I, Shreve S (1997) *Brownian Motion and Stochastic Calculus* Springer, New York, second edition.

Kloeden P, Platen E (1999) *Numerical Solution of Stochastic Differential Equations* Springer, New York, third edition.

Kolmogorov A (1931) Über die analytischen Methoden in der Wahrscheinlichkeitstheorie (On analytical methods in probability theory). *Mathematische Annalen* 104:415–458.

Kolmogorov A (1937) Zur umkerhbarkeit der statistichen naturgesetze (On the reversibility of the statistical laws of nature). *Mathematische Annalen* 104:415–458.

Kolmogorov A (1933) *Grundbegriffe der Wahrsheinlichkeitsrechnung* Springer, Berlin Translated into English in 1950 as "Foundations of the theory of probability".

Krainski E, Gómez-Rubio V, Bakka H, Lenzi A, Castro-Camilo D, Simpson D, Lindgren F, Rue H (2019) *Advanced Spatial Modeling with Stochastic Partial Differential Equations Using R and INLA* CRC Press.

Kramers H (1940) Brownian motion in a field of force and the diffusion model of chemical reactions. *Physica VII* 4:284–304.

Kristensen K, Nielsen A, Berg CW, Skaug H (2016) TMB: Automatic differentiation and Laplace approximation. *Journal of Statistical Software* 70.

Kushner H (1967) *Stochastic Stability and Control*, Vol. 33 of *Mathematics in Science and Engineering* Academic Press, New York.

Kushner H, Dupuis P (2001) *Numerical Methods for Stochastic Control Problems in Continuous Time*, Vol. 24 of *Applications of Mathematics* Springer-Verlag, New York, second edition.

Lamba H (2003) An adaptive timestepping algorithm for stochastic differential equations. *Journal of computational and applied mathematics* 161:417–430.

Lanchier N (2017) *Stochastic modeling* Springer, New York.

Langevin P (1908) Sur la théorie du mouvement brownien. *C.R. Acad. Sci.* 146:530–533.

Lasry JM, Lions PL (2007) Mean field games. *Japanese journal of mathematics* 2:229–260.

Liggett TM (1985) *Interacting Particle Systems* Springer, New York.

Lions P (1982) Optimal stochastic control of diffusion type processes and hamilton-jacobi-bellman equations In *Advances in filtering and optimal stochastic control*, pp. 199–215. Springer, New York.

Madsen H (2007) *Time series analysis* Chapman & Hall/CRC, London.

Mao X (2008) *Stochastic Differential Equations and Applications* Woodhead Publishing, Cambridge, UK.

Maruyama G (1955) Continuous Markov processes and stochastic equations. *Rendiconti del Circolo Matematico di Palermo* 4:48–90.

Merton RC (1969) Lifetime portfolio selection under uncertainty: The continuous-time case. *The Review of Economics and Statistics* 51:247–257.

Metzler R, Klafter J (2000) The random walk's guide to anomalous diffusion: A fractional dynamics approach. *Physics Reports* 339:1–77.

Milstein G, Tretyakov M (2004) *Stochastic Numerics for Mathematical Physics* Springer, New York.

Moler C, Van Loan C (2003) Nineteen dubious ways to compute the exponential of a matrix, twenty-five years later. *SIAM Review* 45:3–49.

Muntean A, Rademacher JD, Zagaris A, editors (2016) *Macroscopic and Large Scale Phenomena: Coarse Graining, Mean Field Limits and Ergodicity* Springer.

Øksendal B (2010) *Stochastic Differential Equations - An Introduction with Applications* Springer-Verlag, New York, sixth edition.

Okubo A, Levin S (2001) *Diffusion and Ecological Problems: Modern Perspectives* Springer, New York.

Pardoux É (2021) *Stochastic Partial Differential Equations: An Introduction* Springer, New York.

Pavliotis G (2014) *Stochastic Processes and Applications: Diffusion Processes, the Fokker-Planck and Langevin Equations* Springer, New York.

Pawitan Y (2001) *In all likelihood: Statistical modelling and inference using likelihood* Oxford University Press, Oxford, UK.

Pham H (2009) *Continuous-time Stochastic Control and Optimization with Financial Applications* Springer, New York.

Puterman ML (2014) *Markov Decision Processes.: Discrete Stochastic Dynamic Programming* John Wiley & Sons, New York.

Rey-Bellet LR (2006) Ergodic properties of Markov processes In *Open quantum systems II*, pp. 1–39. Springer, New York.

Rogers L, Williams D (1994a) *Diffusions, Markov processes, and martingales.*, Vol. 1: Foundations Cambridge University Press, Cambridge.

Rogers L, Williams D (1994b) *Diffusions, Markov processes, and martingales.*, Vol. 2: Itô calculus Cambridge University Press, Cambridge.

Rue H, Held L (2005) *Gaussian Markov random fields: theory and applications* Chapman and Hall/CRC, London.

Rümelin W (1982) Numerical treatment of stochastic differential equations. *SIAM Journal on Numerical Analysis* 19:604–613.

Schaefer MB (1957) Some considerations of population dynamics and economics in relation to the management of the commercial marine fisheries. *Journal of the Fisheries Board of Canada* 14:669–681.

Simon D (2006) *Optimal State Estimation - Kalman, $\mathcal{H}_\infty$, and Nonlinear Approaches* John Wiley & Sons, Hoboken, New Jersey.

Stratonovich R (1966) A new representation for stochastic integrals and equations. *J. SIAM Control* 4:362–371.

Särkkä S, Solin A (2019) *Applied Stochastic Differential Equations* Cambridge University Press, Cambridge.

Touzi N (2013) *Optimal Stochastic Control, Stochastic Target Problems, and Backward SDE* Springer, New York.

Uhlenbeck G, Ornstein L (1930) On the theory of Brownian motion. *Phys.Rev.* 36:823–841.

Versteeg H, Malalasekera W (1995) *An Introduction to Computational Fluid Dynamics: The Finite Volume Method* Prentice Hall, Harlow, England.

Walsh JB (1986) An introduction to stochastic partial differential equations In *École d'Été de Probabilités de Saint Flour XIV-1984*, pp. 265–439. Springer, New York.

Williams D (1991) *Probability with Martingales* Cambridge University Press, Cambridge.

Zhou K, Doyle J, Glover K (1996) *Robust and Optimal Control* Prentice Hall, Hoboken, New Jersey.

Zucchini W, MacDonald IL (2009) *Hidden Markov models for time series: an introduction using R* CRC Press, London.

Index

σ-algebra
    of events, $\mathcal{F}$, 34
    as a model of information, 40
Absorbing boundaries, 269
Adapted process, 77
Additive noise, 105, 153
Advection vs diffusion, 22
Advection-diffusion equation, 20,
    203, 204
Advective transport, 19
Algebraic Lyapunov equation, 106,
    113, 208
Almost surely (a.s.), 35
Arcsine law, 88
Attainable boundary point, 278
Autocovariance function, 226

Backward Kolmogorov equation, 199
    for linear systems, 228
Bayes' rule, 48
Bellman, Richard E., 323
Bessel process, 152
Black-Scholes model, 200, 225
Borel algebra, 34
Borel measurable function, 37
Borel's paradox, 61
Borel, Émile, 35
Borel-Cantelli lemmas, 56
Boundary
    Absorbing, 269
    Attainable or repelling, 278
    Reflecting, 217
Boundedness, stochastic sample
    path, 314
Brown, Robert, 25
Brownian bridge, 68, 84, 114
Brownian motion, 23
    as a Markov process, 192
    as a martingale, 78

as a transformed
    Ornstein-Uhlenbeck
    process, 158
Definition, 64
exit time of sphere, 265
Finite-dimensional distributions,
    66
Hitting time distribution, 73
Lévy characterization, 79
Maximum, 71
on the circle, 153, 162
on the sphere, 165
Physical unit, 26, 66
Recurrence, 73
Recurrence in $\mathbf{R}^n$, 273
Scaling, 68
Simulation, 67, 84
Standard, 26
Total variation, 87
vs white noise, 103
w.r.t. a filtration, 79
Wiener expansion, 84
Brownian motion with drift, 85, 120
    Extremum, 289
Brownian unicycle, 189, 228

Central limit theorem, 59
Chain rule
    in Itô calculus, 143
    in Stratonovich calculus, 159
Commutative noise, 188
Complementary error function, 29
Conditional distribution, 45
    for Gaussians, 62
Conditional expectation, 42, 62
    Graphically, 61
    Properties of, 45
Conditional independence, 47, 62
Conditional variance, 45, 61